Wheel and Pinion Cutting in Horology

Wheel and Pinion Cutting in Horology

A historical and practical guide

J Malcolm Wild FBHI

The Crowood Press

First published in 2001 by
The Crowood Press Ltd
Ramsbury, Marlborough
Wiltshire SN8 2HR

enquiries@crowood.com

www.crowood.com

This impression 2022

British Library Cataloguing-in-Publication Data
A catalogue record for this book is available from the British Library.

ISBN 978 1 86126 245 5

Photography and line illustrations by the author

Typeset by Florence Production Ltd, Stoodleigh, Devon

Printed and bound in India by Parksons Graphics Pvt. Ltd.

CONTENTS

PREFACE

In the early sixties when I first became interested in clockmaking and restoring, there was very little information available in horological books about wheel and pinion cutting. There were numerous technical books on the machining of involute gears for general engineering based on the involute system, but clock and watch gears have a different arrangement altogether, known as the cycloidal system.

Articles in *Model Engineer* by Claude B. Reeve gave basic information and also referred to Philip Thornton, a very good restorer at Great Haywood, Stafford. He also manufactured cutters for cutting clock wheels and pinions. Philip Thornton was very helpful with technical information, and this enabled me to collate the background that forms this book.

The information I gathered was originally put together for an article published in *Model Engineer*. This was well received and was eventually revised and published in book form. With the passing of time it was felt that some of the detail could benefit from updating, and that much more additional information could be included: hence this revised edition.

Where possible, actual set-ups have been used to show the type of work that can be achieved. Obviously where extremely small components are produced, photography is more difficult and line drawings are used instead.

Whilst some of the equipment shown is sophisticated and can be expensive and quite difficult to find on the secondhand market, it is not essential. Much useful work can be carried out on quite simple equipment as described, that can be constructed by the reader in his own workshop.

Having been trained in the age when the imperial system of measurement dominated, I hope I will be forgiven for including this as well as the metric system. In any event, I have an excellent case for including this for our many clockmaking friends in the USA, who still hold on strongly to the good old English system.

I hope both amateur and professional clockmakers will find something that will help them, and if they have not carried out wheel and pinion cutting before, will be encouraged to do so, as it is extremely satisfying and rewarding. If the book does that, then writing it will have been worthwhile.

The methods shown are those I have used over many years, but they are not definitive.

ACKNOWLEDGEMENTS

My thanks are due to my friends and colleagues in horology, many of whom are experts in their field, for providing assistance, whether it were advice or the loaning of either tools or catalogues. Also a special thanks to those who have helped in checking the calculations and text, and to the various museums and the British Horological Institute Library who have provided photographs and information, all of which has helped in the writing of this book. These people include Alan Bennett, David Burton, Peter Clark, Barry Corbishley, Jeremy Evans, Frank Gartside, John Griffiths, Jim Hancock, John Hatt, Charles Haycock, Rodney Law, Chris Lowe, Tony Marks, Alec Marsden, Arthur McDonald, H. Parry, Brian Raggett, Graham Schofield, Prof. Alan Smith, Roger Stevenson, Peter White, Ben Wright.

From overseas I would like to thank Ray Bates (Canada); Ian Fowler (Germany); David Lindow (USA); Archie B. Perkins (USA); Derek Pratt (Switzerland); Mike Simpkins (South Africa).

1 EARLY METHODS OF CUTTING WHEELS

Methods for cutting clock and watch wheel teeth have been available since the seventeenth century, and possibly during the century before, although only one or two wheel-cutting engines have survived. Robert Hooke, a brilliant scientist of the latter part of the seventeenth century, is often credited – though probably not correctly – with the invention of the wheel-cutting engine, due to various references in his diary. Hooke was deeply involved in the horological problems of his day. It was Hooke who invented the balance spring, and he was a close friend and associate of Thomas Tompion, the father of English clockmaking.

In 1672 he mentions a meeting with a Lancashire watchmaker's son about wheel-cutting engines. A further entry on 18 March 1673 states 'Harry cleansd lathe, began wheel cutting engine'.[1] If he were making his own engine, was this the type with a rotary cutter, or some method of guiding a file similar to later machines used for finishing teeth? What a pity his machine did not survive! On Saturday 2 May 1674, the following extract was noted:

> Slept till 7. At Garaways with Lem Oliver, Godfry, and new Carpenter, at mercers. Agreed with Lem about the Theater. To Thomkin in Water Lane [he is referring to Thomas Tompion]. Much discourse with him about watches. Told him the way of making an engine for finishing wheels, and how to make a dividing plate; about the forme of an arch; about another way of teeth work; about pocket watches and many other things.

Anyone who has the chance of reading a copy of Robert Hooke's diary should do so. It is extremely interesting, and remarkable that such an important record of seventeenth-century life has survived; also that Hooke, who was possibly the leading scientist of the day, managed to be involved in so many projects. He was not a healthy man, and a number of his extracts explain his many illnesses – in quite graphic detail.

A manuscript written by John Carte *circa* 1708 survives in the Bodleian Library. Carte, a Coventry watchmaker, had been working in London since 1698 and he states:

> I shall now say something of a peculiar improvement the English have made in their engines for the working part whereby their works are performed with the most quickness & exactness. They have invented that curious engine for the cutting the teeth of a wheel whereby that part of the work is done with an exactness which far exceeds what can be performed by hand: Then there is the engine for equalling the balance wheel the old name for a verge watch escape wheel]. Likewise the engine for cutting the turns of the fusee; and lastly the instrument for drawing of steel pinion wire: All which ingenious inventions were first conceived and made at Liverpool in Lancashire in England.[2]

John Carte also mentions the production of steel pinions. This will be covered later.

All these early citations are important. However, an earlier reference has been noted from the will of Peter Dehind, an early seventeenth-century clockmaker, written in February 1608/1609, which states:

> Item I geve will & bequeath unto my brother Jeromy the lease of the shopp wherein he dwelleth wch I hold of Robert Lancaster.
>
> Item 1 I geve unto him also ffyve poundes in money. One instrument to cutt wheeles, one anvild, certaine files and other working tooles so many as wilbe fitting and necessarie for him to make up a peece of work wthall. And to John Lampard thother instrument to cut wheeles.[3]

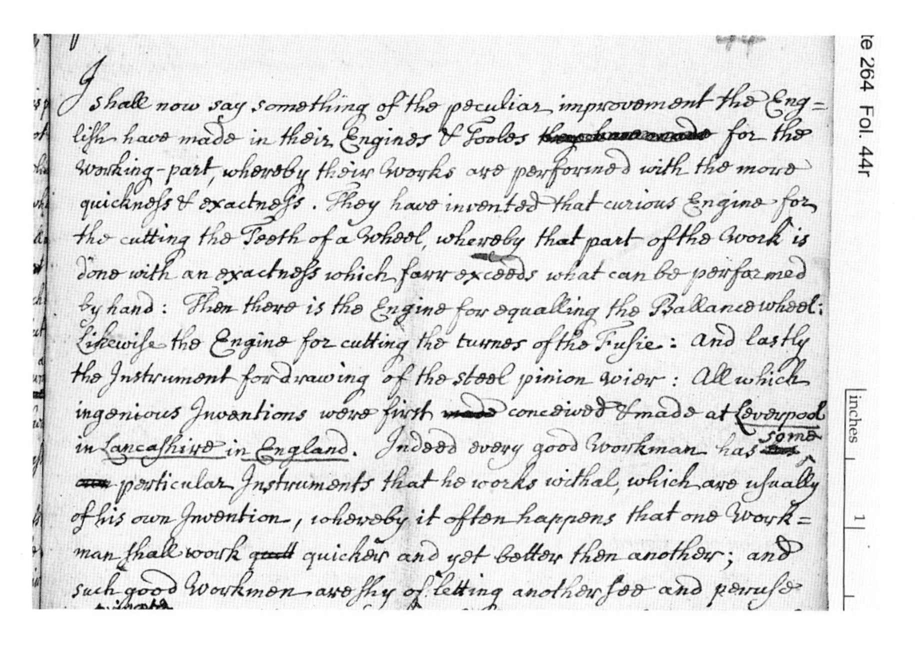
I shall now say something of the peculiar improvement the English have made in their Engines & Tooles for the working-part, whereby their works are performed with the more quicknesse & exactnesse. They have invented that curious Engine for the cutting the Teeth of a Wheel, whereby that part of the work is done with an exactnesse which farr exceeds what can be performed by hand: Then there is the Engine for equalling the Ballance wheel; Likewise the Engine for cutting the turnes of the Fusie: and lastly the Instrument for drawing of the steel pinion wier: All which ingenious Inventions were first conceived & made at Liverpool in Lancashire in England. Indeed every good Workman has some particular Instruments that he works withal, which are usually of his own Invention, whereby it often happens that one Work=man shall work quicker and yet better then another; and such good Workmen are shy of letting another see and peruse

Fig. 1 The Carte manuscript referring to an early wheel-cutting machine.

Clearly, methods for wheel cutting were being used around the beginning of the seventeenth century.

An early reference that Juanelo, engineer and clockmaker to Emperor Charles V, cut toothed wheels on a machine was recorded by Ambrosio de Morales, professor at the University of Alcala Records in *Las Antiguedades de Las Ciudades de Espana*, Alcala 1575. The fact that he had invented a new type of wheel-cutting machine was obtained direct from Juanelo. This machine made it possible to produce about three toothed wheels a day, all of them different and without failures. The machine was used *c.* 1540 to make the wheels for the great astronomical clock for Charles V, which contained 1,800 toothed wheels.[4]

Whilst wheel-cutting engines may have been used in certain parts of Europe, not all clockmakers would be aware of this and would still have been using quite crude methods, by marking out the wheel blanks individually and hand filing the teeth. Clocks have been noted where the radial scribe marks on the wheel are still visible, also centre punch marks. Fairly accurate work could be carried out using this method, but considerable skill must have been involved.

In fact the art of dividing the circle into equal parts has occupied the minds of astronomers, scientists and instrument makers from the earliest times and is mentioned in many early manuscripts. The same problem would apply when constructing a division plate because this would have been the most difficult part of the machine to construct.

EVOLUTION OF THE DIVISION PLATE

Early methods used to divide the circle were by geometry. Henry Hindley, 1701–71, described a method that still holds good today and is used by model engineers to construct an accurate division plate: a simple jig is produced with two holes at a suitable pitch. One hole is for a locating pin, the other for the drill. A thin strip of metal is drilled using the jig, by drilling one hole and then using the pin to locate from each subsequent

Fig. 2 Wheel from an early iron clock, clearly showing the centre punch marks.

Fig. 3 Division plate signed Hans Göbe, Dresden 1564.

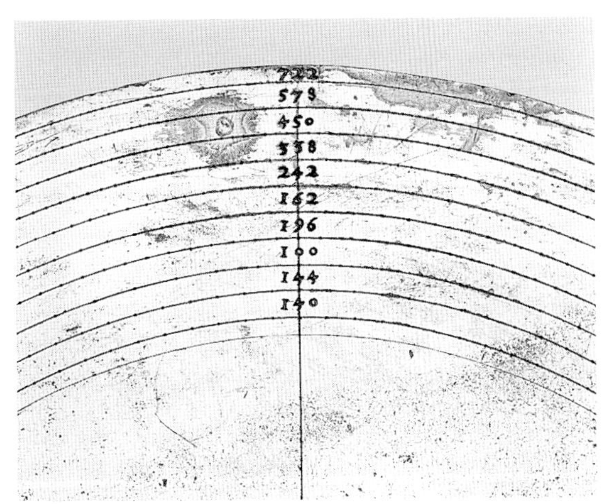

Fig. 4 Details of counts on the Göbe division plate.

hole drilled until the required number of holes have been completed for the divisions required. Each end of the strip is pegged to form a circle, and a disc of wood is turned to suit the circumference of the drilled metal strip. When a division plate is drilled from this master, the strip can be reduced in length for the next lower number of divisions required.[5]

The oldest surviving example of a division plate is that signed by Hans Göbe or Gebe dated 1564, which is in the Mathematisch-Physikalischer Salon in the Zwinger in Dresden. Hans Göbe was a clockmaker by appointment to the Saxon Court in Dresden from 1558, presumably until his death in 1574. He produced astronomical instruments such as calendars, diptych and horizontal dials (a horizontal dial is in the possession of the MPS in Dresden). From the high numbers on the division plate it is clear he produced complicated instruments. Obviously it was used in combination with a pair of dividers. The blank to be calibrated was fixed in the centre by means of three dogs (now missing) connected through the slots to the winged screws in the manner of a mandrel lathe. The divisions would have been scratched onto the blank and subsequently engraved or, in the case of a gear wheel, the teeth would have been filed out.

It has been erroneously assumed by a number of authors including Klaus Maurice (*Deutsche Raderuhr*, 1976), Ted Crom (*Horological Shop Tools*, 1980), Fritz Weger (*Zur Geschichte des Uhrenzahnrads in Alte Uhren* 1991/94) *et al.* that this division plate was made by Eberhard Baldewein of Marburg and accompanied the huge astronomical clock made by him for the Elector of Saxony between 1563 and 1567, because both are contemporary. However, when one contemplates the fine gearing on Baldewein's clocks and those of Jost Burgi, his successor at the Court of the Landgrave of Hesse-Kassel, one can only assume that they must have had access to some form of wheel-cutting engine. Intellects capable of producing such complex mechanisms must have developed a more rational method of gear cutting considering the huge number of teeth to be cut for the astronomical clocks.

The illustration below is from Nicholas Bion's book of 1709 *Traité de la Construction et des Principaux Usages des Instruments de Mathematique*. This is the earliest known illustration of a wheel-cutting machine.

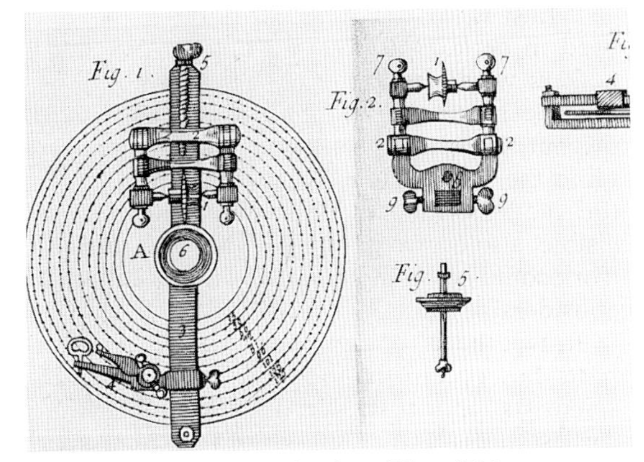

Fig. 5 Wheel-cutting engine from Bion, 1709.

Anyone interested in the history of the wheel-cutting engine should obtain Theodore R. Crom's book *Horological Wheel Cutting Engines 1700–1900*, published in 1970.

EARLY WHEEL-CUTTING MACHINES

Undoubtedly the first of the wheel-cutting machines cut straight teeth with a straight cutter like a file in some form of guide. Later the cutter was probably more like a rotary file than the normal multi-tooth cutter we use today. The first illustration shows a wheel-cutting engine from Ferdinand Berthoud *Essai sur l'horlogerie* 1763, which distinctly shows the slitting cutter and its cutting operation on a wheel blank. After the teeth had been cut, then the radius at the tip of the tooth would have had to have been filed; still a tedious task. Later, machines were constructed for that sole purpose, as the one shown from *A Catalogue of Tools for Watch & Clock Makers* by John Wyke of Liverpool.[6]

William Derham, in his book *The Artificial Clockmaker, a Treatise on Watch and Clock-work* (1696), made reference to Dr Hooke and his invention of cutting engines, which were never thought of until towards the end of Charles II's reign (King Charles II reigned from 1660–1685). A hundred years had therefore elapsed since Juanelo had cut his wheels, which would cast doubt on Robert Hooke being the inventor of the wheel-cutting engine.

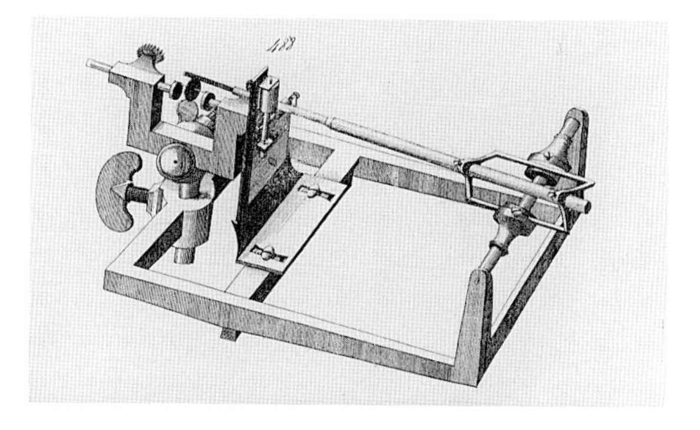

Fig. 7 Machine for rounding watch and clock teeth – Wyke catalogue.

Basically, the wheel-cutting machine consists of a rigid frame, on which is mounted a vertical spindle. At the lower end of this is mounted a large brass indexing plate, drilled with rows of equally spaced holes, the numbers being those most commonly required for clock wheels. The wheel blank is mounted at the upper end of the vertical spindle. In some cases the spindle is supported by an additional arm to give added rigidity when a wheel is being cut. The cutter is mounted on a horizontal spindle set in cone bearings with screw adjustment. The frame, which supports the cutter spindle, is either mounted on a vertical slide attached to the main frame of the wheel-cutting engine, or pivoted radially. Both methods give the necessary move-

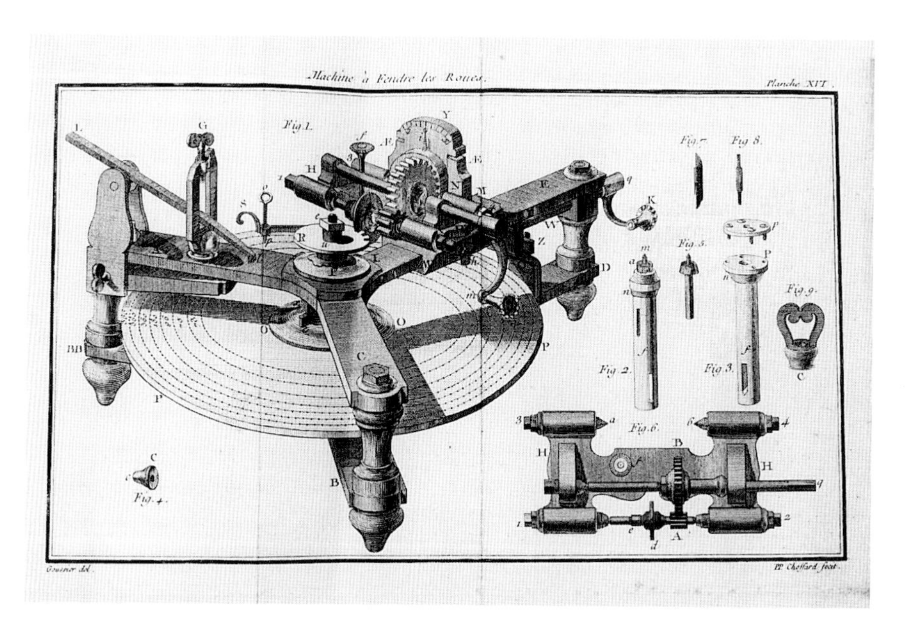

Fig. 6 Early wheel-cutting engine, from Ferdinand Berthoud's Essai sur l'horlogerie *(1763).*

Fig. 8 Early wheel-cutting engine, c. *1670.*

Fig. 9 Underside of the division plate showing the engraving of Gunters Quadrant.

ment for the cutter to be traversed through the wheel blank, although the radial cutting frame does not give a perfect tooth shape. In use, the blank is held in place and the index pin arranged to locate a hole in the particular circle of holes being used. The cutter is then rotated by either a bow or small handwheel pulley-and-gut drive; when one tooth has been cut, then the index plate is moved one space and the index pin relocated. This procedure continues until all the teeth have been cut.

Figure 8 shows what is possibly one of the oldest surviving wheel engines of the time, when Hooke and Tompion were discussing wheel-cutting engines. This machine was originally thought to be *c.* 1670, but it is more likely to be early eighteenth century; it is on display in the Science Museum, London. As can be seen from the underside in the next picture, the dividing plate is signed Humphrey Marsh Highworth, and the engraving is of Gunters quadrant.[7]

Figures 10–13 show a typical engine of the early eighteenth century, taken from *Les Sciences Les Arts Libéraux et Les Arts Mechaniques*, 1765. This machine was attributed to Sully and is shown in many of the books of the early eighteenth century. Note the quadrant that enables the cutting head to be tilted for angular cutting.

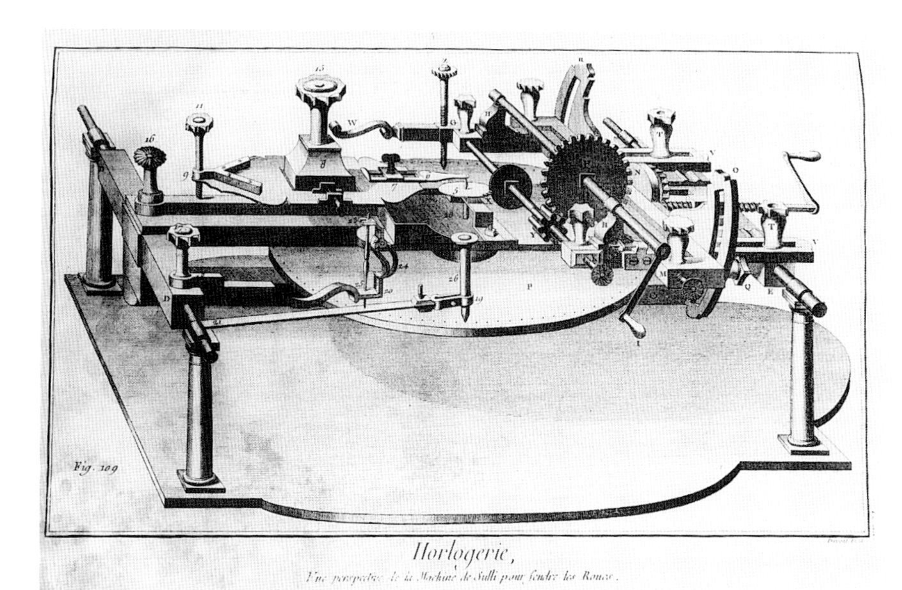

Fig. 10 Eighteenth-century wheel-cutting engine attributed to Sully.

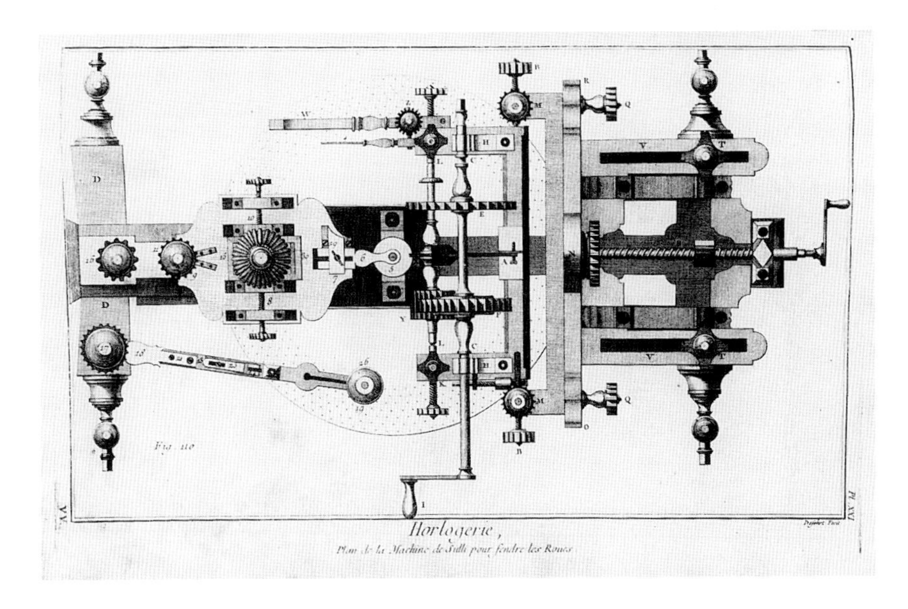

Fig. 11 Sully engine showing the feed screw and crank handle for the rotating cutter.

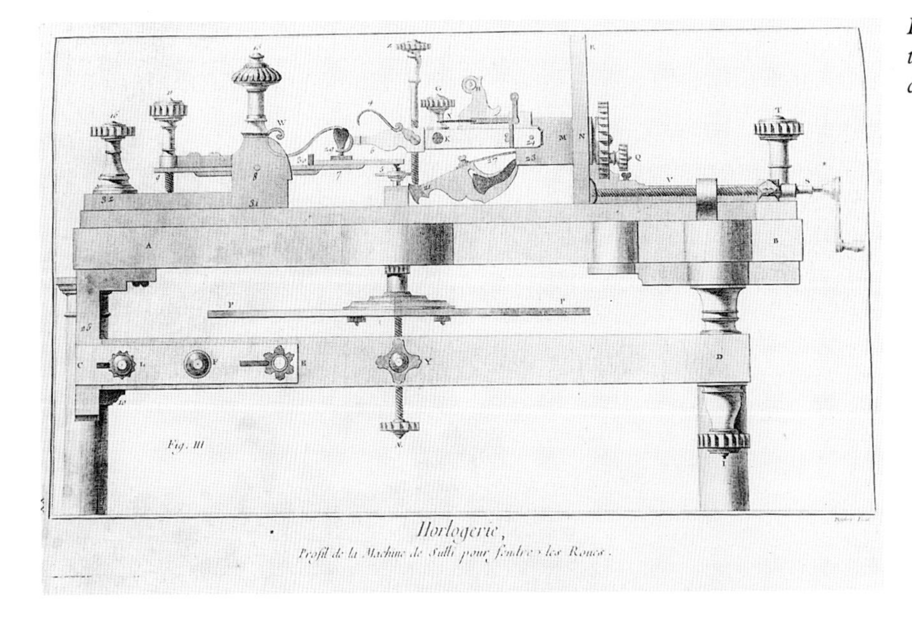

Fig. 12 Side elevation of the Sully engine; note the decorative handles.

The fine early wheel-cutting engine shown in Fig. 14 is European, and is on display in the Main Frankisches Museum, Wurzburg, Germany; it is signed Johann Melchior Wetschgi, Augsburg, *c.* 1710. As can be seen, the winged screws are beautifully shaped and are reminiscent of machines shown in the early Continental books on horology. The division plate has 33 concentric holes covering counts from 3 to 160, including prime numbers – some 2,462 holes!

The construction of these machines is quite detailed, and they would have been capable of extremely accurate work. Figure 15 is a nicely constructed English wheel-cutting engine with an iron frame and brass division plate. It is signed 'I. Waite Fecit' and also by the clockmaker Joseph Jackman, which is most unusual. On searching through records, no mention was found of 'I. Waite'; he was obviously a toolmaker or manufacturer and was proud of his work to have

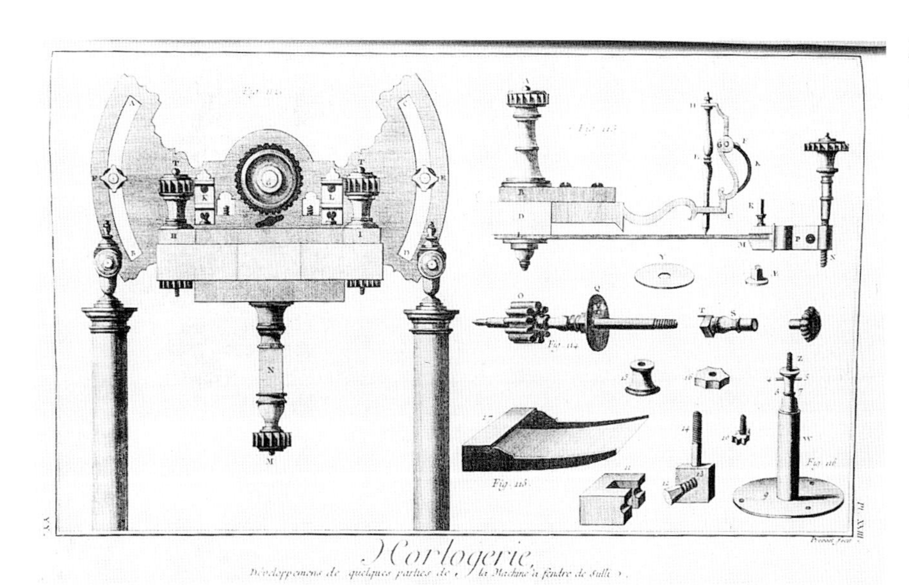

Fig. 13 Quadrant arrangement for angular cutting – Sully.

Fig. 14 German wheel-cutting machine, signed Johann Melchior Wetschgi Augsburg, c. 1710.

Fig. 16 Signature 'I Waite Fecit'.

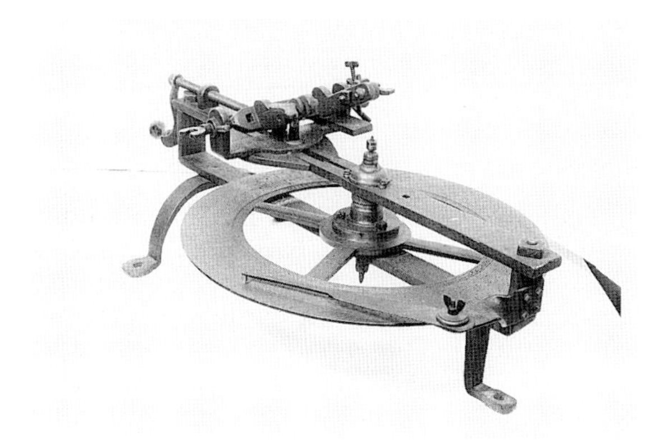

Fig. 15 English wheel-cutting machine, possibly seventeenth century.

Fig. 17 Joseph Jackman's signature; he was working in 1683.

Fig. 18 Heavy duty wheel-cutting machine by Rehi.

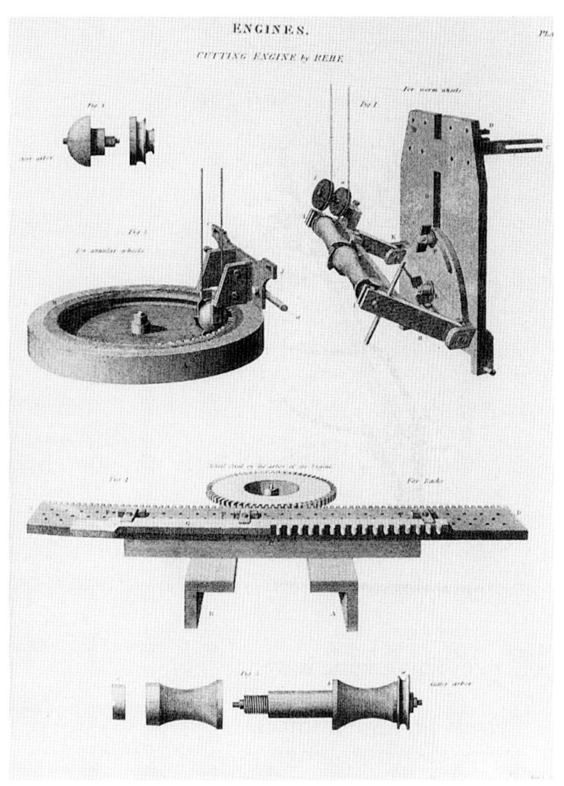

Fig. 19 Wheel-cutting frame of the Rehi machine.

engraved his signature on the division plate. Joseph Jackman is listed in Loomes' *Watchmakers & Clockmakers of the World* as working in London in 1683. This could, therefore, be one of the earliest wheel-cutting machines extant.

The above two illustrations (Figs 18 and 19) are from *Rees Cyclopaedia*, 1819–20. *Rees* was an important technical encyclopaedia running to some thirty-nine volumes and six volumes of plates; the horology section was reprinted in 1970.

The very heavy duty machine by Rehi shown here has a large division plate and an extremely heavily constructed bed, and what appears to be a cranked handle to activate the cutting frame.

Illustrated to the right is an engine said to have been constructed by Henry Hindley of York, a regional clockmaker of repute. It is known Hindley made tools for sale, and a wheel-cutting engine by him was sold by John Smeaton FRS to Thomas Reid of Edinburgh, the author of *A Treatise on Clock and Watch Making*, first published

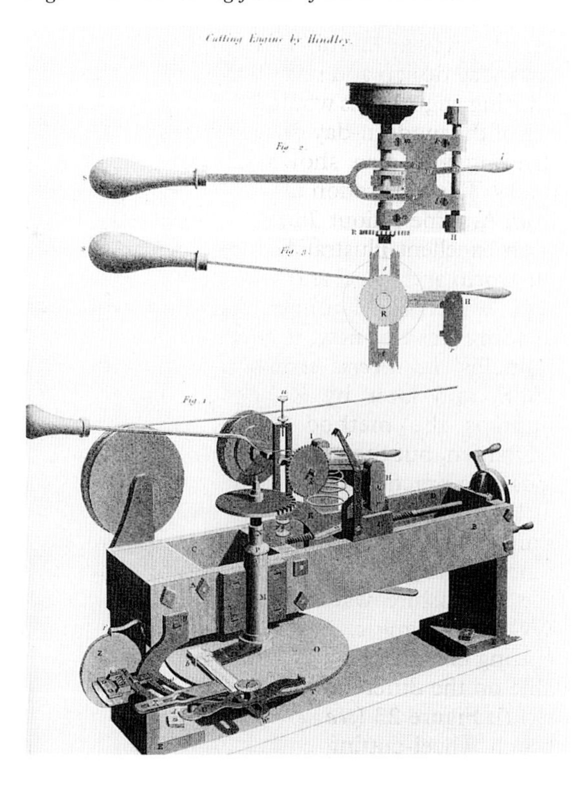

Fig. 20 Henry Hindley wheel-cutting machine showing worm-and-wheel dividing.

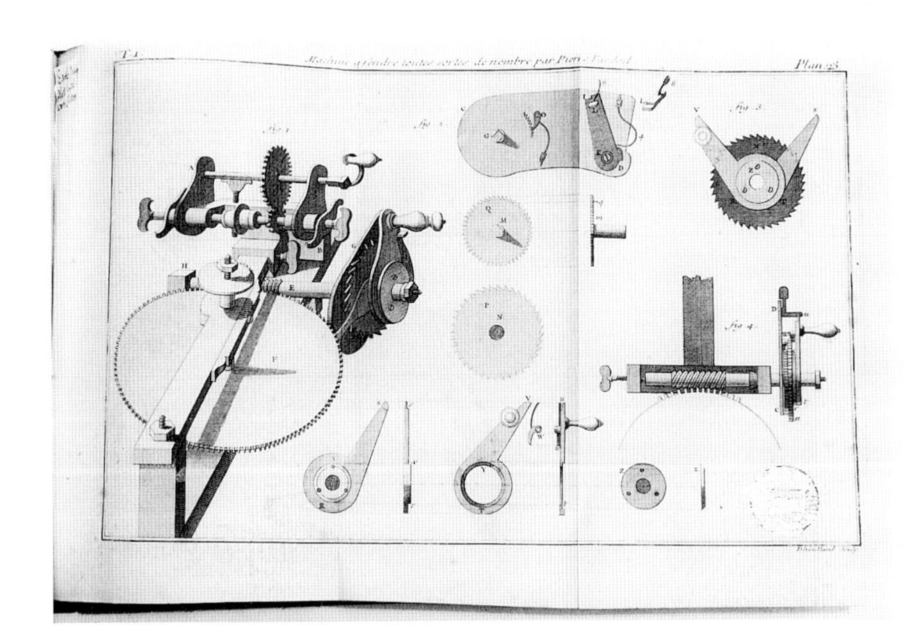

Fig. 21 Wheel-cutting machine showing worm wheel and quadrant for dividing.

in 1826. This engine is well documented in Reid's book – which, incidentally, is one of the best books on clock- and watch-making written in English.

Whereas most engines of the period had direct indexing methods from a division plate, Hindley's engine was operated by a worm and wheel, so giving infinitely variable numbers of wheel counts. At the time, this was quite an advanced design and revolutionized the use of the dividing engine; this would have been the forerunner of the modern-day dividing head. This type of dividing was also shown in the early French books. The illustration above shows an engraving from Antoine Thiout *Traité de l'horlogerie*, 1741; it is an excellent illustration also showing detail of the worm and wheel and sector arms.

The method of using a worm and wheel for dividing was not new: it had been mentioned in Ramellis' *Le diverse et artificiose machine*, Paris 1588, and later by Robert Hooke in 1674. Perhaps the method was known by a few craftsmen, but they had not been able to put the idea into practice successfully.

Firms such as Wyke & Green produced Lancashire wheel-cutting engines in great number; they were prolific tool manufacturers for the Lancashire watch and clock industry. The Wyke catalogue *c.* 1770 shows a watch wheel-cutting engine with a table of wheel counts available on the different division plates available (*see right*). Figure 23 (*see next page*) shows a very fine watch wheel-cutting engine by Wyke & Green, now in the Prescot Museum.[8] According to Professor Alan Smith in his introduction to the reprint of the Wyke catalogue, these engines were probably produced for jobbing work, not the production cutting of wheels; the more heavily constructed machines previously shown by Rehi and Hindley would have been more suited to production work. Whilst the constructional materials used in this model are mainly brass and perhaps more suitable for watch work, the Wyke engine

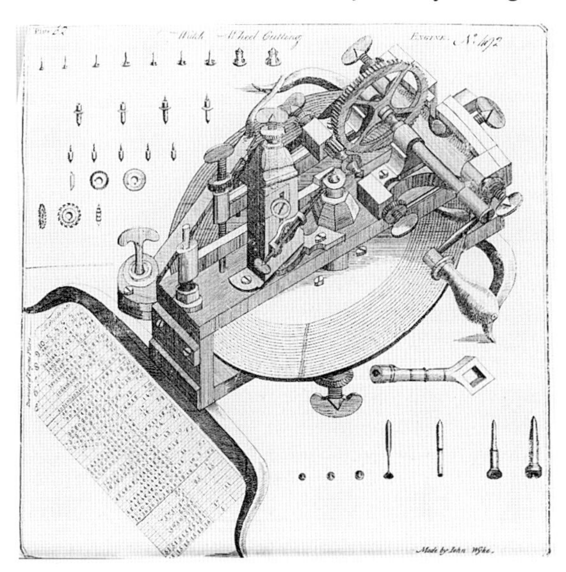

Fig. 22 Wheel-cutting engine c. *1770 from the Wyke catalogue.*

Fig. 23 Fine Wyke wheel-cutting engine – Prescot Museum.

Fig. 24 Lancashire wheel-cutting engine by Wyke.

shown below it, on display in the British Horological Institute, is primarily constructed of iron and consequently much stronger.

In the nineteenth century when the Swiss watch industry started to flourish, there was an increased demand for more tools. Consequently, the Swiss produced wheel engines to satisfy the market. These must have been produced in considerable quantity, as many are still in existence. Illustrated below is a very elegant machine *c.* 1860: the construction is predominantly brass, with a large brass division plate. The cutter frame is mounted on a vertical dovetailed slide and operated by a lever feed. A similar machine offered for sale in Grimshaw, Baxter & J. J. Elliott's tool catalogue of 1910 is shown (right). A number of accessories were available with the engine, as seen in the illustration. An enlarged illustration of the same engine from another tool catalogue of similar period is also illustrated (bottom right). The construction of the machine is extremely solid and capable of good work.

Whilst this design carried on from the nineteenth into the twentieth century, much heavier wheel-cutting machines were then being offered.

A fine, large engine *c.* 1760 is shown overleaf; the construction is mainly of iron. This is in the workshop of William Haycock of Ashbourne, and it is still in use today. It was previously in the Whitehurst Workshop in Derby, and possibly made by John Whitehurst himself. The engine

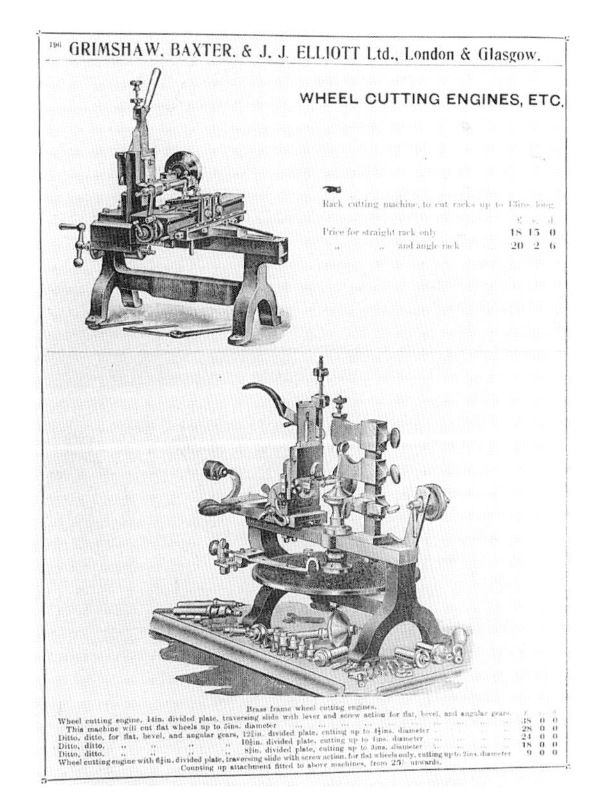

196 GRIMSHAW, BAXTER, & J. J. ELLIOTT Ltd., London & Glasgow.

WHEEL CUTTING ENGINES, ETC.

Rack cutting machine, to cut racks up to 13ins. long.

	£	s.	d.
Price for straight rack only ...	18	15	0
" " and angle rack	20	2	6

Brass frame wheel cutting engines.

	£	s.	d.
Wheel cutting engine, 14in. divided plate, traversing slide with lever and screw action for flat, bevel, and angular gears. This machine will cut flat wheels up to 5ins. diameter ...	38	0	0
Ditto, ditto, for flat, bevel, and angular gears, 12¾in. divided plate, cutting up to 4½ins. diameter ...	28	0	0
Ditto, ditto, " " " " 10½in. divided plate, cutting up to 4ins. diameter ...	24	0	0
Ditto, ditto, " " " " 8¼in. divided plate, cutting up to 3ins. diameter ...	18	0	0
Wheel cutting engine with 6¼in. divided plate, traversing slide with screw action, for flat wheels only, cutting up to 2ins. diameter	9	0	0

Counting up attachment fitted to above machines, from 25/- upwards.

Fig. 26 Illustration from Grimshaw, Baxter & J. J. Elliott's catalogue of 1910.

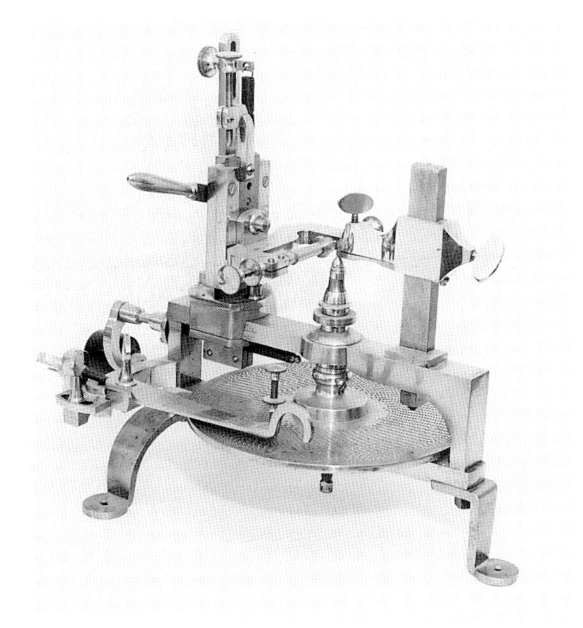

Fig. 25 Swiss wheel-cutting engine.

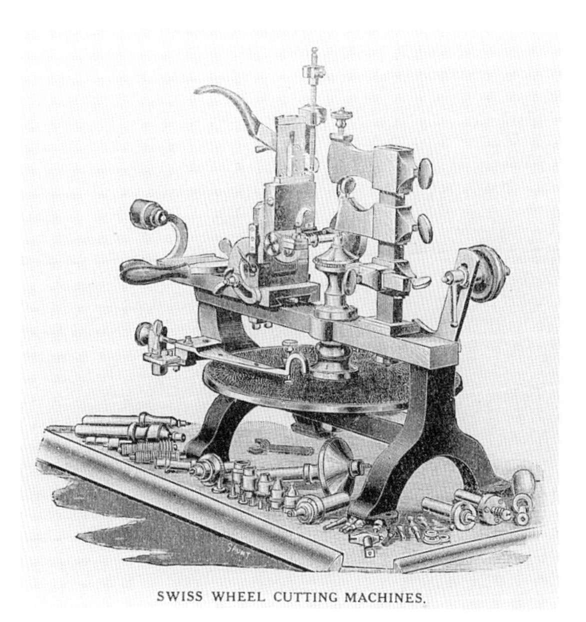

Fig. 27 Swiss wheel-cutting engine with its many accessories.

Fig. 28 The Haycock wheel-cutting machine, still in use today.

BELOW:
Fig. 29 Engraving from Horologia Ferrea, *depicting an early clockmaker's workshop.*

Fig. 30 Clockmaker's workshop from Diderot, c. *1765.*

was acquired by Haycock some time around 1859, together with various skeleton clock patterns, at the Whitehurst closing down sale.

The Whitehurst family were noted clockmakers, and produced clocks and watches during the eighteenth and early nineteenth centuries. An excellent book is available, written by Maxwell Craven, entitled *John Whitehurst, Clockmaker & Scientist 1713–1788* and published in 1996. The main contents of Whitehurst's company were sold to Robert Roskell of Liverpool. According to Charles Haycock, the wheel-cutting engine was used by his grandfather to cut wheels for the addition of a quarter chime mechanism to St Oswald's parish church in Ashbourne in 1897. Obviously the machine would have to be substantial to cope with the cutting of turret clock wheels.

EARLY WORKSHOPS

Very few engravings exist showing early workshops and the machines used. One of the earliest is the print of Stradanus's design of a clockmaker's workshop *c.* 1600, engraved by Philip Galle;[9] this shows the construction of various clocks, with a forge and only hand tools. No machines are in evidence.

In the *Encylopédie de Diderot et d'Alembert Paris, 1751–1772* is a plate depicting a clockmaker's workshop. In the left foreground a clockmaker can be seen using a bow with his turns or small lathe. There is no evidence of any wheel-cutting facilities, although wheel-cutting machines are depicted in Diderot's encyclopaedia. The horological section of this was reprinted in 1971, and the engravings that have been reproduced are excellent.

Fig. 31 The Samuel Deacon workshop as discovered.

Fig. 32 The reconstruction of the Samuel Deacon workshop.

Fig. 33 Deacon wheel-cutting machine.

A fine reconstruction of an eighteenth- to nineteenth-century workshop exists at the Newark House Museum in Leicester;[10] this was discovered virtually intact in 1951. It had originally been set up in 1771 by Samuel Deacon, a skilled craftsman in Barton in the Beans. The workshop contents were acquired by Leicester City Museum, and it is complete with tools, benches, a wheel and pinion cutting machine and turns. Many of the business records were also found. Deacon made many clocks and watches, including musical clocks and turret clocks, and many of his own tools. He was also a skilled engraver.

We have now seen illustrations of typical workshops over the last 300 years. The type of workshop in use today for the making and restoring of clocks and watches will be shown in later chapters.

EXAMPLES OF SEVENTEENTH-/ EARLY EIGHTEENTH-CENTURY WORK

The following illustrations give some indication of the type of work being carried out in the

Fig. 35 Movement from an early longcase clock, with musical train.

Fig. 34 Seventeenth-century winged lantern clock.

seventeenth and early eighteenth century where, in most instances, the wheel-cutting machine has been used.

Figure 34 shows a seventeenth-century winged lantern clock. Whilst the movement of this type of clock was fairly basic by today's standards, considerable skill would have been involved in making such a clock, and wheel-cutting facilities would more than likely have been employed. Figures 35 and 36 are photographs of a very rare thirty-hour, three-train longcase clock *c.* 1700 which, when restored, showed clear evidence that some of the wheels had been marked out and filed to shape by hand! Workmen in different areas, therefore, could still have been using both methods, as their skill and knowledge were closely guarded trade secrets and would not readily be passed on to their fellow tradesmen.

The following are photographs (*see* Figures 37–39) of a Joseph Knibb longcase clock *c.* 1690. Knibb, along with Tompion, was one of the finest English clockmakers of the late seventeenth

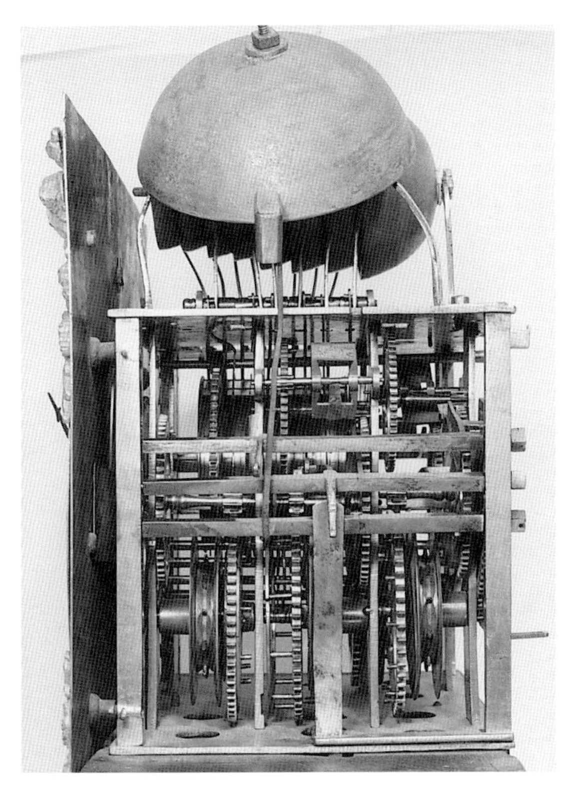

Fig. 36 Another view of the musical three-train lantern clock.

century. Proportions are extremely important, and it is difficult to imagine that any improvements could be made to the longcase shown. The case is veneered walnut with beautifully inlaid marquetry work. The movement is of eight day duration, has nicely cut wheels, and is generally well constructed. When the clock was stripped prior to cleaning, the wheels were examined and the teeth were most definitely cut by machining methods. An enlarged view of the great wheel and barrel shows how nicely cut and evenly spaced the teeth were.

Many of the best English clocks were manufactured in the early eighteenth century. A typical example is this small ebonized bracket clock (*see* opposite) by Robert Bumstead, London, *c.* 1730, who was a member of the Clockmakers Company. The movement is a timepiece with pull repeat. Also shown is the under-dial work and the quarter repeating rack. This clock is very nicely made.

As skills were being developed in the seventeenth and early eighteenth centuries, one workshop stood head and shoulders above the

Fig. 37 Joseph Knibb seventeenth-century longcase clock.

ABOVE LEFT: *Fig. 38 Movement of the Joseph Knibb longcase clock.*

ABOVE RIGHT: *Fig. 39 The great wheel from the Knibb longcase movement.*

BELOW LEFT: *Fig. 40 Small ebonized bracket clock by Robert Bumstead.*

BELOW RIGHT: *Fig. 41 Timepiece movement with pull quarter repeat – Robert Bumstead.*

Fig. 42 Complicated longcase movement by Thomas Tompion.

Fig. 43 Another view of Tompion's excellent workmanship.

Fig. 44 Tompion's longcase movement, showing the wheel work.

rest, and that belonged to Thomas Tompion. As already noted, he was a friend of Robert Hooke and the two were exchanging ideas. Judging from the work carried out by Tompion and his workmen,[11] wheel-cutting machines must have been employed. Whilst it was known that slitting cutters were used and the addendum curves were produced by filing, one wonders why they did not think of one cutter with the radius already formed, as we are looking at two geniuses with extremely fertile minds who were developing methods to improve the quality of their products. Also their output, especially in Tompion's case, was quite prolific.

Perhaps we will never know who invented the wheel-cutting machine, but it is possible that further research in the years to come may produce new evidence.

2 PINIONS AND PINION WIRE

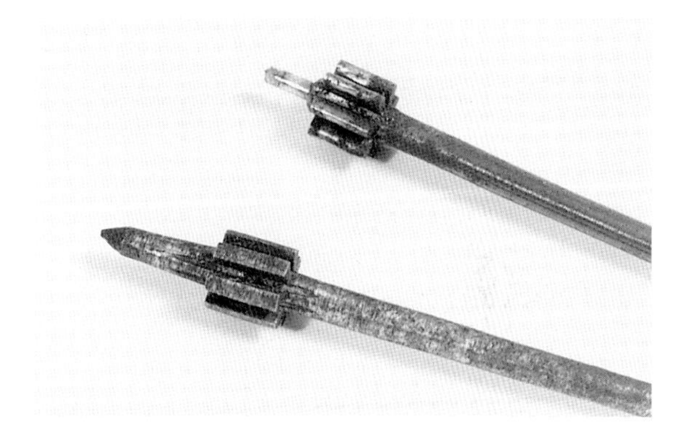

Fig. 45 Pinions machined using slitting method.

As anyone who has cut wheels and pinions will verify, it is comparatively easy to cut wheel teeth in brass, but an entirely different matter to cut steel pinions. Don't let this put you off, however, as methods will be shown later that make this job well within the capacity of a small workshop. Early pinions were filed from the solid, not an easy task. Later they were produced by dividing out and then slitting the blank with a circular cutter, having teeth similar to a file; the teeth or leaves were then shaped with special files. The illustration *(right)* shows an early pinion blank produced by the slitting method; also shown is a finished pinion.

Cut pinions were used from the early seventeenth century, perhaps for higher grade work only, as they would have been more difficult to produce with the machines available at the time. The first known reference to cut pinions is in a letter from a Wigan clockmaker to his customer in Lancashire, dated 1756. The Wigan clockmaker wrote: 'these pinions I send you on tryol as are made by a tool maker of the best steel and cut down in an engine and everyone who has of 'em likes them.'[12]

Some later wheel-cutting machines had attachments for also cutting pinions, but these

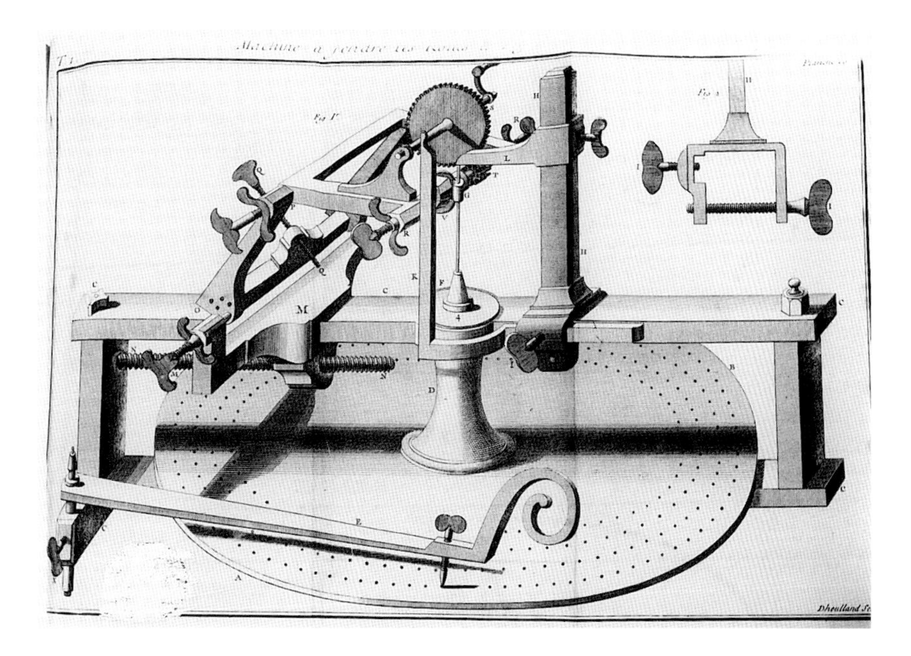

Fig. 46 Pinion-cutting machine from Antoine Thiout 'Traite de l'Horlogerie Mecanique et Pratique' 1741.

could not have been successful, as great rigidity was required to cut the steel to the depth required. Very few early pinion-cutting machines have survived, so one can only assume that throughout the eighteenth and into the early nineteenth century the slitting method for producing pinions was still employed. The method was laborious and time-consuming, and speedier methods of producing pinions were required. Whilst pinion wire had been known a hundred years earlier, it had not been used on the production of pinions on clocks or watches throughout the eighteenth century. Liverpool and the Prescot area is credited with the introduction of pinion wire.

PINION WIRE

To produce the shape of the pinion, the wire was drawn through a series of hardened steel dies. This was necessary to produce a pinion of acceptable proportions. Once the pinion wire was available in the required size, the surplus leaves had to be removed by either filing or breaking away. The leaves still had to be finished and often deepened by filing. The arbor could then be turned true, and the final polishing with abrasive could take place.

Anyone interested in the history of early horological tools and the development of pinion wire should read E. Surry Dane *Peter Stubs and the Lancashire Hand Tool Industry* (1973). Stubs were major tool and file manufacturers, and produced large quantities of pinion wire throughout the eighteenth and well into the nineteenth century. Another very informative book is *An 18th Century Industrialist – Peter Stubs of Warrington* by T. S. Ashton, published in 1939 and reprinted in 1961. Fig. 47 shows pinion wire, and from the sectional piece, it can be seen how thick the leaves are. They require both deepening and thinning.

Fig. 47 Pinion wire.

Thomas Hatton, in his book *An Introduction to the Mechanical Part of Clock and Watchwork* (1773), states:

> . . . small tools, and particular files are best made by the workman, the files for watch pinions are easily made for the centre pinion with curved edges, equal to that of the body of the pinion.

Hatten mentions watch pinions, but the same methods would apply for larger files used in finishing pinions for clocks. He also states: 'The tools made in Lancashire are best executed' and, as we have seen from illustrations shown previously from the Wyke catalogue, this is particularly true of, and is confirmed by, the examples still in existence.

PINION ENGINES

A fine pinion engine in the collection of the Prescot Museum (*see* opposite)[13] is typical of the machines made in Lancashire in the nineteenth century. A small index plate is used to locate the blank, and the depth of cut is adjusted by means of a vertical threaded rod with a winged lock nut. The pinion blank is traversed under the rotating cutter by lead screw, whilst the cutter is rotated by means of the wheel and pinion and the cranked handwheel drive.

Machines were developed to finish pinion leaves. Illustrated is an engraving from Thiout (*see* page 30). It is obvious from the engraving that the pinion has already been slit, and one can only assume that this machine is shaping the addendum curves, the file being operated by hand in guides. An index plate is fitted to the machine spindle to control the necessary divisions.

It is quite surprising that the official delegate of the Swiss Confederation to the Chicago Exhibition of 1893 reported as follows on the pinion-cutting machines:

> The pinions are cut from the solid steel whereas we use pinion wires, which means that the pinions cost less to produce. With one or two

ABOVE: Fig. 48 Pinion-cutting machine – Prescot Museum.

Fig. 49 Pinion with index plate – Prescot Museum.

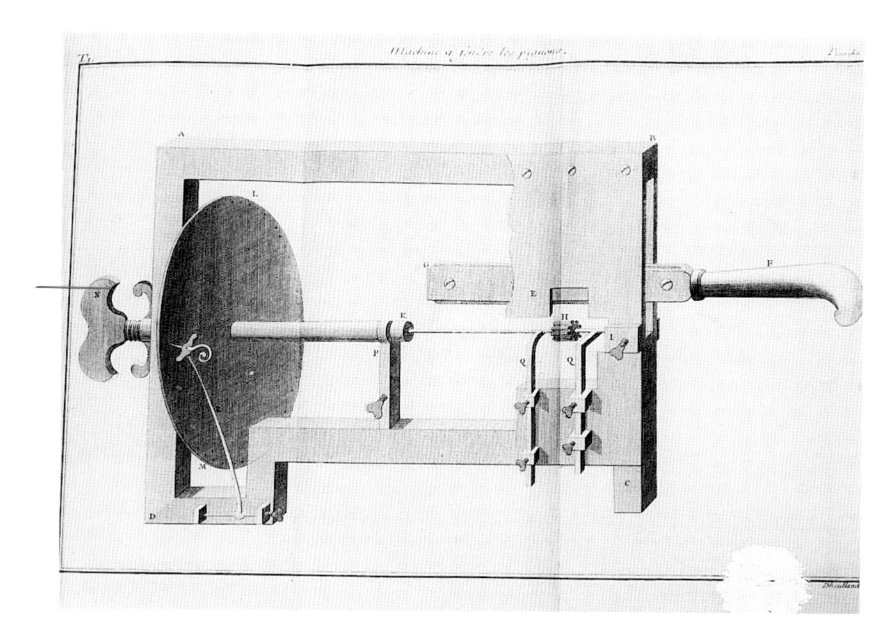

Fig. 50 Pinion-finishing machine from Antoine Thiout 'Traite de l'Horlogerie Mecanique et Pratique' 1741.

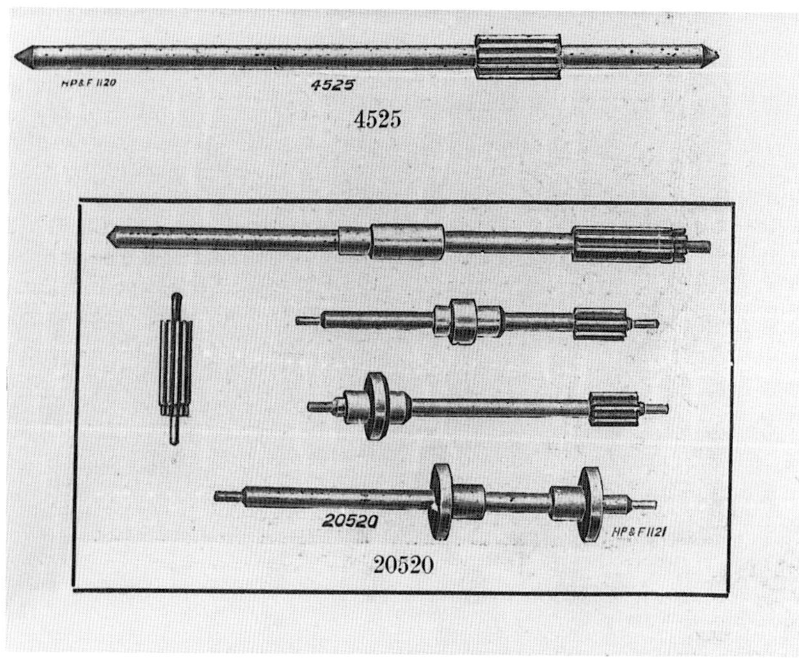

Fig. 51 Cut pinions available in a horological material supplies catalogue circa *1900.*

exceptions, the American method has not been adopted in Switzerland for this reason.

One would have thought that at this time pinions cut from the solid would have completely superseded the use of pinion wire.

Horological material supplies catalogues at the end of the nineteenth century were offering cut pinions for both clocks and watches complete on their arbors, with just the pivots requiring finishing. A number of suppliers were still offering pinions and pinion wire up until the 1930s. For the clock repairer who did not have pinion-cutting facilities, pinion wire was an acceptable means of replacing a worn pinion on longcase and similar clocks.

3 LATER WHEEL- AND PINION-CUTTING MACHINES

At the end of the nineteenth century, tools being produced were taking on a more sophisticated appearance. Such companies as Holtzapffel of London were producing extremely fine lathes and tools, and it was only natural that this development which took place in the Victorian era should include horological tools. Samuel Smiles, in his *Industrial Biography* on Richard Roberts, a famous engineer of the nineteenth century, states:

> Then Roberts trudged all the way on foot to the great hiding place (London) and first tried Holtzapffels, the famous tool makers, but failing in his application he next went on to Maudsleys and succeeded in getting employment.

This shows the determination of workmen eager to acquire the skills and to receive training in the best possible engineering companies of the day.

Shown below is a general view of an Holtzapffel ornamental turning lathe manufactured in 1870. The overall appearance is extremely pleasing, and this was the type and quality of work being produced at that time. Figure 53 shows a cutting frame in use: whilst this was used mainly for ornamental turning, most probably it could and would have been made for cutting wheel teeth.

German Designs

Joseph Koepfer of Germany had a very comprehensive catalogue showing many types of wheel- and pinion-cutting machines, some completely automatic in operation. The company was started around 1850. Figure 55 shows a machine from their 1927 catalogue: it has worm and wheel dividing, giving a greater number of wheel counts than direct indexing. Note the number of mandrels, cutters and backing plates to support the work offered with the machine. There is a facility for tilting the cutter head to enable bevel gears to be cut. A similar machine is also shown (*see* page 33), the conventional direct indexing model 'K'. It cost £38 10s whilst the worm dividing machine shown cost £50 when new – not cheap by any means. To give some indication

Fig. 52 A Holtzapffel ornamental turning lathe.

Fig. 53 Cutting a wheel on a Holtzapffel lathe.

Jos. Koepfer & Söhne
G. m. b. H.
Furtwangen (Baden)

Gegründet 1867 – Telegramme: Koepfer Söhne Furtwangen – Telefon No. 33
Code 5th Edition A B C – Mosse Code

Spezial-Fabrik
für moderne Präzisions-
Zahn-Fräs-Maschinen / Fräser
für die Fabrikation von
Uhr-, Lauf-, Zählwerken, Gas-, Wassermessern, Druckmessern,
Telegrafen-, Kinoapparaten, Sprech-, Schreib-, Rechenmaschinen
u. s. w.
Massenfabrikation
von
Zahnrädern – Trieben – Schnecken – Zahnstangen

1927

Fig. 54 Jos. Koepfer & Söhne catalogue.

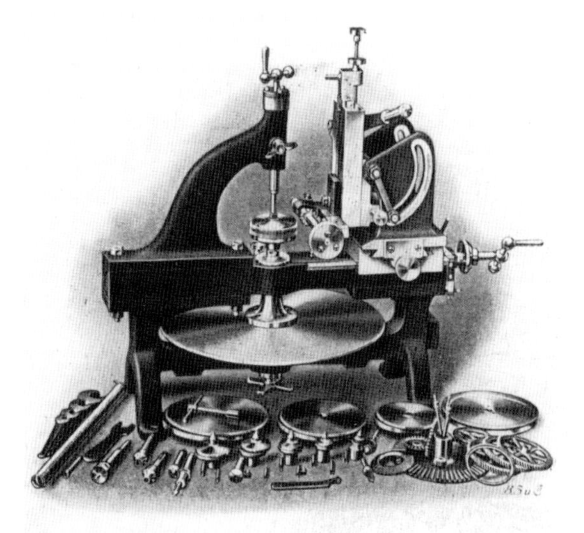

Fig. 55 Koepfer wheel-cutting machine shown in their catalogue 1927.

ABOVE: *Fig. 56 Cutting a wheel.*

LEFT: *Fig. 58 The heavy construction of the Model 'K'.*

TOP LEFT: *Fig. 57 Model 'K' wheel-cutting machine.*

Fig. 59 Automatic Koepfer wheel- and pinion-cutting machine.

Fig. 60 Close-up of the work-head.

of the overall size of the machine, the division plate is 16in (40.5cm) in diameter. The machine was constructed of cast iron and steel, and was extremely sturdy and rigid.

As mentioned previously, Koepfer also supplied automatic wheel- and pinion-cutting machines. Figure 59 is an actual machine still in use at William Haycock's workshop in Ashbourne, Derbyshire. The cutting of the wheel or pinion is by the gashing process, and the indexing of the blank is entirely automatic. Figure 60 shows a close-up of the cutter head.

The carriage or slide for the work holding is traversed by a worm drive whilst cutting takes place. This is a controlled, slow speed. The return feed is rapid, by a skew gear when the automatic clutch operates. Indexing is carried out by a segment of a gear operating a rack; the actual wheel count is achieved by two ratchet wheels. There is a complete set of ratchet wheels supplied with each machine – that is, one count for each wheel that is required to be cut. A number of wheel blanks could be mounted on a single arbor, thus enabling a small batch to be machined at one setting.

Swiss Designs

A very versatile Swiss wheel- and pinion-cutting machine is shown in Figures 61 and 62, manufactured *c.* 1880 by Maxthurm, Geneva. It has a number of interesting features: notched index plates are used, there is provision for helical gears

Fig. 61 Maxthurm wheel- and pinion-cutting machine c. *1880.*

Fig. 62 Cutting a barrel.

to be cut, and the head can be set over to cut bevel gears. An attachment facilitates the cutting of racks. The machine is very rigid in construction and quite compact, being bench mounted, and is still in daily use today. Whether many more machines were made at the time is not known, but none appears to have been advertised in early horological tool catalogues.

American Designs

With the American watch industry growing in the mid-nineteenth century, there was a requirement for precision lathes and wheel- and pinion-cutting machines. A number of catalogues offered these items, one of the earliest noted being F. W. Gesswein. The company was situated in New York, and the catalogue is dated 1883. It is extremely comprehensive, with 140 pages covering all types of hand tools, lathes, and the wheel- and pinion-cutting machines shown. Also in the illustration is an automatic pinion leaf polisher, an automatic balance staff, and a pivot-finishing machine.

One of the most important companies producing small precision machines was the American Watch Tool Company, born from Ambrose Whitcomb of Waltham, Mass. joining forces with the previous firm of J. E. Whitcomb & Company in 1876. This was to be a successful partnership for nearly forty years, producing all types of precision machines for the thriving watchmaking industry; one of their most successful designs was the Webster-Whitcomb, or W. W. watchmaker and instrument lathe. The design of the bed cross-section was extremely rigid, and was far more substantial than the Swiss or German lathes available at the time. The W. W. lathe revolutionized the small lathe market and is still popular today, most watch- and clockmakers having a preference for this design of lathe with its extra rigidity.

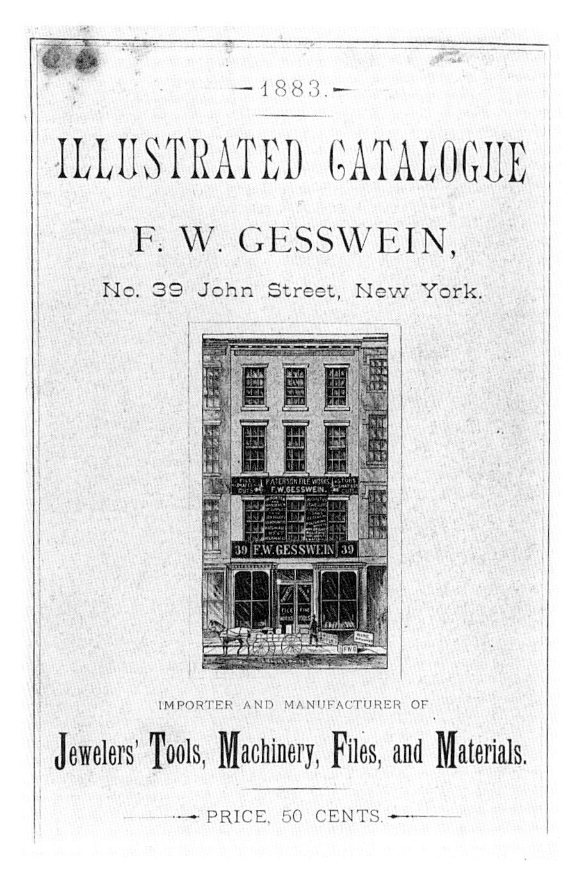

Fig. 63 F. W. Gesswein catalogue 1883.

Fig. 64 Machines illustrated in the Gesswein catalogue.

As with most companies who produced a range of items, a catalogue was essential to promote the range of tools and machines, and the American Watch Tool Co. was no exception. Fortunately some of their catalogues have survived, and a copy of the 1890 edition has been reproduced by the Ken Roberts Publishing Company for the Adams Brown Company. This is an excellent publication of some fifty pages, many illustrated with engravings of lathes, accessories and wheel- and pinion-cutting machines.[14]

Reproduced here are pages showing a small milling machine: this can be used for both wheel and pinion cutting and general light milling work. There is lever feed with rack and pinion to

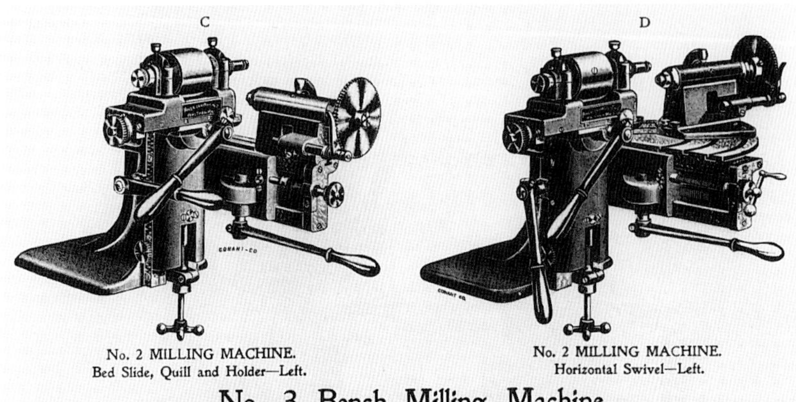

No. 3 Bench Milling Machine.

The work is held in a No. 3 chuck, on centers, or in a vise. Both screw and lever feeds are provided for each slide, and there are adjustable stops for each direction. The cutter slide can be set to any angle from 0 to 35 degrees. The following interchangeable slides and attachments are made:—

1. Bed slide, A, B and C, which takes Quill Holder carrying a quill fitted with a No. 3 chuck, index and pawl, also a Tailstock. We also make the following attachments which interchange with the bed slide: Vise, Spiralling attachment for making twist drills, spiral gears, etc., also a Dividing Engine A and B by which

Fig. 65 American Watch Tool Co. No. 3 bench milling machine – from 1890 catalogue.

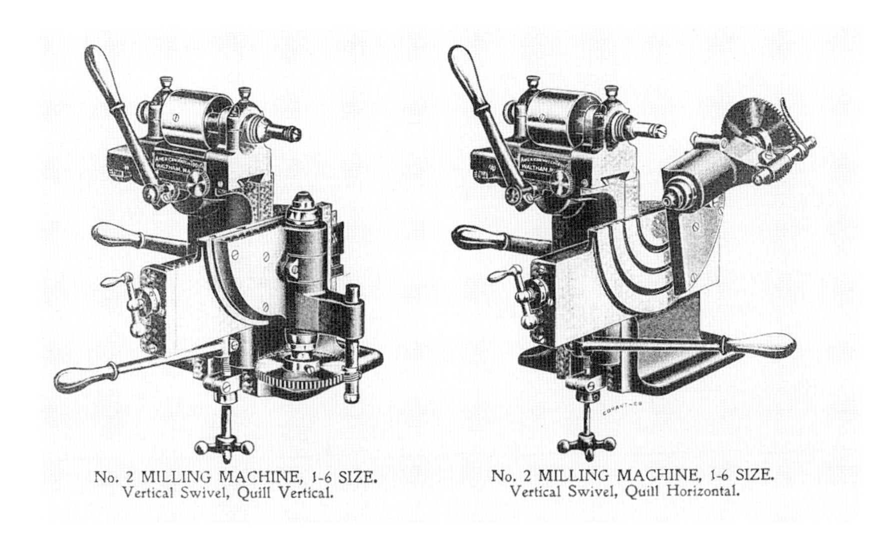

Fig. 66 American Watch Tool Co. No. 2 bench milling machine – from 1890 catalogue.

Fig. 67 Actual wheel-cutting machine.

Fig. 68 Name plate from American Watch Tool Company machine.

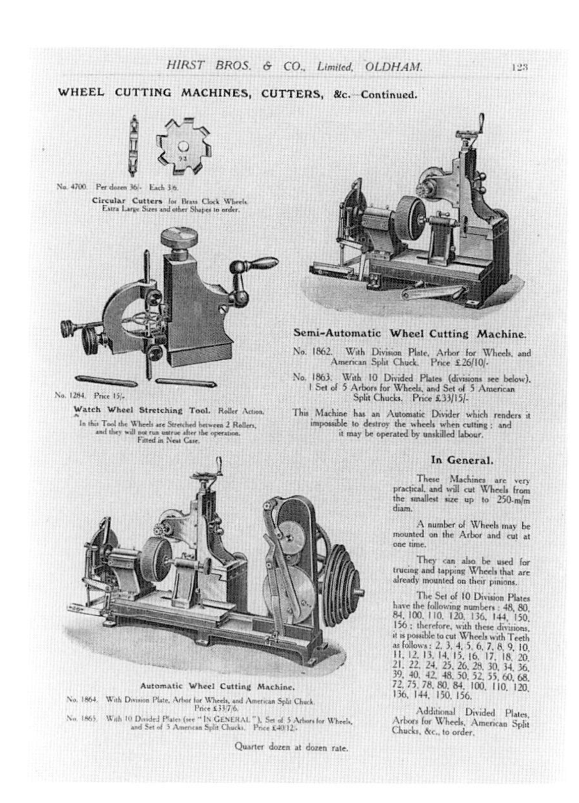

HIRST BROS. & CO., Limited, OLDHAM. 123

WHEEL CUTTING MACHINES, CUTTERS, &c.—Continued.

No. 4700. Per dozen 36/- Each 3/6.

Circular Cutters for Brass Clock Wheels. Extra Large Sizes and other Shapes to order.

No. 1284. Price 15/-

Watch Wheel Stretching Tool. Roller Action. In this Tool the Wheels are Stretched between 2 Rollers, and they will not run untrue after the operation. Fitted in Neat Case.

Semi-Automatic Wheel Cutting Machine.

No. 1862. With Division Plate, Arbor for Wheels, and American Split Chuck. Price £26/10/-

No. 1863. With 10 Divided Plates (divisions see below), 1 Set of 5 Arbors for Wheels, and Set of 5 American Split Chucks. Price £33/15/-

This Machine has an Automatic Divider which renders it impossible to destroy the wheels when cutting; and it may be operated by unskilled labour.

In General.

These Machines are very practical, and will cut Wheels from the smallest size up to 250-m/m diam.

A number of Wheels may be mounted on the Arbor and cut at one time.

They can also be used for trueing and tapping Wheels that are already mounted on their pinions.

The Set of 10 Division Plates have the following numbers: 48, 80, 84, 100, 110, 120, 136, 144, 150, 156; therefore, with these divisions, it is possible to cut Wheels with Teeth as follows: 2, 3, 4, 5, 6, 7, 8, 9, 10, 11, 12, 13, 14, 15, 16, 17, 18, 20, 21, 22, 24, 25, 26, 28, 30, 34, 36, 39, 40, 42, 48, 50, 52, 55, 60, 68, 72, 75, 78, 80, 84, 100, 110, 120, 136, 144, 150, 156.

Additional Divided Plates, Arbors for Wheels, American Split Chucks, &c., to order.

Automatic Wheel Cutting Machine.

No. 1864. With Division Plate, Arbor for Wheels, and American Split Chuck. Price £33/7/6.

No. 1865. With 10 Divided Plates (see "IN GENERAL"), Set of 5 Arbors for Wheels, and Set of 5 American Split Chucks. Price £40/12/-

Quarter dozen at dozen rate.

Fig. 69 Semi-automatic wheel-cutting machine – Hirst Bros. catalogue.

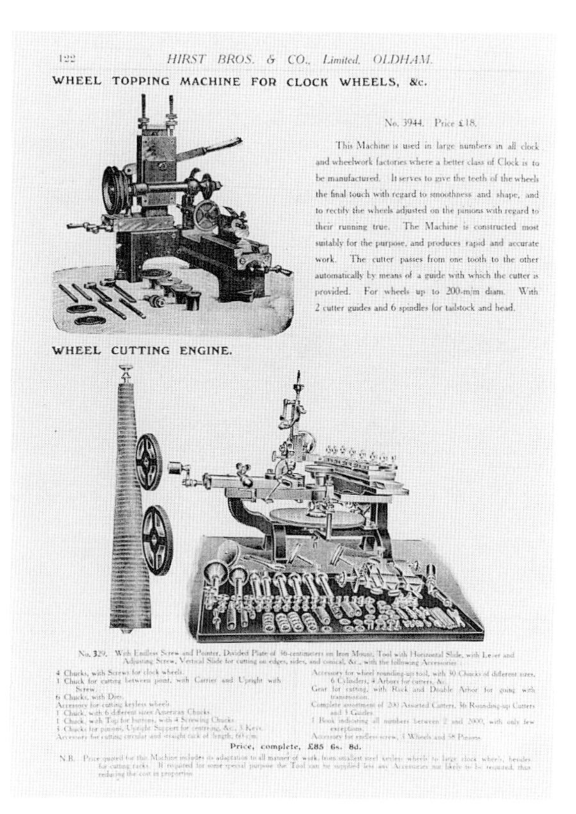

122 HIRST BROS. & CO., Limited, OLDHAM.

WHEEL TOPPING MACHINE FOR CLOCK WHEELS, &c.

No. 3944. Price £18.

This Machine is used in large numbers in all clock and wheelwork factories where a better class of Clock is to be manufactured. It serves to give the teeth of the wheels the final touch with regard to smoothness and shape, and to rectify the wheels adjusted on the pinions with regard to their running true. The Machine is constructed most suitably for the purpose, and produces rapid and accurate work. The cutter passes from one tooth to the other automatically by means of a guide with which the cutter is provided. For wheels up to 200-m/m diam. With 2 cutter guides and 6 spindles for tailstock and head.

WHEEL CUTTING ENGINE.

No. 329. With Endless Screw and Pointer, Divided Plate of 36-centimeters on Iron Mount, Tool with Horizontal Slide, with Lever and Adjusting Screw, Vertical Slide for cutting on edges, sides, and conical, &c., with the following Accessories:

4 Chucks, with Screws for clock wheels.
1 Chuck for cutting between point, with Carrier and Upright with Screw.
6 Chucks, with Dies.
Accessory for cutting keyless wheels.
1 Chuck, with 6 different sizes American Chucks.
1 Chuck, with Top for buttons, with 4 Screwing Chucks.
3 Chucks for pinions, Upright Support for centring, &c., 3 Keys.
Accessory for cutting circular and straight rack of length, 60 c/m.
Accessory for wheel rounding-up tool, with 30 Chucks of different sizes, 6 Cylinders, 4 Arbors for cutters, &c.
Gear for cutting, with Rack and Double Arbor for going with transmission.
Complete assortment of 200 Assorted Cutters, 36 Rounding-up Cutters and 3 Guides.
1 Book indicating all numbers between 2 and 2000, with only few exceptions.
Accessory for endless screw, 1 Wheels and 58 Pinions.

Price, complete, £85 6s. 8d.

Fig. 70 Wheel-cutting and wheel-topping machine – Hirst Bros. catalogue.

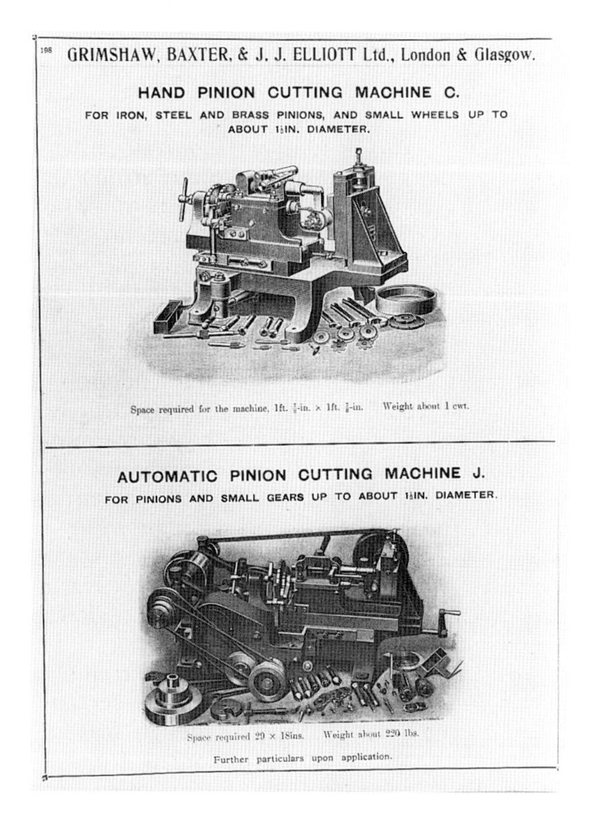

198 GRIMSHAW, BAXTER, & J. J. ELLIOTT Ltd., London & Glasgow.

HAND PINION CUTTING MACHINE C.

FOR IRON, STEEL AND BRASS PINIONS, AND SMALL WHEELS UP TO ABOUT 1½IN. DIAMETER.

Space required for the machine, 1ft. ¾-in. × 1ft. ¾-in. Weight about 1 cwt.

AUTOMATIC PINION CUTTING MACHINE J.

FOR PINIONS AND SMALL GEARS UP TO ABOUT 1½IN. DIAMETER.

Space required 29 × 18ins. Weight about 220 lbs.

Further particulars upon application.

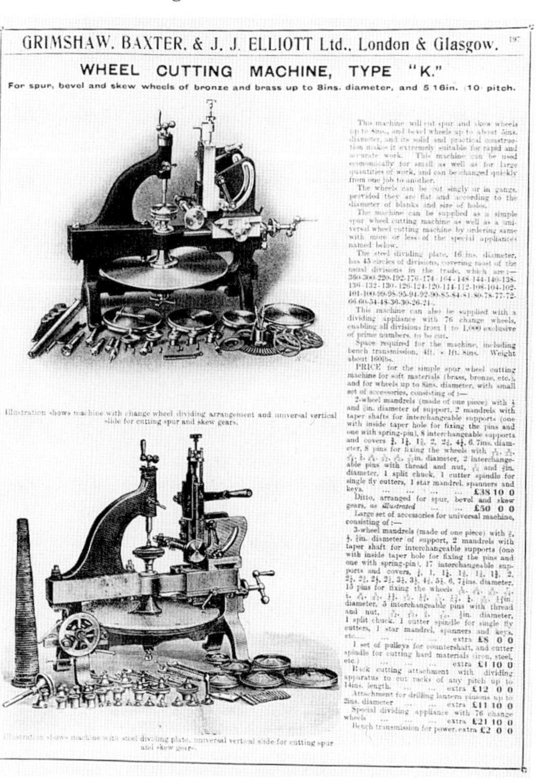

GRIMSHAW, BAXTER, & J. J. ELLIOTT Ltd., London & Glasgow. 197

WHEEL CUTTING MACHINE, TYPE "K."

For spur, bevel and skew wheels of bronze and brass up to 8ins. diameter, and 5/16in. (10) pitch.

This machine will cut spur and skew wheels up to 8ins., and bevel wheels up to about 5ins. diameter, and its solid and practical construction makes it extremely suitable for rapid and accurate work. This machine can be used economically for small as well as for large quantities of work, and can be changed quickly from one job to another.

The wheels can be cut singly or in gangs, provided they are flat and according to the diameter of blanks and size of holes.

The machine can be supplied as a simple spur wheel cutting machine as well as a universal wheel cutting machine by ordering same with more or less of the special appliances named below.

The steel dividing plate, 16 ins. diameter, has 45 circles of divisions, covering most of the usual divisions in the trade, which are:—360-300-220-192-176-174-164-148-144-140-138-136-132-130-126-124-120-114-112-108-104-102-101-100-99-98-95-94-92-90-85-84-81-80-78-77-72-66-60-54-48-36-30-26-24.

This machine can also be supplied with a dividing appliance with 76 change wheels, enabling all divisions from 1 to 1,000 exclusive of prime numbers, to be cut.

Space required for the machine, including bench transmission, 4ft. × 1ft. 8ins. Weight about 160lbs.

PRICE for the simple spur wheel cutting machine for soft materials (brass, bronze, etc.), and for wheels up to 8ins. diameter, with small set of accessories, consisting of:—

2-wheel mandrels (made of one piece) with [illegible] and [illegible]in. diameter of support, 2 mandrels with taper shafts for interchangeable supports (one with inside taper hole for fixing the pins and one with spring-pin), 8 interchangeable supports and covers ¾, 1¼, 1[illegible], 2, 2½, 4½, 6, 7ins. diameter, 8 pins for fixing the wheels with [illegible]in. diameter, 2 interchangeable pins with thread and nut, [illegible] and [illegible]in. diameter, 1 split chuck, 1 cutter spindle for single fly cutters, 1 star mandrel, spanners and keys. £38 10 0

Ditto, arranged for spur, bevel and skew gears, *as illustrated* £50 0 0

Large set of accessories for universal machine, consisting of:—

3-wheel mandrels (made of one piece) with [illegible]in. diameter of support, 2 mandrels with taper shaft for interchangeable supports (one with inside taper hole for fixing the pins and one with spring-pin), 17 interchangeable supports and covers, [illegible], 1, 1¼, [illegible], 2, [illegible], 3½, [illegible], 4½, 5½, 6, 7½ins. diameter, 15 pins for fixing the wheels [illegible]in. diameter, 5 interchangeable pins with thread and nut, [illegible]in. diameter, 1 split chuck, 1 cutter spindle for single fly cutters, 1 star mandrel, spanners and keys, etc.... ... ... extra £8 0 0

1 set of pulleys for countershaft, and cutter spindle for cutting hard materials (iron, steel, etc.) ... ... extra £1 10 0

Rack cutting attachment with dividing apparatus to cut racks of any pitch up to 14ins. length. ... ... extra £12 0 0

Attachment for drilling lantern pinions up to 2ins. diameter ... ... extra £11 10 0

Special dividing appliance with 76 change wheels ... ... extra £21 10 0

Bench transmission for power, extra £2 0 0

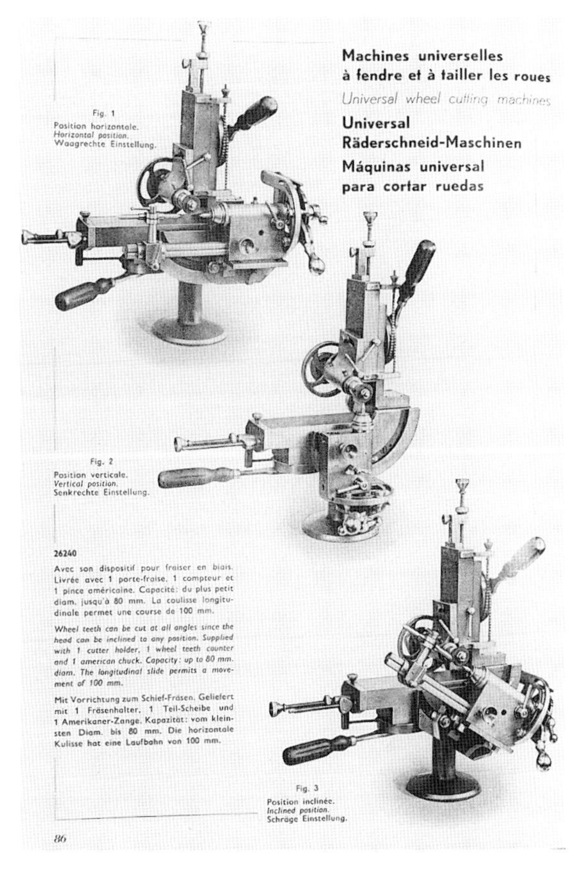

Fig. 73 'Star' catalogue showing wheel and pinion machine.

Fig. 74 General view showing drive to cutter spindle.

the machine slides or, alternatively, a screw feed can be engaged. Many collets and wheel arbors were supplied with the machine.

EARLY TWENTIETH-CENTURY MACHINES

In the early twentieth century such companies as Grimshaw & Baxter, Hirst Bros and Henri Picard & Frère were major suppliers to the horological trade, their catalogues being full of accessories and materials, and all types of machine for cutting wheels and pinions. An interesting wheel-cutting machine was manufactured by Star, a Swiss company, one of the few still remaining who manufacture watchmakers' lathes. This is a small machine but quite strong in construction. It appears to have been available seventy or eighty years ago, as it appeared in early tool catalogues and was still listed in the Star catalogue of 1992. The spindle is of quite light construction, similar to a watchmaker's lathe. Whilst it is possible to cut pinions, the machine is more suitable for wheel cutting. The work-holding head can be swung through 90 degrees to enable wheels to be cut in both planes – and, of course, bevel and crown wheels can be accommodated. When the machine was purchased, a small pulley was fitted to the spindle, but a heavy duty flywheel has now been fitted that gives a much smoother finish to the work when cutting takes place. This will be covered in a later chapter.

The operation of the slides is extremely smooth due to the chain drive, which is a section of small-linked pin chain. This design feature would appear to have been copied from a straight-line ornamental chuck, as shown in *Holtzapffel's Turning & Mechanical Manipulation*

FACING PAGE, BELOW LEFT: *Fig. 71 Hand and automatic pinion-cutting machine – Grimshaw, Baxter & J. J. Elliott catalogue.*

FACING PAGE, BELOW RIGHT: *Fig. 72 Wheel-cutting machines – Grimshaw, Baxter & J. J. Elliott catalogue.*

Fig. 75 Wheel cutting on the 'Star'.

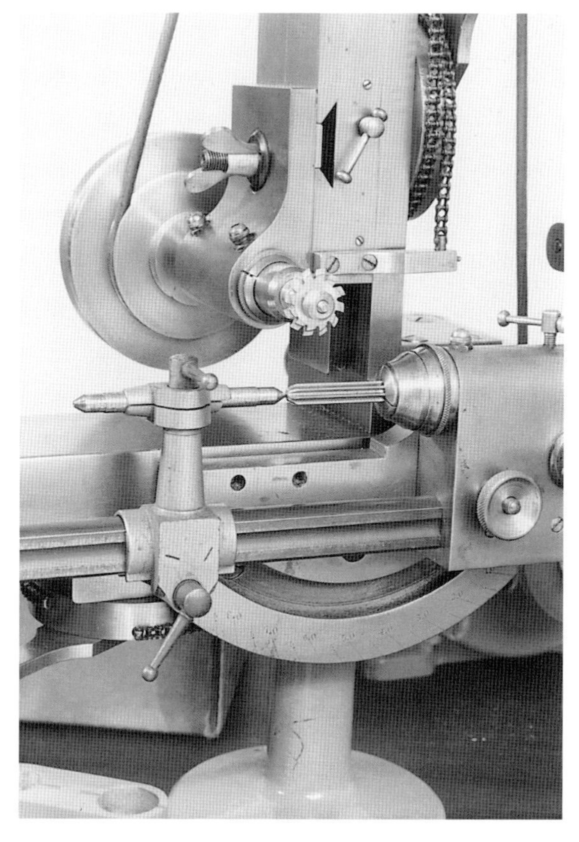

Fig. 76 Pinion cutting on the 'Star' – note the position of the work-holding slide.

Vol 5. Stops are provided for all the slides, and in place of the normal division plate, notched plates are used. This is a quicker method of indexing. There is also a device to enable the rapid counting of intermediate divisions. This is based on an invention by T. J. Ashton, his *Automatic Counting Apparatus*, used mainly by J. H. Evans on the ornamental lathes he produced. This device was discussed in Evans' book *Ornamental Turning*, dated 1886.

Only a small number of wheel-cutting machines were manufactured after the 1920s, as requirements for this type of work had diminished. Very few clocks were being produced, and the wheels and pinions required for repair or restoration were cut by the remaining specialist out-workers. Glimpsing through the trade magazines of the day, in *The Practical Clock & Watchmaker* of 1928 and the *Horological Journal* of 1936, there is only one company offering the service of wheel cutting. That company was Biddle & Mumford, who were still working in Clerkenwell until 1986.

Another excellent machine is the one manufactured by Dixi. This was advertised in catalogues around 1920. Again, direct indexing by the use of notched plates gives a rapid machine operation, although the current owner has designed and manufactured a dividing head that can easily be fitted to the machine. This method will give an infinite number of counts, prime numbers being possible. Also built into the attachment is a micro device that enables compound indexing to be carried out. Another feature of the machine is that the work-holding quill can be removed and fitted into another machine: for example, if a wheel blank is machined using the quill, this can be transferred direct to the Dixi and wheel cutting commenced, knowing that extreme accuracy is ensured.

MID-TWENTIETH-CENTURY MACHINES

The resurgence of amateur clockmaking in the late 1950s was due mainly to the writings of Claude B. Reeve of Hastings. Claude was a prolific builder: he completed some fifty clocks over as many years, and his writings in the *Model Engineer* magazine had many enthusiastic followers. This resurgence developed through the sixties and into the seventies. When John Wilding commenced his instructional articles in the *Horological Journal*, he simplified the art of clockmaking to enable many who had not built a clock before to be completely successful.

There then developed a demand for clockmaking tools once again. Chronos of Northampton produced a wheel-cutting engine: this was not constructed from castings, but fabricated to enable the tool to be manufactured at a competitive price. It proved so successful that a pinion mill was also produced. This was based on the design of a small horizontal mill and fabricated in mild steel; simple vee-notched plates were used for indexing.

A number of other machines were produced,

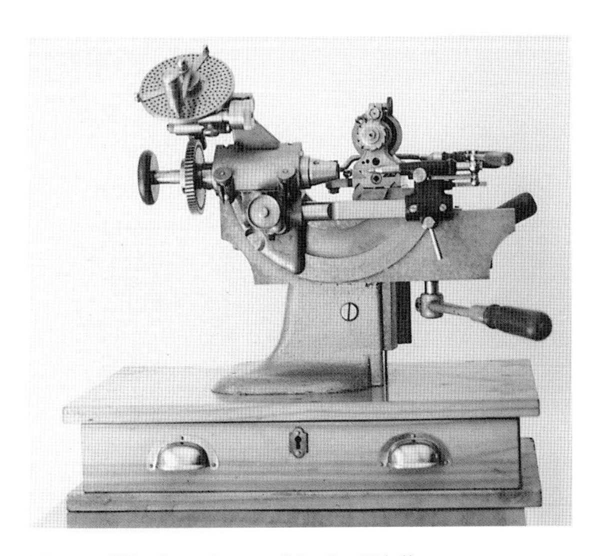

Fig. 77 Wheel-cutting machine by 'Dixi'.

Fig. 78 'Dixi' machine showing addition of a worm and wheel dividing head.

Fig. 79 Chronos wheel-cutting engine.

Fig. 80 Chronos pinion mill.

Fig. 81 Corry wheel-cutting engine.

Fig. 82 Wheel-cutting machine by John Wardle.

Fig. 83 Wheel engine described in a constructional article – Horological Journal.

Fig. 84 Close-up of wheel-cutting spindle.

Fig. 85 Wheel-cutting machine with automatic indexing.

Fig. 86 Close-up showing gallows for checking depthing.

Fig. 87 Pinion mill.

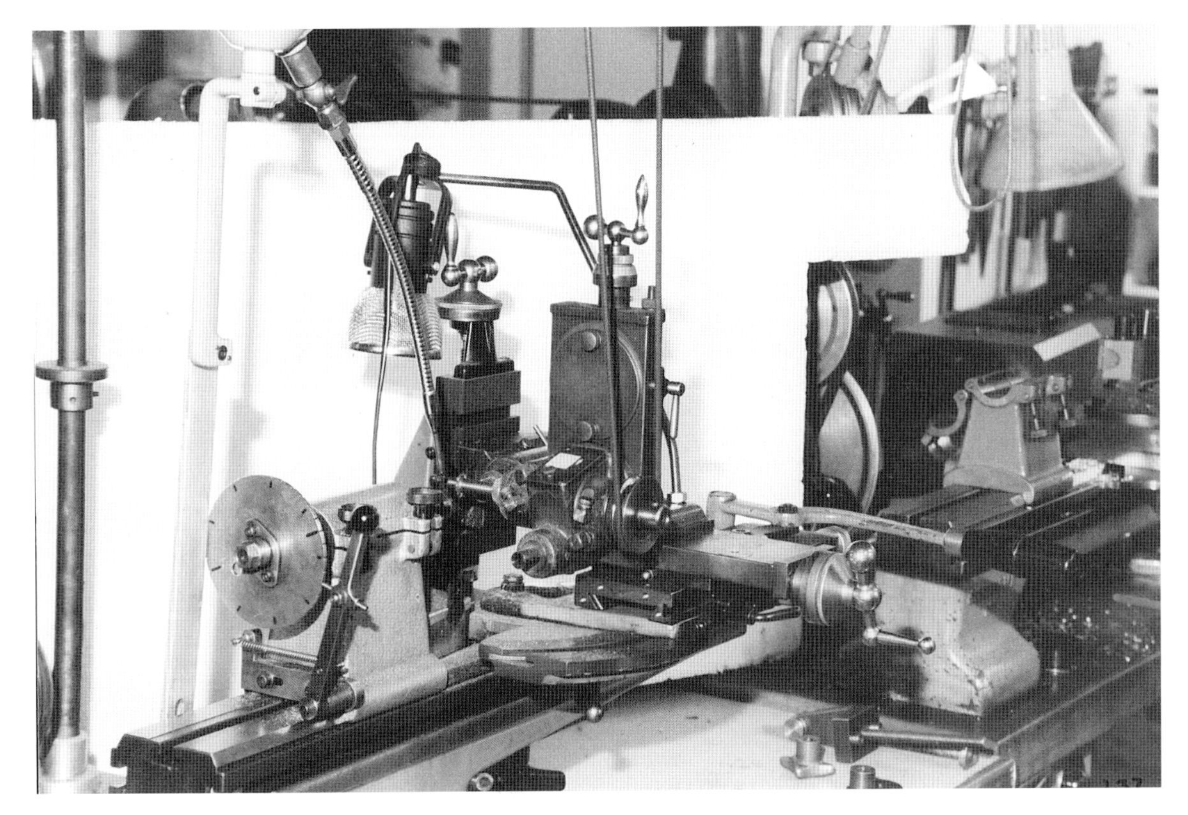

Fig. 88 Schaublin 102 adapted for wheel and pinion cutting.

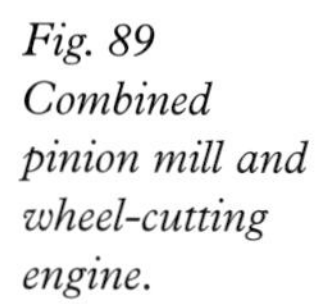

Fig. 89 Combined pinion mill and wheel-cutting engine.

Fig. 90 Combination machine showing dividing head and accessories.

namely the Boyle and Corry. The latter was a very robust machine, and used the Myford vertical slides and an excellent ball-bearing spindle. Another is illustrated here and was manufactured by the late John Wardle of Derby. A small batch was produced in the early 1980s, constructed on traditional lines, with a large brass division plate. John was a skilled clock-maker and manufactured parts for longcase and bracket clocks in large batches. His grandson Steve Bevan is still carrying on the business near Derby.

For those who may still be interested in building their own wheel-cutting machine, an excellent detailed constructional article by G. E. Lloyd-Jones appeared in the *Horological Journal*; the series began in February 1979 and ran for some six issues. The machine shown here was constructed by Arthur McDonald, and as can be seen from the photographs, it is an extremely nice wheel-cutting engine.

An excellent machine for wheel cutting was constructed by Brian Ragget, a clockmaker in the Midlands. It is ingenious in its design, being very quick to operate. It is self-indexing: as soon as the cutter handle is returned, indexing automatically takes place. The rotation of the division plate is powered by a clock spring which is

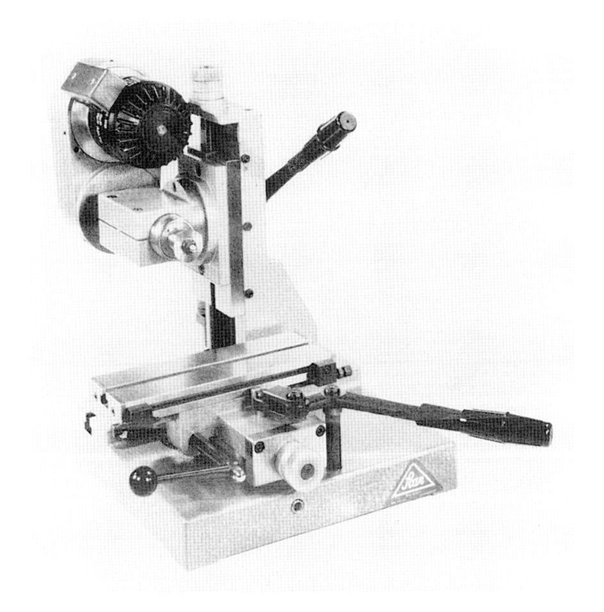

Fig. 91 'Star' wheel and pinion mill.

Fig. 92 'Star' direct indexing dividing head.

wound up a few times before wheel cutting commences. Sector arms mounted on the division plate can be set to determine the actual number of divisions being selected. Note the gallows for depthing whilst the wheel is still mounted in the machine; then, any adjustments can be made before dismounting the wheel.

Also constructed by the same clockmaker is a very robust and compact pinion-cutting machine. It is in constant use in his workshop. Note the large pulley that acts as a fly wheel and speed reducer.

Both amateur and professional watch- and clockmakers design and manufacture their own machines, often to embrace features that will suit their own particular requirement or type of work. Derek Pratt, an excellent watchmaker in Switzerland, has constructed such a machine using Schaublin lathe parts. It is essential that the machine is rigidly constructed when cutting pinions. With the machine shown in Figure 88, it has been possible to cut six-leaf pinions as small as 0.5mm in diameter.

Another fine machine that accommodates both wheel and pinion cutting was made by Dennis Harris of Oundle, Northants; it was his solution when finding out that two separate machines were traditionally used for the two functions concerned. This machine accomplishes both operations and is based on the castings of the well known Kennet Tool & Cutter Grinder. For clock wheel cutting, the lever at the right-hand side of the bed is used to traverse the blank through the revolving multi-toothed cutter. The speed of this operation is decided by the operator according to how fast the cutter revolves.

When pinions are to be cut, the lever referred to is pulled out of engagement with the bed (but still retained by the bed). A locking pin, seen to the left of the centring wheel in front of the bed, is screwed in, engaging the lead screw, and the feed is put on by means of the handle.

There are two headstocks, one for cutting pinions and the other wheel teeth, and each has its own set of division plates. The reduction ratio of the dividing gear is 60:1 for wheel cutting, and for pinions the detent engages directly with the divisions in the edge of the plates. Both headstocks were made to accept Myford 'Series 7' standard collets so that the wheel or pinion blanks can be transferred from the lathe to the machine with no loss of concentricity.

PRESENT-DAY CUTTING MACHINES

At the present time there appear to be very few wheel- and pinion-cutting machines currently available. A machine by the Star Watch Lathe Company was shown in their catalogue until recently, but this has now been discontinued. It was an extremely compact machine and was the successor to the Star machine described previously. A view of the direct indexing dividing head is also shown. This has the means of rapid counting when indexing takes place.

Another Swiss machine is shown in Figure 93: this is by Bergeon and is currently available. The machine is extremely well finished, and the cutting head can be tilted, which is a very useful feature.

The importance of rigidity when wheel cutting has been mentioned, and this is extremely important when pinion cutting. The Lindow machine illustrated here is designed with this in mind. Manufactured in Pennsylvania, USA, it is of a fabricated construction, using large-section material, and is capable of cutting both wheels and pinions. Notched plates are provided or, alternatively, division plates can be fitted. The headstock and cutter spindle are machined to accept 8mm collets or arbors, and the motor drive is by steel pulleys, either with polyurethane belts for wheel cutting or miniature vee-belts when pinion cutting. When cutting wheels, a lever feed is employed and production is extremely rapid. For pinion cutting, a screw feed attachment is fitted.

Although possibly outside the scope of the typical amateur clockmaker, included are photographs of an automatic wheel- and pinion-cutting machine. This is a SAFAG Model 11A, of

Fig. 93 Bergeon wheel-cutting machine.

Fig. 94 Lindow machine.

Fig. 95 Lindow machine with screw feed fitted for pinion cutting.

Fig. 96 Safag IIA Swiss automatic pinion-cutting machine.

Swiss manufacture. The cutter traverses along the work, and then the blank is automatically indexed. As with all the gear-cutting machines being covered in this chapter, the method of machining or cutting the tooth is a 'gashing' process, and not a generated form. This type of machine is more suited for batch work, as the setting up involved for a 'one-off' is not really economic.

Another automatic machine is illustrated from the Petermann catalogue. This is a very versatile machine for wheels and pinions and is capable of being fitted with a magazine feed. The front cover shows a general view of the machine, and a further view shows the magazine feed used when cutting batches of pinions. There is also a facility for tilting the cutter head.

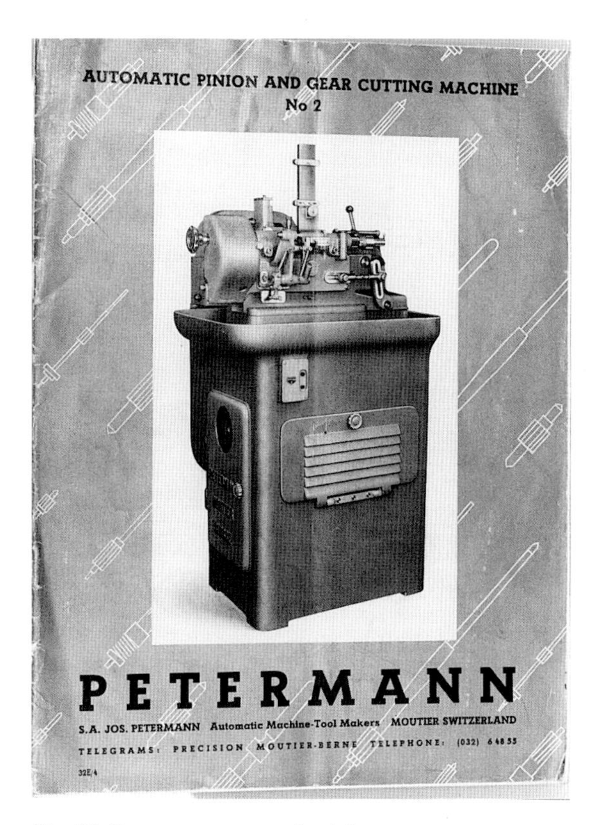

Fig. 97 Petermann automatic pinion cutter.

THE TOPPING TOOL

Another tool that is quite interesting and should be covered is the topping, or rounding-up tool. This is a useful small machine for the workshop, used mainly on watch wheels for either correcting the form of the teeth or reducing the full form of the teeth to improve the depthing in a train of wheels.

Larger machines were manufactured and these are most useful for correcting wheel teeth in carriage clocks and other small clocks. They will also correct wheels that are slightly out of round. These machines were still shown in horological tool catalogues up to the 1930s.

The tool is normally mounted on a mahogany base, under which is fitted a drawer containing the various arbors for mounting the wheels, together with a range of cutters to suit the different sizes of wheel teeth. The cutters are serrated for approximately half their circumference. The remainder of the circle acts as a guide, and its section is similar to the shape of the cutting portion. This guide is adjustable to form a helix around the periphery of the cutter. When the wheel to be topped or rounded is mounted between the runners, the cutter is presented to a tooth and the guide adjusted to accommodate the pitch of one tooth. The cutter is then rotated by the large handwheel at the rear of the tool. The cutting edge, which takes up some 180 degrees of the circumference, profiles the tooth, and the guide positions the wheel

Fig. 98 Close-up of the work-head.

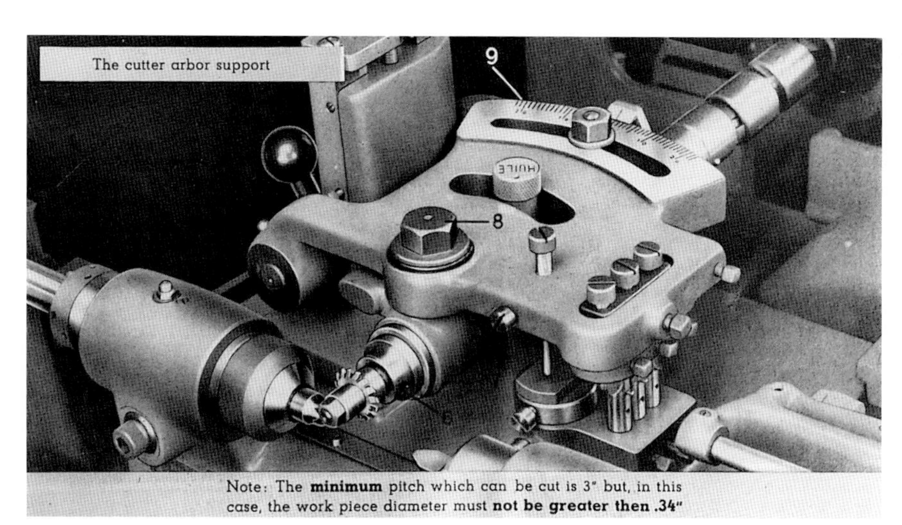

Fig. 99 The cutter head showing angular setting.

Fig. 100 Swiss wheel-topping or rounding-up tool with cutters.

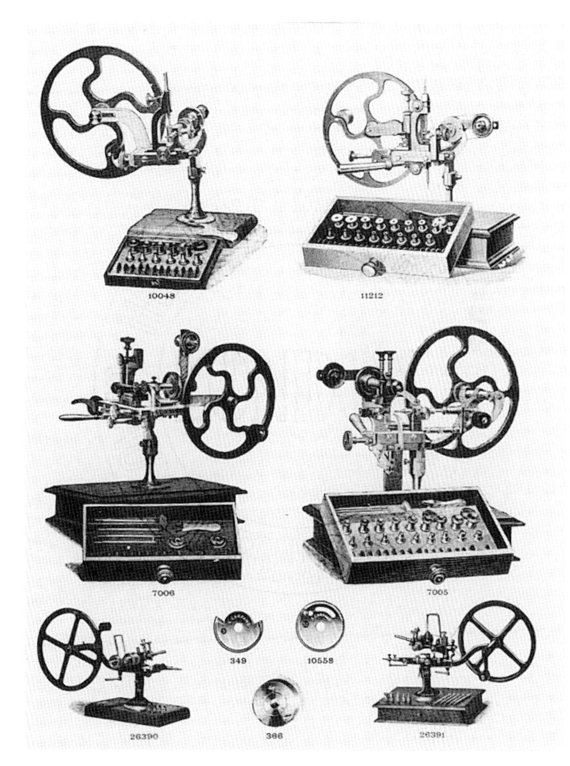

Fig. 101 Range of wheel-topping tools from early twentieth-century tool catalogue.

ready for the next cut. This is, of course, a continuous process.

Figure 104 illustrates a heavy-duty tool from the Koepfer catalogue. This will accommodate wheels up to 75mm (3in). A range of accessories is also shown.

The wheel-topping machine shown in Figure 105 is by Morat and is similar to the Koepfer machine. Both were of German manufacture and constructed in cast iron, whereas the Swiss machines were generally brass. The company of Morat was established in 1840, although the machine from its design could have been manufactured some seventy-five years before this. There is a large range of cutters in a two-drawer

Fig. 102 Topping or rounding-up cutters.

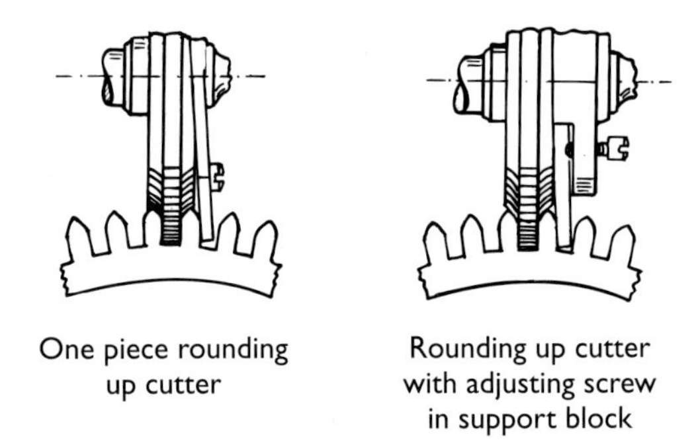

Fig. 103 Topping cutter in action.

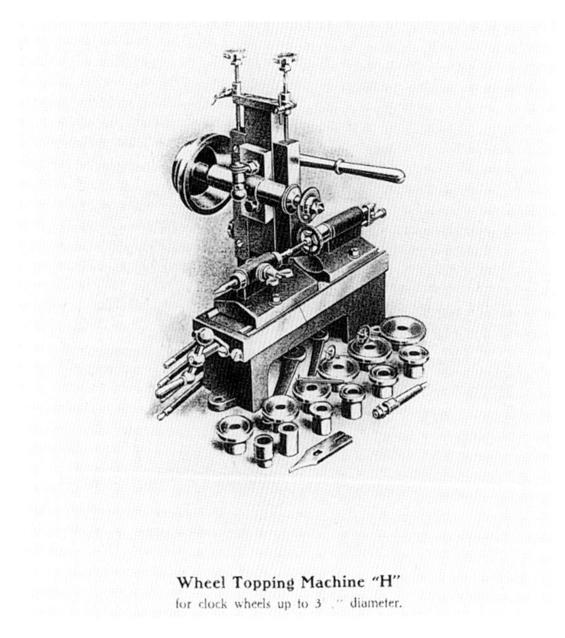

Fig. 104 Koepfer heavy-duty wheel-topping machine.

Fig. 105 Morat heavy-duty wheel-topping machine.

cabinet, and all the larger cutters have faces that are radially ground. This method of grinding did not appear until around 1860, which again dates the machine. The largest of the cutters could accommodate a 2.0 module wheel.

Very little historical information is available for either Koepfer or Morat due to records being destroyed in two world wars. Both companies produced many types of machines for the Black Forest clock industry, and the catalogue that Koepfer produced was quite extensive. The companies are still in existence today, manufacturing components for the automotive and engineering industries. The current Morat catalogue gives details of a special-purpose machine developed for the Black Forest and described as a 'trundle drill' for producing lantern pinions. This was described in *Dinglers Polytechnical Journal* of 1840, but no mention is made of wheel or pinion machines.

Watchmakers' lathes were sometimes supplied with a topping tool as a separate accessory, which could be mounted on the lathe cross-slide or tailstock by direct fixing with a small bolt, or, more often, fitted in the tee-rest holder. An illustration from an early tool catalogue shows two mountings for a topping or rounding-up attachment.

Fig. 106 Morat machine showing cutter spindle.

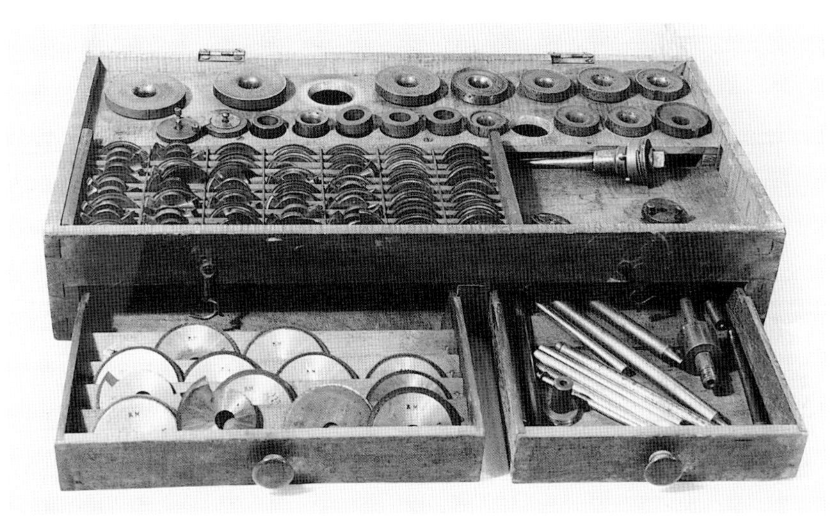

Fig. 107 Extensive range of topping cutters for the Morat machine.

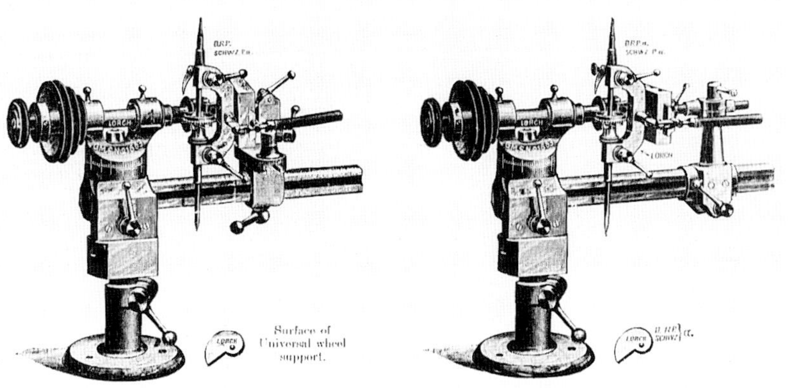

Fig. 108 Topping or rounding-up attachments shown in early Lorch catalogue.

Fig. 109 Topping attachment in fitted box.

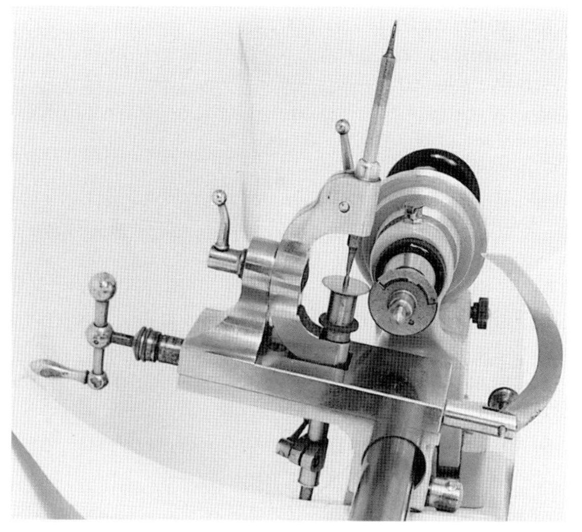

Fig. 110 Lorch design of topping attachment fitted to the lathe bed.

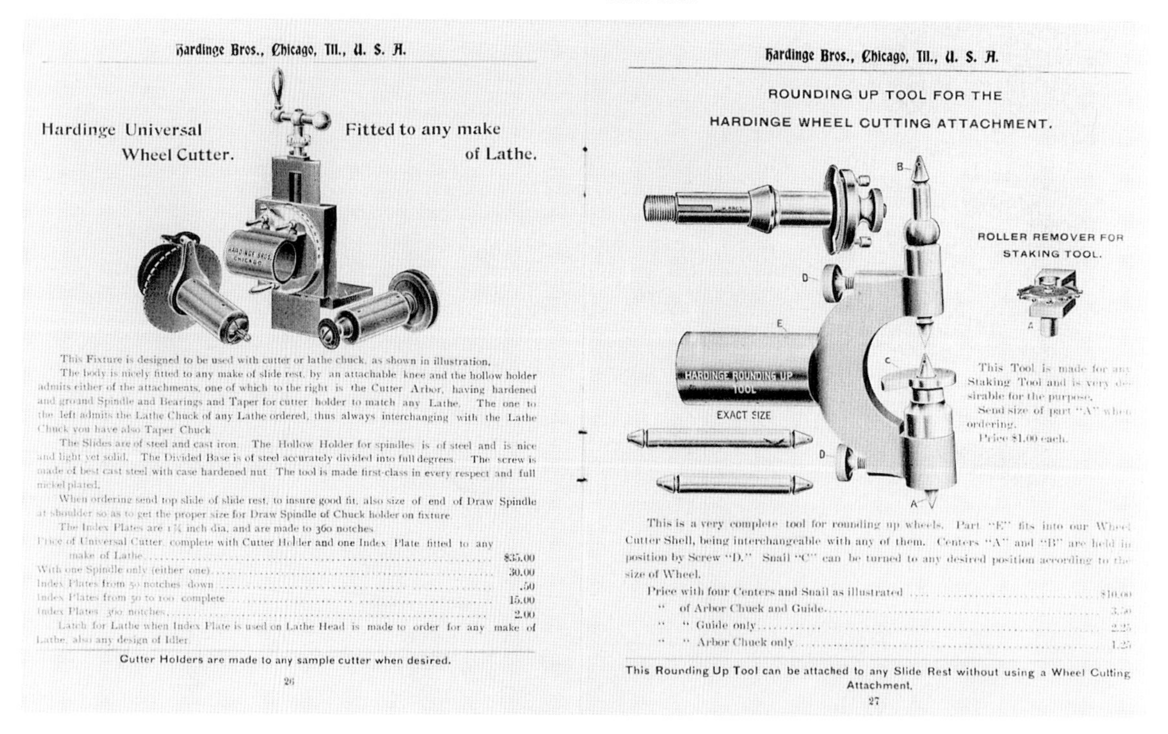

Hardinge Bros., Chicago, Ill., U. S. A.

Hardinge Universal Wheel Cutter.

Fitted to any make of Lathe.

This Fixture is designed to be used with cutter or lathe chuck, as shown in illustration.

The body is nicely fitted to any make of slide rest, by an attachable knee and the hollow holder admits either of the attachments, one of which to the right is the Cutter Arbor, having hardened and ground Spindle and Bearings and Taper for cutter holder to match any Lathe. The one to the left admits the Lathe Chuck of any Lathe ordered, thus always interchanging with the Lathe Chuck you have also Taper Chuck

The Slides are of steel and cast iron. The Hollow Holder for spindles is of steel and is nice and light yet solid. The Divided Base is of steel accurately divided into full degrees. The screw is made of best cast steel with case hardened nut The tool is made first-class in every respect and full nickel plated.

When ordering send top slide of slide rest, to insure good fit, also size of end of Draw Spindle at shoulder so as to get the proper size for Draw Spindle of Chuck holder on fixture.

The Index Plates are 1¾ inch dia, and are made to 360 notches

Price of Universal Cutter, complete with Cutter Holder and one Index Plate fitted to any make of Lathe $35.00
With one Spindle only (either one) 30.00
Index Plates from 50 notches down .... .50
Index Plates from 50 to 100 complete 15.00
Index Plates 360 notches 2.00

Latch for Lathe when Index Plate is used on Lathe Head is made to order for any make of Lathe, also any design of Idler.

Cutter Holders are made to any sample cutter when desired.

26

Hardinge Bros., Chicago, Ill., U. S. A.

ROUNDING UP TOOL FOR THE HARDINGE WHEEL CUTTING ATTACHMENT.

ROLLER REMOVER FOR STAKING TOOL.

This Tool is made for any Staking Tool and is very desirable for the purpose.

Send size of part "A" when ordering.

Price $1.00 each.

HARDINGE ROUNDING UP TOOL

EXACT SIZE

This is a very complete tool for rounding up wheels. Part "E" fits into our Wheel Cutter Shell, being interchangeable with any of them. Centers "A" and "B" are held in position by Screw "D." Snail "C" can be turned to any desired position according to the size of Wheel.

Price with four Centers and Snail as illustrated $10.00
" of Arbor Chuck and Guide 3.50
" " Guide only 2.25
" " Arbor Chuck only 1.25

This Rounding Up Tool can be attached to any Slide Rest without using a Wheel Cutting Attachment.

27

Fig. 111 Hardinge topping tool.

The units were usually supplied in a fitted box with runners and a range of cutters, though occasionally the unit was complete and mounted directly onto the lathe bed itself. The example shown is a Lorch rounding-up tool. It is a beautifully made piece, as were all Lorch lathes, in its own fitted box with runners, cutters and arbors.

A tool from the 1901 Hardinge catalogue is shown in Figure 111. This is an excellent design. The unit fits into the vertical slide of a wheel-cutting attachment, the quills and topping-tool shank being interchangeable.

Topping tools were still being manufactured until quite recently, and they must have been manufactured in large numbers as many are still around today. If one is purchased, it is important that a good range of cutters is available with the machine, as it is almost impossible to find these separately.

THE INGOLD FRAISE

Another tool for correcting wheel teeth was the ingold fraise. Unlike the rounding-up machine, this is an extremely rare tool and it is unlikely that the reader will come across one of these. As can be seen from the photograph, the tool is similar in appearance to a depthing tool. There is a series of steel rollers that are grooved longitudinally, and these rollers fit onto the runners provided. The wheel to be shaped is mounted on one of the runners, whilst the cutter is mounted on the adjacent one. The two are brought into contact with one another, and the fraise is rotated by a bow; this in turn rotates the wheel and impacts an extremely fine finish to the wheel teeth.

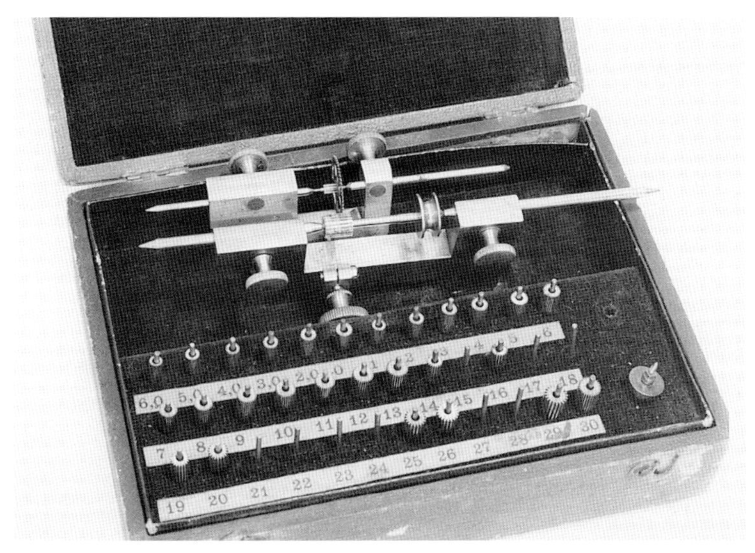

Fig. 112 Ingold fraise tool for shaping wheel teeth.

Fig. 113 Ingold fraise in use with the bow.

4 LATHE ATTACHMENTS FOR WHEEL AND PINION CUTTING

It is unlikely that everyone will be able to obtain a wheel- or pinion-cutting machine, or have the time to construct such an item. Methods will now be described where all this type of work can be carried out on the lathe.

Most workshops will have a medium-sized bench lathe, which, with adaptation, is ideal for the purpose of wheel and pinion cutting; the lathe lends itself most admirably for this purpose, due mainly to its rigidity. With this method, one main advantage over a wheel-cutting machine is that the work can quite often be machined, and then the wheel teeth or pinion leaves cut all at one setting – and this, of course, ensures greater accuracy of concentricity. Also, it is easier to see the work in progress as the cutter is acting on top of the work.

Whilst many of the photographs of set-ups are shown using the Myford Super 7, 3½in lathe, as this is probably the most popular small lathe found in a clockmaker's or model engineer's workshop, the methods apply equally to similar machines.

As early tool catalogues show, many accessories were available for wheel and pinion cutting. A number of illustrations are included here from Grimshaw, Baxter & J. J. Elliott, Hirst Bros. and the Lorch catalogue *c.* 1920. A heavy duty milling spindle is illustrated, in the general view, from Hirst Bros. catalogue; another view is shown giving close-up details of the bevel gear drive. Vertical slides with milling spindles are shown, also cutter frames and dividing plates. An excellent Lorch universal dividing head for mounting directly on to the lathe spindle at the rear of the lathe is shown. Lorch claimed that any prime number could be accommodated. There is a large wheel divided into 500 divisions fitted to the worm wheel, and to assist in viewing the small divisions, a magnifying glass is fitted to the main casting.

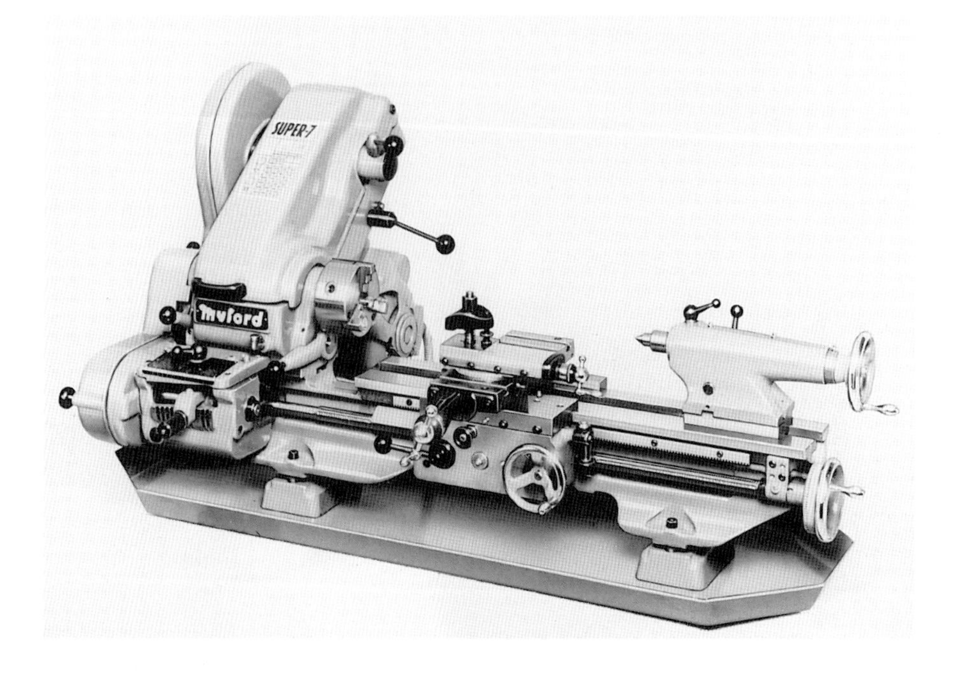

Fig. 114 Myford Super 7, 3½in bench lathe.

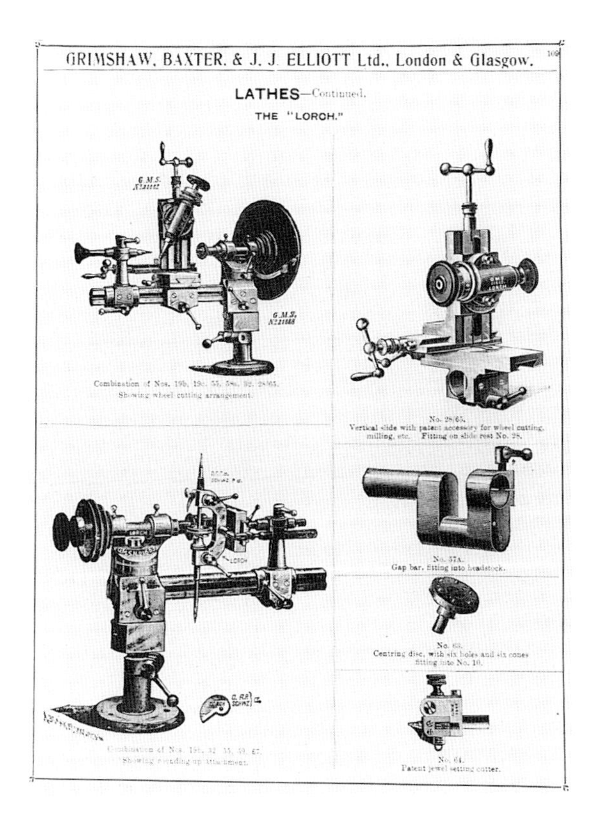

Fig. 115 Grimshaw, Baxter & J. J. Elliott catalogue showing lathe attachments.

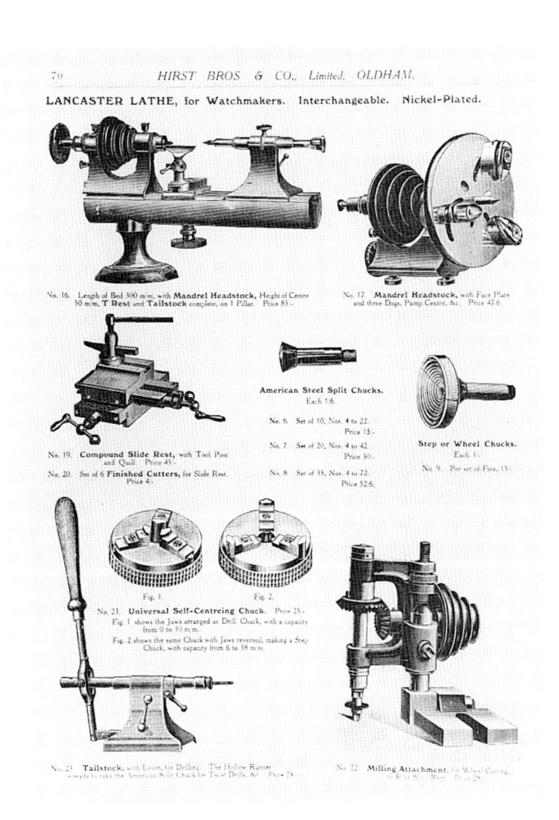

Fig. 117 Hirst Bros. catalogue with various accessories for wheel cutting.

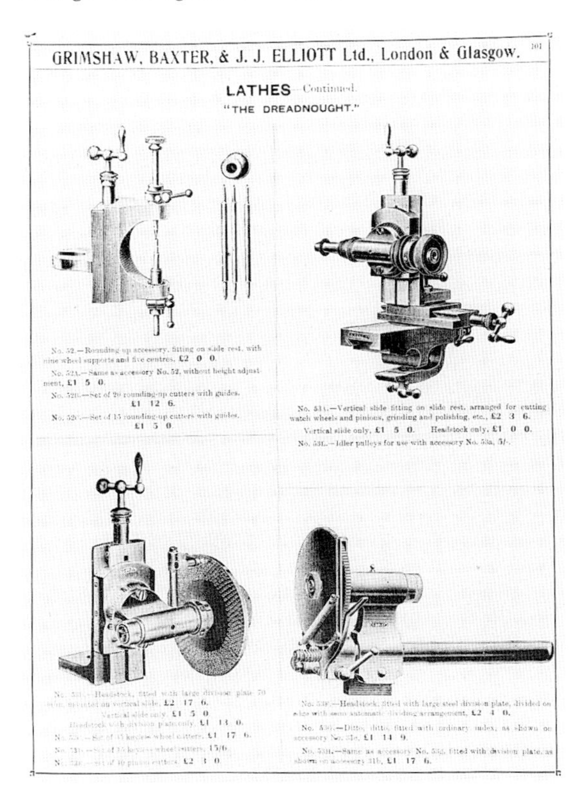

Fig. 116 Grimshaw, Baxter & J. J. Elliott catalogue showing lathe attachments.

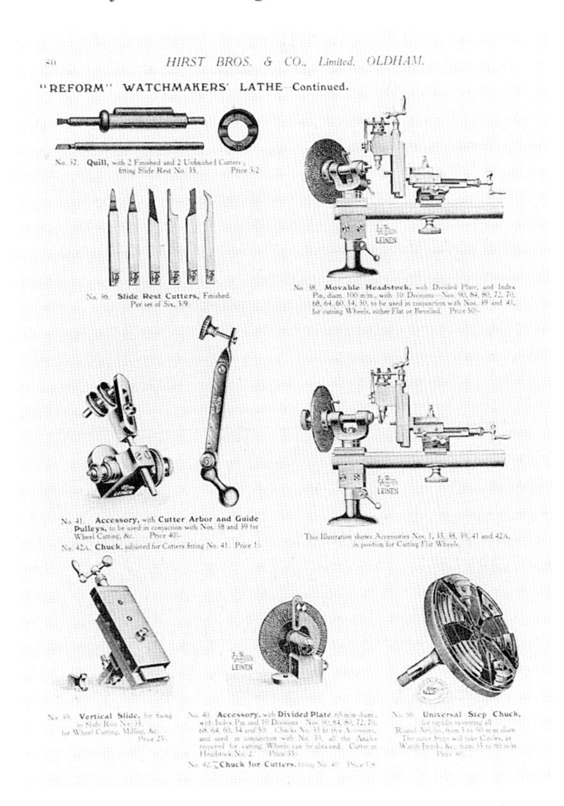

Fig. 118 Hirst Bros. catalogue with various accessories for wheel cutting.

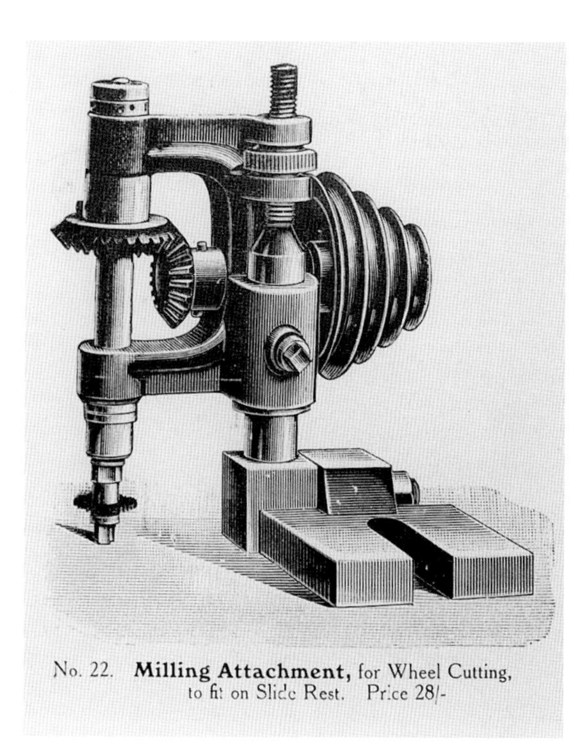

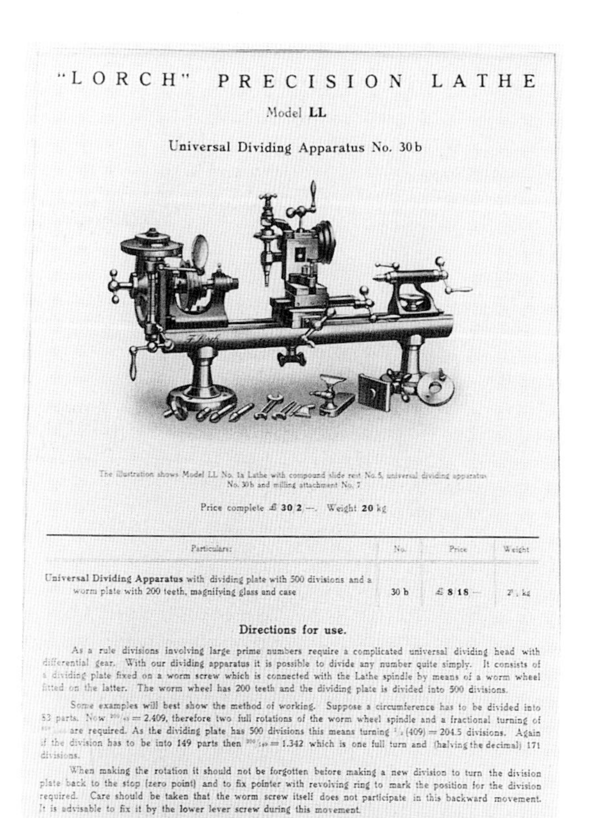

"LORCH" PRECISION LATHE

Model **LL**

Universal Dividing Apparatus No. 30 b

The illustration shows Model LL No. 1a Lathe with compound slide rest No. 5, universal dividing apparatus No. 30 b and milling attachment No. 7

Price complete £ **30**/**2**/—. Weight **20** kg

Particulars:	No.	Price	Weight
Universal Dividing Apparatus with dividing plate with 500 divisions and a worm plate with 200 teeth, magnifying glass and case	30 b	£ **8**/**18**/—	2½ kg

Directions for use.

As a rule divisions involving large prime numbers require a complicated universal dividing head with differential gear. With our dividing apparatus it is possible to divide any number quite simply. It consists of a dividing plate fixed on a worm screw which is connected with the Lathe spindle by means of a worm wheel fitted on the latter. The worm wheel has 200 teeth and the dividing plate is divided into 500 divisions.

Some examples will best show the method of working. Suppose a circumference has to be divided into 83 parts. Now $^{200}/_{83} = 2.409$, therefore two full rotations of the worm wheel spindle and a fractional turning of $^{409}/_{1000}$ are required. As the dividing plate has 500 divisions this means turning ½ (409) = 204.5 divisions. Again if the division has to be into 149 parts then $^{200}/_{149} = 1.342$ which is one full turn and (halving the decimal) 171 divisions.

When making the rotation it should not be forgotten before making a new division to turn the division plate back to the stop (zero point) and to fix pointer with revolving ring to mark the position for the division required. Care should be taken that the worm screw itself does not participate in this backward movement. It is advisable to fix it by the lower lever screw during this movement.

For numbers which can be exactly divided into 200 only full rotations of the worm spindle are required so that the division plate will not be moved.

SET-UPS FOR WHEEL CUTTING

The process of wheel cutting is relatively simple once the necessary tools are to hand, and it is extremely satisfying. A lot of would-be constructors are put off by the thought of gear cutting or wheel cutting (as clockmakers call it) but a lot of the interest in all forms of making and restoring comes from tackling the many variations in machining operations. If a wheel-cutting machine is available some constructors will prefer this, as it leaves their lathe free for other operations. The methods to be described can be adopted for either.

The set-up in Figure 122 shows the division plate mounted at the rear of the lathe mandrel together with an index pin. A cutting frame is clamped in the lathe tool post, and this is driven by a small, fractional horsepower motor mounted on a hinged board. A light spring is attached to the outer edge of the motor mounting board to ensure that there is always tension on the driving belt. This method was used by the author for a number of years, and proved entirely satisfactory for cutting wheels; it was also the method used by the late C. B. Reeve. Pinions, however, were cut on the milling machine, or pinion wire was

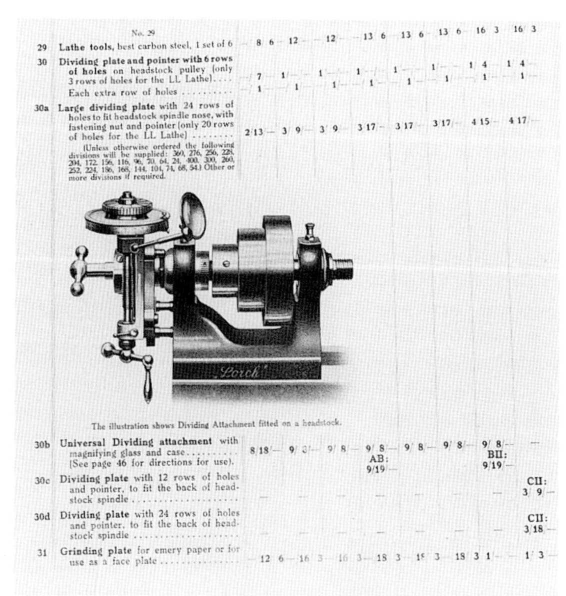

No.	Particulars								
29	**Lathe tools,** best carbon steel, 1 set of 6 (No. 29)	8/6	12/—	12/—	13/6	13/6	13/6	16/3	16/3
30	**Dividing plate and pointer with 6 rows of holes** on headstock pulley (only 3 rows of holes for the LL Lathe)	7	1/—	1/—	1/—	1/—	1/—	1/4	1/4
	Each extra row of holes	1	1/—	1/—	1/—	1/—	1/—	1/—	1/—
30a	**Large dividing plate** with 24 rows of holes to fit headstock spindle nose, with fastening nut and pointer (only 20 rows of holes for the LL Lathe) (Unless otherwise ordered the following divisions will be supplied: 360, 276, 256, 228, 204, 172, 156, 116, 96, 70, 64, 24, 400, 300, 260, 252, 224, 186, 168, 144, 104, 74, 68, 54.) Other or more divisions if required.	2/13/—	3/9/—	3/9/—	3/17/—	3/17/—	3/17/—	4/15/—	4/17/—
30b	**Universal Dividing attachment** with magnifying glass and case (See page 46 for directions for use).	8/18/—	9/3/—	9/8/—	9/8/— AB: 9/19/—	9/8/—	9/8/—	9/8/— BII: 9/19/—	—
30c	**Dividing plate** with 12 rows of holes and pointer, to fit the back of headstock spindle	—	—	—	—	—	—	—	CII: 3/9/—
30d	**Dividing plate** with 24 rows of holes and pointer, to fit the back of headstock spindle	—	—	—	—	—	—	—	CII: 3/18/—
31	**Grinding plate** for emery paper or for use as a face plate	12/6	16/3	16/3	18/3	18/3	18/3	1/—/—	1/3/—

The illustration shows Dividing Attachment fitted on a headstock.

TOP LEFT: *Fig. 119 Milling spindle and vertical slide with bevel gear drive.*

BOTTOM LEFT: *Fig. 120 Lorch L L lathe with dividing head.*

BOTTOM RIGHT: *Fig. 121 Lorch dividing head.*

Fig. 122 Division plate mounted on rear of lathe mandrel.

Fig. 123 Cutting frame and motor drive.

used, as the cutting frame lacks the rigidity for pinion cutting, where slow speeds, high torque and positive drive are essential.

A similar arrangement is shown, but in place of the division plate a dividing head has been substituted. The dividing-head spindle is attached to the lathe mandrel by an expanding bolt, and the head is prevented from rotating by a short stay attached to the end of the lathe bed. This is a home-made dividing head, made some thirty years ago.

A similar arrangement to the previous photo is shown in Figure 125, but in this instance a Myford dividing head and a proprietary mounting kit is being used to attach this to the lathe.

The dividing head with its worm and wheel arrangement and one plate will cover a far wider range of divisions, although more care is required in its use, as it is easier to miscount than with a simple direct index plate. The only problem with this arrangement is the engagement and disengagement of the dividing head if further turning operations are to be carried out to the blank.

Fig. 124 Home-made dividing head mounted on rear of lathe mandrel.

Fig. 125 Myford dividing head.

Figure 126 gives a view of the headstock dividing attachment mounted on the Myford S7, designed by the late George Thomas. This is an excellent arrangement as it can be mounted and removed from the lathe easily, in a matter of seconds; work can be machined and then cut whilst *in situ*. Another advantage with this tool is the micro attachment, a simple method of differential indexing using an extra worm and wheel; it enables every wheel count to be accommodated, including prime numbers.

The cutting frame shown was made many years ago. This is ideal for cutting wheels, but is not suitable for pinion cutting. It is perhaps more elaborate than is actually required as it had a dual purpose, having been designed with a view to using it on the ornamental turning lathe; but anyone can simplify the construction if necessary. The main frame was cut from mild steel in one piece, using a bandsaw to reduce the hard work of hacksawing, drilling and filing. Fabricating the frame by using mild steel sections and silver sol-

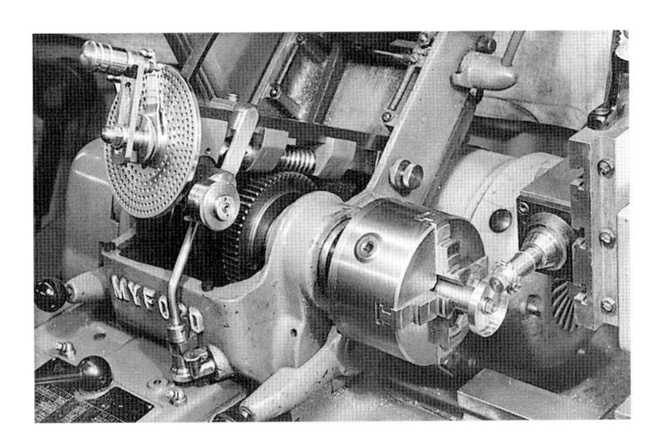

Fig. 126 Headstock dividing attachment.

Fig. 127 Wheel-cutting frame.

dering would be quite acceptable. The adjusting screws are made from silver steel, and the cone point is hardened for maximum resistance to wear. A drawing has been included to enable anyone to make the tool if they so wish.

MILLING SPINDLES

As rigidity is the key to satisfactory wheel and pinion cutting, a good milling spindle is necessary, particularly for cutting pinions. The model shown has been designed for maximum rigidity, with pre-loaded angular contact bearings; this gives an excellent finish when cutting wheel teeth, and particularly pinion leaves. Note also the flywheel pulley that helps to damp out the intermittent cutting forces caused by multi-tooth cutters when cutting takes place and consequently improves the finish. The spindle is bored No. 1 Morse taper, and accepts arbors for mounting the various cutters. Several cutter spindles are shown, and a drawing is included giving suitable dimensions. The arbors are made from No. 1 Morse taper blanks.

Two wheel cutter arbors are required, 3.5mm and 7mm diameter, and if other bore cutters are used, then a suitable arbor will have to be made. Also useful is a fly cutter arbor: this takes a 5mm (3/16in)-diameter round tool steel and any unusual

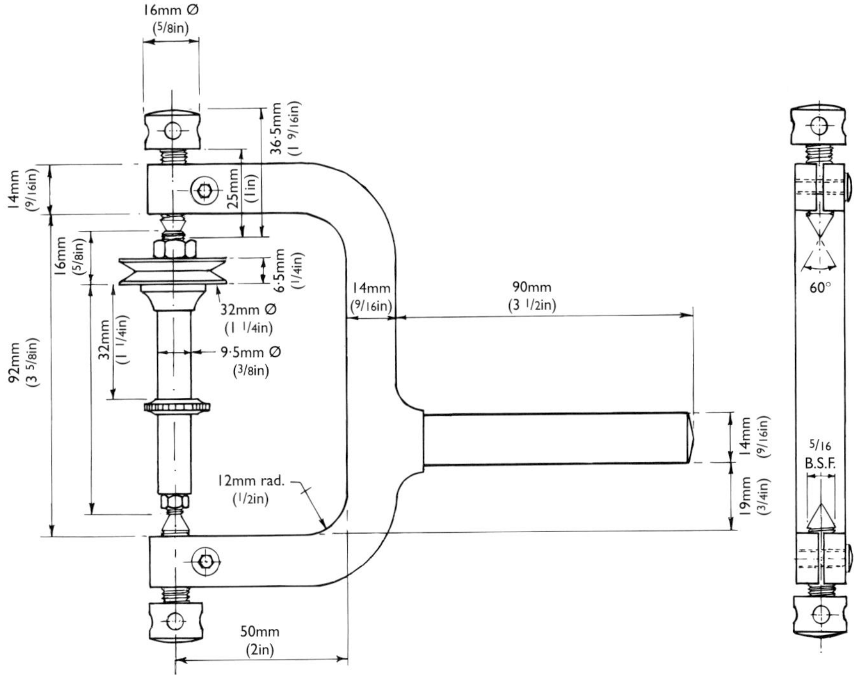

Fig. 128 Dimensioned drawing of wheel-cutting frame.

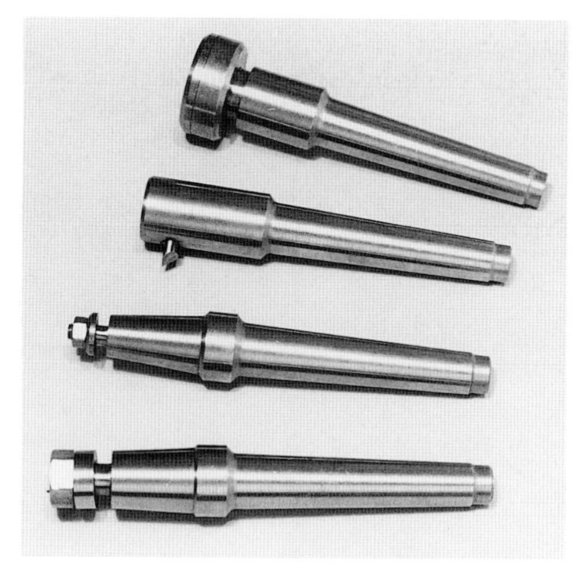

ABOVE: *Fig. 129 Lathe milling spindle.*

RIGHT: *Fig. 130 Range of cutter arbors.*

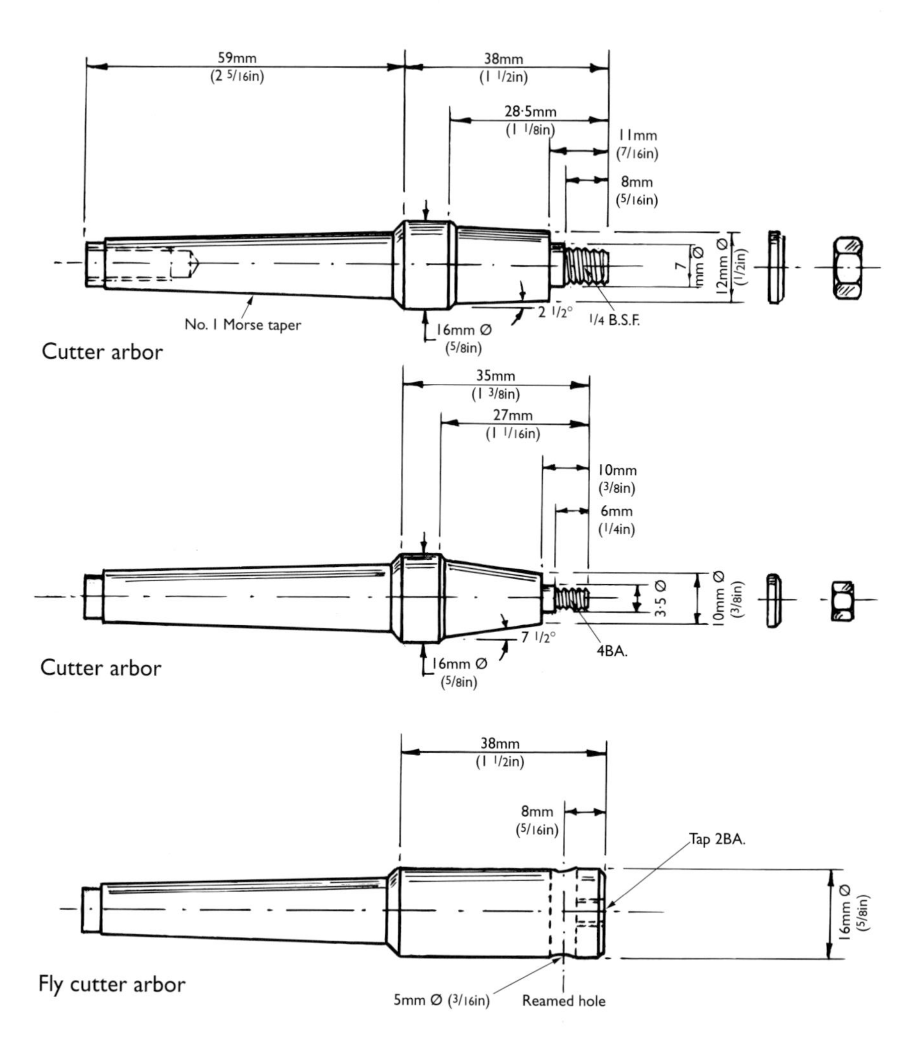

Fig. 131 Dimensioned drawing of cutter arbors.

Fig. 132 Milling spindle mounted on the vertical slide.

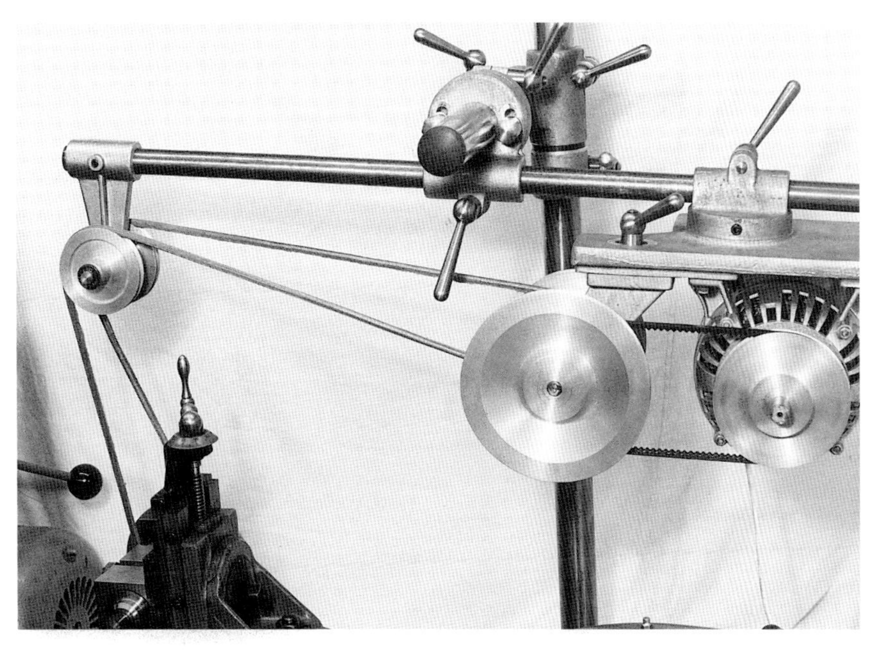

Fig. 133 Overhead drive unit.

shapes for escape wheels and ratchet wheels and so on. A 6mm (¼in) capacity drill chuck on a No. 1 Morse taper arbor is useful when drilling pin wheels and lantern pinions.

Figure 132 shows the milling spindle mounted on the vertical slide. This can be driven by a motor fitted to the rear of the lathe bed, although some form of countershaft speed reduction is necessary, even if a DC motor and variable speed control is used. A more sophisticated method to drive the milling spindle is the overhead drive unit shown, which is available in kit form. This gives a good variation in speed range, having a built-in countershaft. Cutter speeds can be varied from as little as 150rpm, up to 4,000rpm. Pinion cutters require to be driven at speeds from the lower end of the range, whereas wheel cutters cutting brass can be run at up to 4,000rpm.

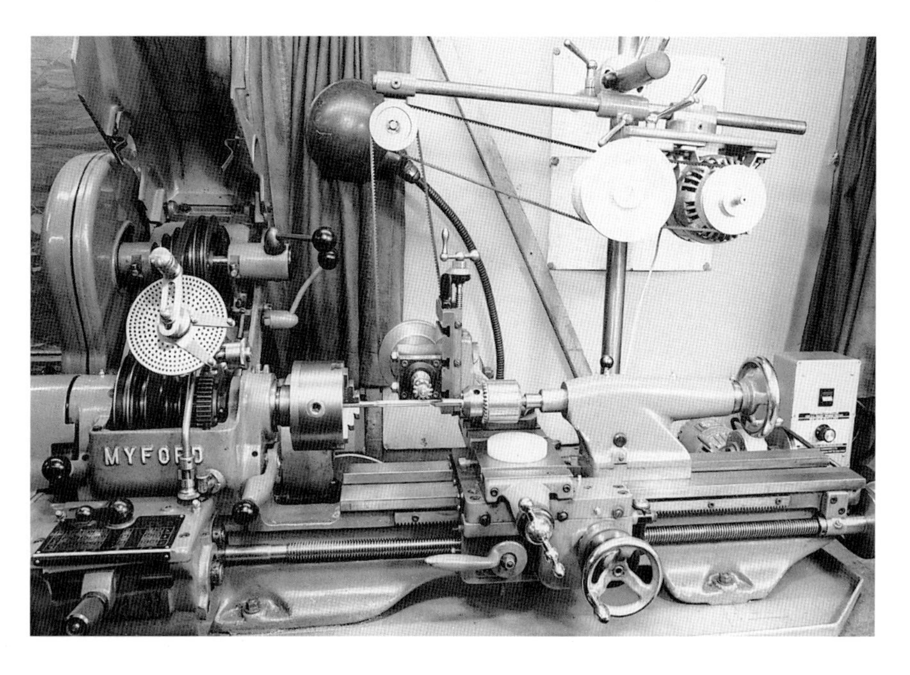

Fig. 134 General view of lathe with overhead drive, milling spindle and dividing head.

Fig. 135 Milling spindle machined to accept 8mm watchmaker's lathe collets.

The design of this particular overhead drive enables the saddle to be traversed a good 150–200mm (6in–8in). This is more than is actually required for wheel and pinion cutting, but is useful for other milling work.

Figure 135 shows a similar milling spindle, but in place of the No. 1 Morse taper bore, the main spindle is machined to take 8mm watchmakers' collets. Often these lathes have 8mm arbors with threaded shanks to accept wheel and pinion cutters.

The description of equipment is given as a guide, and individual clockmakers will develop the methods best suited to their own demands.

Various methods of cutting wheels and pinions in the lathe have been used, and from experience, the method of using a milling spindle and separate drive has proved to be the most convenient. It is possible to use the lathe headstock drive for the cutter, and mount the dividing head on the vertical slide. The advantage with this method is the positive drive from the main lathe motor to the headstock; the disadvantage is

Fig. 136 Myford dividing head mounted on the vertical slide.

Fig. 137 Wheel-and pinion-cutting attachment by Colin Walton.

Fig. 138 Wheel cutting on the Cowells lathe.

that it is not very easy to see the work being machined.

Another attachment similar to the previous arrangement is the device designed by Colin Walton. This has a large brass division plate, similar to a wheel-cutting engine, and is mounted on a right-angle bracket fitted directly onto the lathe cross-slide; the unit will accommodate both wheels and pinions. Cowells have produced a small lathe aimed specifically at the clockmaker. It can be supplied with milling attachment and individual drive to the cutting spindle; also available are division plates that mount directly onto the lathe spindle.

ELECTRONICALLY CONTROLLED ROTARY INDEXER

Finally we come to the ultimate in dividing in the lathe. Both methods described previously can raise problems with counting: mistakes can be made and care is required, as those who have previously carried out wheel cutting can verify. It is quite disconcerting after spending time setting and then cutting a wheel, to find on the last cut that you have miscounted, and have to start again. The device developed by Bryan Mumford for Sherline will provide any number of divisions required between 1 and 999: the unit is a 100mm

Fig. 139 C.N.C. rotary indexer.

Fig. 140 Electronically controlled rotary indexer fitted to the Myford Super 7 lathe.

(4in) rotary table driven by a stepping motor, controlled by a computer; this is mounted on the rear of the lathe mandrel with an adapter, similar to the Myford dividing head, previously shown. A keypad provides the means of entering the required number of divisions or number of degrees. The unit can be set manually with the rotary table handwheel. Backlash on the worm gear is adjustable, and a programmable backlash feature in the computer allows the remainder to be compensated for electronically. Previous positions can be recalled with complete accuracy, which is a most useful feature.[15]

Having now considered the necessary equipment required for wheel and pinion cutting, this will be dealt with in more detail later.

5 WHEEL AND PINION THEORY

Having briefly described the equipment required, it is time to look at the general design of tooth form, together with some of the basic formulae used. The theory and its application are quite simple as all formulae have been transposed, so no mathematics are required, only the ability to insert numbers into a calculator. In the case where a constructor is following an existing design of clock, then all the calculations will already have been completed and he can disregard this section.

WHEEL TRAINS OF EARLY CLOCKS

In the eighteenth and nineteenth centuries there was very little knowledge of the theory of gearing, the methods used being 'rule-of-thumb' – and this could be taken quite literally. Many years ago a retired workman from the old Prescott works of the Lancashire Watch Company gave a lecture to a local branch of the Antiquarian Horological Society. When asked by a member of the audience what shape he made his pinions, he held up his thumb and said 'that shape'! Therefore, as late as the end of the nineteenth century and into the early twentieth century, no official standard was available for wheel and pinion teeth.

Consequently, when one has to measure and check the tooth size of wheels on antique clocks, there is a wide variation and the modern cutters often do not match exactly. This calls for some intelligent guesswork at times. The country clockmakers who made the majority of the clocks still in existence, whilst undoubtedly being good craftsmen, did not have the grasp of mathematics required for the various simple formulae used for the gearing of wheels and pinions in clocks.

If the wheel trains of early clocks are studied, the variation of wheel teeth proportions is quite dramatic. Often the teeth are quite thin, particularly at the top end of the train, but when the gaps between adjacent teeth are measured, it is quite apparent that the same cutter has been used throughout the train. Even in the mid-nineteenth century, there was no standard for the shape of wheel teeth or pinion leaves. An English edition of a famous work on gearing presented various questions to the horological trade regarding the basis on which they shaped their wheel teeth. One reply stated:

> In Lancashire they make the teeth of watch wheels of what is called the bay leafe pattern; they are formed altogether by the eye of the workman; and they would stare at you for a simpleton to hear you talk about the epicycloidal curve.

It was the Frenchman M. Camus, 1699–1768, who first applied theory to horological gearing and provided the basis for a standard for a gear tooth form. His work *Cours de Mathématiques*, published in Paris in 1749, was subsequently translated into English by John Isaac Hawkins in 1806; it proved so popular that at least three further editions were printed. Camus was so advanced in science and mathematics that at the age of twelve he gave lectures in Paris, and at an early age he attained the highest academic honours in his own country of France and in most other foreign academies. In 1765 he was elected a Fellow of the Royal Society of London. It required someone of his genius to grasp the problems of the day: his treatise, entitled *A TREATISE ON THE TEETH OF WHEELS, demonstrating the best forms that can be given to them for the purpose of machinery such as MILL WORK AND CLOCK WORK*, covered cycloidal tooth forms, and even covered the design of lantern pinions.

The Sector, or Proportional Gauge

One method that simplified the sizing of wheel and pinion teeth was the Sector, or proportional gauge. Basically, this is made from brass with two engraved scales, hinged at one end and adjustable at the other by a quadrant and locking

Fig. 141 Sector or proportional gauge.

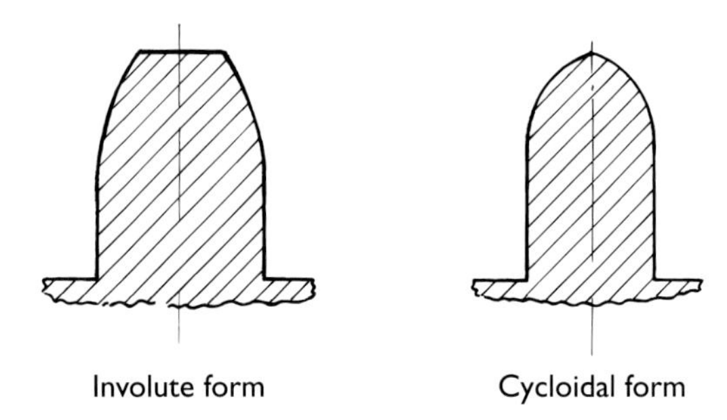

Fig. 143 Typical gear tooth forms.

Fig. 142 Sector arm from Treatise on Clock and Watchmaking.

screw, the inside edges of both scales being engraved into 120 parts. When used for sizing wheels and pinions, let us assume a pinion of eight leaves is required for a wheel of seventy-two teeth. The wheel is placed between the two scales at the seventy-two mark, then the distance is measured between the scales at the eight mark for the diameter of the pinion. There is, of course, an addition required for the addendum. In modern-day practice a Sector would not be required, due to the tooth shapes now being manufactured to approved standards and all data being easily calculated. An illustration of a Sector from Thos. Reid's book *Treatise on Clock & Watchmaking* is shown in Figure 142.

Generally speaking, wheels and pinions in clockwork differ from the normal types of involute form of gearing found in the gearbox of a modern motor car. In nearly all cases in clocks, the wheel drives the pinion and this is termed 'gearing up'; whereas in engineering gearing the reverse normally takes place. Consequently a different tooth shape is required, and this is known as the cycloidal form. It helps to reduce friction and improve the efficiency of the drive between the motive force through the train to the escapement.

Involute Gearing

There are, of course, exceptions to every rule. One is in the motion work for the hands of the clock, particularly in modern clocks, where it is quite normal to use involute gearing for the 12:1 ratio required from the centre pinion to the hour wheel, as in this instance the pinion drives the wheel. Also in winding wheels, where intermediate wheels are used either to relocate the winding square in a more convenient position, or to gear down to reduce the torque required when winding.

There is a company in existence in the UK who manufactures fine clocks and cuts all their wheels and pinions using the involute system. With this system, the smallest practical number of leaves on a pinion is eight, and a correction factor is always required for the wheel and pinion centre distance to improve the rolling action.

When matching wheel work in the restoration of old clocks, it is essential that cutters are used that are manufactured to the cycloidal form.

Driving Action Explained

Referring now to Figure 144, this shows on the right-hand side two circles, representing the pitch circles of a wheel and pinion. If disc 'B' revolves in the direction of the arrow – clockwise – then disc 'A' will revolve anticlockwise,

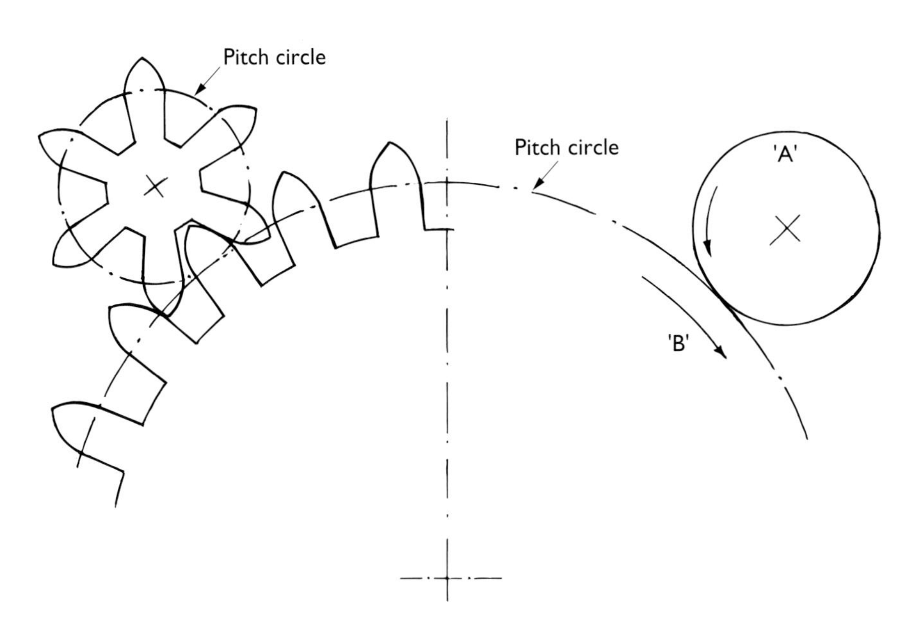

Fig. 144 Wheel and pinion engagement.

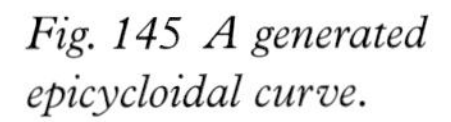

Fig. 145 A generated epicycloidal curve.

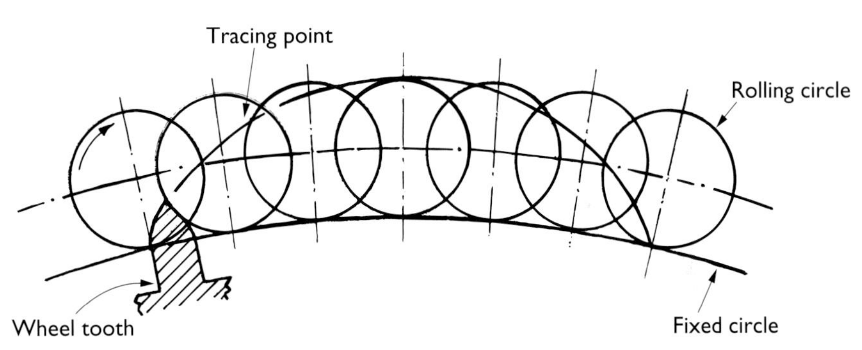

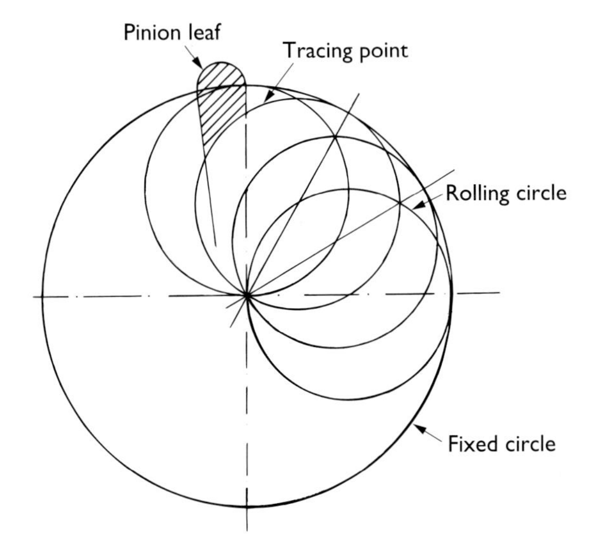

Fig. 146 A generated hypocyloidal curve.

assuming the discs revolve without slipping. Also, if the pitch circle of 'B' is twice that of 'A', then for every revolution of 'B', 'A' will revolve twice. This is the *velocity ratio* and is the basis on which gearing is worked. However, two discs revolving in this manner could not be relied upon to transmit equal motion without slipping; therefore, teeth are projected upwards above pitch circle 'B', and downwards for pitch circle 'A'. The extension of the wheel and pinion outside the pitch circle is known as the 'addendum'. Conversely, it is necessary for the gap between the teeth to project downwards inside the pitch circle, and this is known as the 'dedendum'. This will then give the necessary uniform angular motion, as shown on the left-hand side of Figure 144.

We have seen that without teeth on the circumference of two circles there would be no positive drive. Furthermore if the teeth were just straight with matching slots, jamming would occur. Projections or teeth should, therefore, be shaped to form a rolling action where one is driving the other, although some sliding action does take place, particularly with low-numbered pinions. Also, in clockwork it is essential that friction is reduced to a minimum. The radius of the teeth are formed by generated circles: one is the epicycloid, which is a curve generated by a point on the circumference of a circle that is rolling at uniform speed upon the circumference of another larger circle. The other generated curve is the hypocycloid, which is a curve traced by a point on the circumference of a circle rolling on the inside of the circumference of another.

The addenda of the wheel teeth are formed by the epicycloidal curve, and the leaves of the matching pinion that it drives are formed to the hypocycloidal curve. Both are originated from the same rolling circle.

For perfect driving action between wheel and pinion teeth to take place, hundreds of cutters would have to be made available, which is not a practical solution. A compromise has to be made, therefore, to enable a limited range of cutters to be manufactured that are capable of producing acceptable results. The main friction that takes place is engaging and disengaging: engaging friction takes place *before* the line of centres, more prevalent on pinions with low counts, *i.e.* six and seven leaves; this is the reason that pinions of ten and twelve leaves are used on precision clocks such as high grade regulators. Disengaging friction takes place *after* the line of centres, and is a more sliding type of friction.

THE SIZING OF WHEEL AND PINION TEETH

The formulae for modern cutter manufacturing are based on a ratio of 7½ : 1 – that is, a 45-tooth wheel driving a six-leaf pinion. This is a good average, and one that will produce a range of cutters that will produce wheels and pinions with minimum driving friction.

From the diagram (Figure 144) we have seen that pitch diameters have been used to form the basis of the gearing. However, since the addendum is usually quoted as the extra added to one tooth, the full diameter is worked out by adding twice the addendum to the pitch diameter. It will be noticed from the formulae to follow that addenda constants are used for both wheels and pinions.

Essentially, this book is about the practical side of wheel and pinion cutting, and only a brief description of the theory has been touched upon. Anyone interested in the theory of gearing, however, should obtain a copy of *Gears for Small Mechanisms* by W. O. Davis – first edition 1953, revised edition 1970, re-printed 1993.

The measurement and sizing of wheel and pinion teeth is based on one of two systems; the diametral pitch, or DP, with dimensions in inches; and the module system, or M, which is measured in millimetres.

Whilst in England and the USA the DP system has been the one most widely used, this is now changing due to the introduction of the metric system, although this system has always been the predominant one used on the Continent and by professional horologists in the UK.

Fig. 147 Engaging and disengaging friction.

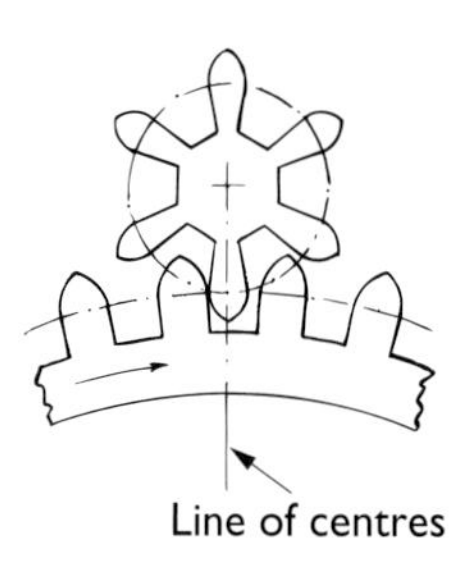

6 leaf pinion – action well before line of centres

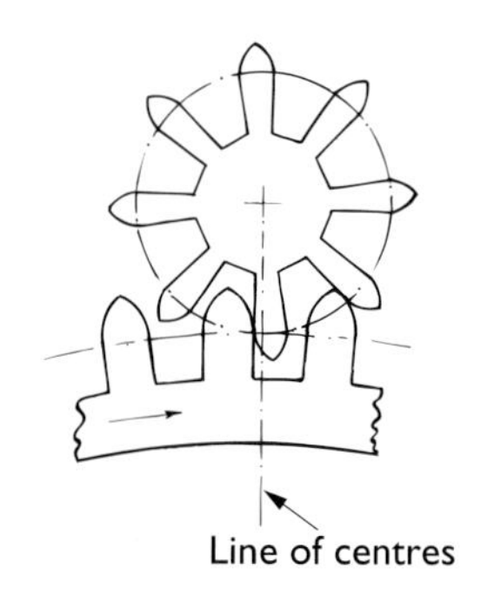

8 leaf pinion – action just before line of centres

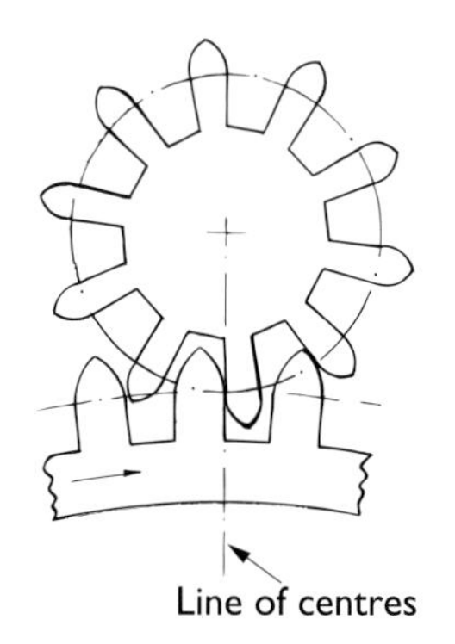

10 leaf pinion – action taking place on line of centres

Therefore

$$DP = \frac{\text{number of teeth}}{\text{pitch dia. in inches}} \text{ or } \frac{N}{PD}$$

$$M = \frac{\text{pitch dia. in mm}}{\text{number of teeth}} \text{ or } \frac{PD}{N}$$

You will notice that the module system is the reverse, or reciprocal, of the DP system.

All the horological cutters available today are based on the metric module system: one can work in either imperial or metric measurements; both methods will be shown. The younger fraternity, who are conversant with the metric system, will already have no trouble in using this method throughout. The standards most commonly used for wheels and pinions with cycloidal form are the Swiss NIHS 20.01, 20.02, 20.10 and 20.25, formerly NHS 56702 and 56703, and BS 978, Part 2 (*see* Appendix). It will be noted that from either of these standards the teeth or leaves are shown to have a radius at the root. This is, of course, good engineering practice, and obviously gives additional strength to the tooth form, but unfortunately it does not match up with the wheels and pinions cut by the old clockmakers. The cutters we generally use are therefore manufactured to cut teeth with square bottoms.

The Module System

Beginning first with the module system *where all dimensions are in millimetres*, the notation and definitions are the ones that will be used throughout:

M = Module
N = Number of teeth in wheel
n = Number of leaves in pinion
OD = Outside dia. of wheel and pinion
PD = Pitch circle dia.
CD = Centre distance of wheel and pinion
2.76 = Addenda constant for wheels
1.71 = Addenda constant for pinions of 6, 7 and 8 leaves, where the wheel drives the pinion
1.61 = Addenda constant for pinions of 10, 12 and 16 leaves, where the wheel drives the pinion

(1) Outside dia. of wheel
= [number of teeth + 2.76] × module dimensions in mm
Abbreviated $OD = [N + 2.76]M$

(2) Outside dia. of pinion
= [number of leaves + addendum constant] × module
Abbreviated $OD = [n + 1.71]\ M$
for 6, 7 and 8 leaves
or $OD = [n + 1.61]\ M$
for 10, 12 and 16 leaves

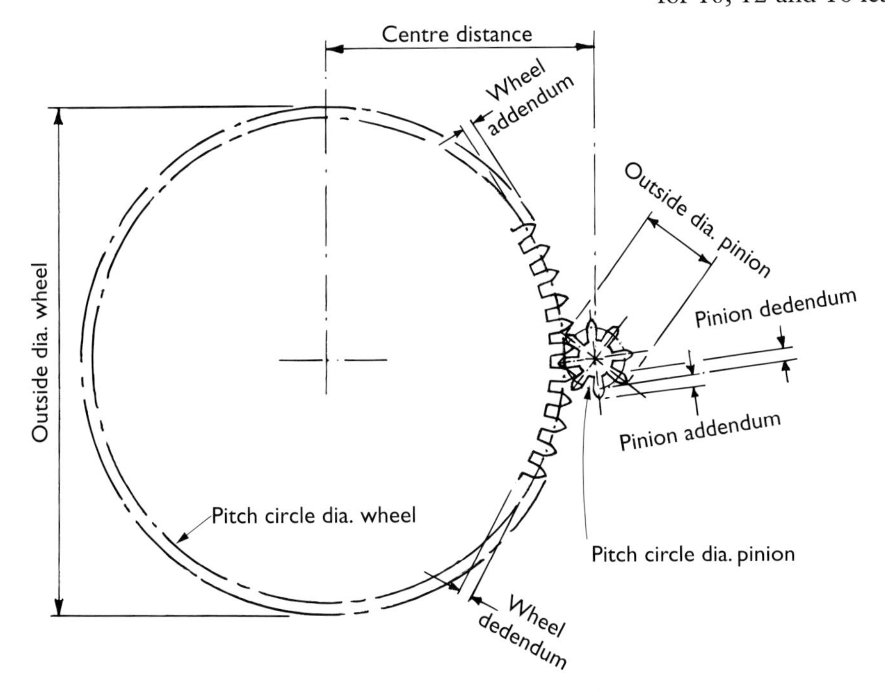

Fig. 148 Wheel and pinion sizing data.

(3) Centre distance of wheel and pinion

$$= \left[\frac{\text{number of teeth in wheel + leaves in pinion}}{2}\right] \times \text{module}$$

Abbreviated $CD = \left[\frac{N + n}{2}\right] M$

The above formulae are all that are required when cutting wheels and pinions when the cutters are readily available. If you wish to make your own cutters, then additional formulae will be required, and these will be covered later, together with information on how to manufacture cutters in your own workshop.

Note that the addenda constants used are for P. P. Thornton's cutters, as these are readily available in the UK. Where cutters from other manufacturers are used, check the recommended data.

The DP English System

Now for the formulae using the diametral pitch (DP) English system of measuring teeth:

(4) Outside dia. of wheel (dimensions in inches) $= \frac{N + 2.76}{DP}$

(5) Outside dia. of pinion (6, 7 and 8 leaves) (dimensions in inches) $= \frac{n + 1.71}{DP}$

The constants used before for 10, 12 and 16 leaves apply just the same.

(6) Centre distance (dimensions in inches) $= \frac{N + n}{2 \times DP}$

Useful conversion formulae are included, together with a table covering all the normal sizes encountered.

To convert module to DP $DP = \frac{25.4}{M}$

To convert DP to module $M = \frac{25.4}{DP}$

Applying the formulae now to a practical example, let us assume that a centre wheel and matching pinion are required for a longcase (grandfather) clock. The size of tooth is the first factor to be decided. From past experience and the study of previous clock movements, a suitable tooth size for a longcase train wheel would be 0.8 module, and a typical number of teeth would be 60. Then, therefore, the outside dia. of the wheel would equal, from formula (1):

$$\text{Outside dia.} = [N + 2.76]\ M$$
$$OD = [60 + 2.76] \times 0.8$$
$$OD = 50.21\text{mm}$$

and if this is required in inches: $\frac{50.21}{25.4}$

$$= 1.977\text{in}$$

A suitable count for the matching pinion would be 8 leaves. Now using formula (2):

$$OD = [n + 1.71]\ M$$
$$\text{Outside dia.} = [8 + 1.71] \times 0.8$$
$$= 7.77\text{mm}$$

If this is required in inches: $\frac{7.77}{25.4}$

$$= 0.306\text{in}$$

Having arrived at the dimensions of the wheel and pinion, the next measurement required is the centre distance that the wheel and pinion mesh together on. Therefore, using formula (3):

$$CD = \left[\frac{N + n}{2}\right] M$$

Centre distance of wheel and pinion

$$= \left[\frac{60 + 8}{2}\right] 0.8$$

CONVERSION TABLE

Module		DP	Module		DP
M0.20	=	127 DP			
M0.25	=	102 DP	M0.60	=	42 DP
M0.30	=	84 DP	M0.65	=	39 DP
M0.35	=	73 DP	M0.70	=	36 DP
M0.40	=	64 DP	M0.75	=	34 DP
M0.45	=	56 DP	M0.80	=	32 DP
M0.50	=	51 DP	M0.85	=	30 DP
M0.525	=	48 DP	M0.90	=	28 DP
M0.55	=	46 DP	M0.95	=	27 DP
M0.575	=	44 DP	M1.00	=	25.4 DP

$$= 27.2\text{mm} = \frac{27.2}{25.4}$$

$$= 1.071\text{in}$$

All the necessary information is now available to go ahead and manufacture the wheel and pinion.

Variations on these Systems

When carrying out restoration work on old clocks, variation on the above methods can be applied. As already stated, the old clockmakers worked using their own addendum constants, consequently wheel and pinion teeth proportions vary tremendously.

When replacing damaged wheels and pinions where only the OD or number of teeth or leaves are known, the following formulae may be found useful:

$$\text{CD} = \frac{\text{wheel PD} + \text{pinion PD}}{2}$$

$$\text{M} = \frac{\text{OD}}{\text{N} + 2.76}$$

Note this constant may have to be adjusted

$$\text{PD} = \frac{\text{OD} \times \text{N}}{\text{N} + 2.76} \text{ and } \text{N} = \frac{\text{PD}}{\text{M}}$$

If a wheel or pinion is missing, calculate the module and the pitch radius PD/2 of the existing wheel/pinion from the tooth count and the OD and also measure the centre distance CD. By subtraction, the pitch radius can be found, and hence the PD of the missing wheel/pinion. Since the module and the PD of the missing gear is now known, the formula N = PD/M will give the number of teeth/leaves required.

However, since old wheels and pinions rarely conformed to theory, it is likely that the answer may have a peculiar number or even a fraction of a number, in which case a study of the rest of the train is necessary in order to choose the nearest suitable number (*see* Chapter 11 for help here). One can also use the fact that the wheel and pinion counts are directly proportional to the pitch radii.

Example:

centre distance = 27mm

pitch radius of pinion = 3mm

∴ pitch radius of wheel = 24mm

∴ ratio pinion : wheel = 3 : 24

leaves on pinion = 8, so multiply 3 by 2.66 to achieve the number 8,

then 3 × 2.66 : 24 × 2.66 = 8 : 63.84

Nearest suitable wheel count = 64

Also listed below are a number of formulae that may be found useful. Some formulae have been transposed for those who do not like maths. All that is required is to slot the numbers into your calculator – always remember the units you are working in.

(7) $\text{M} = \frac{2 \times \text{CD}}{\text{N} + n}$

or

(8) $\text{M} = \frac{\text{OD}}{\text{N} + 2.76}$ for wheels

(9) $\text{M} = \frac{\text{OD}}{n + 1.71}$ for pinions 6, 7 and 8 leaves

(10) $\text{M} = \frac{\text{OD}}{n + 1.61}$ for pinions 10, 12 and 16 leaves

(11) $\text{M} = \frac{\text{PD}}{\text{N}}$ for wheels

(12) $\text{PD} = \text{M} \times \text{N}$ for wheels

(13) $\text{M} = \frac{\text{PD}}{n}$ for pinions

(14) $\text{PD} = \text{M} \times n$ for pinions

(15) $\text{PD} = \frac{\text{OD} \times \text{N}}{(\text{N} + 2.76)}$ for wheels

WHEEL AND PINION CUTTERS

Multi-teeth wheel and pinion cutters are readily available in high-speed steel. The purchase of these is strongly recommended if funds will permit, as they will give good service and last a lifetime if treated with respect.

A wheel cutter will cover teeth from, say, 20–140, whereas a pinion cutter will only deal with one count, i.e. a different cutter would be

Fig. 149 Thornton wheel cutter.

required for a 0.6M, six-leaf pinion and a 0.6M, seven-leaf pinion, and so on.

For those who wish to make their own cutters, details of the various methods employed, together with dimensions of wheel and pinion tooth forms, will be given. Fly cutters are the easiest form to manufacture, but are not suitable for cutting steel. The standard cutter range available in the UK is as follows, and is produced by P. P. Thornton:

- Wheel cutters M0.2–M1.0 in 0.05 steps with the addition of M0.525 and M0.575. The smaller sizes, module 0.2–0.4, are generally more suited to carriage clocks.
- Larger cutters are also available from M1.0 in steps of 0.10 up to 1.50. These are for longcase calendar wheels, Dutch clocks and small turret clocks.
- Pinion cutters in 6, 7, 8, 10, 12 and 16 leaf in the range M0.2–M1.0 in steps of 0.05.

6 WHEEL CUTTING

Now that the basic equipment required for wheel cutting has been covered, the actual job can commence. As brass wheels are far easier to cut than steel pinions, these will be dealt with first. The use of the machinery and accessories will be dealt with in more detail later.

CUTTING BRASS WHEELS

For modern clocks the material most suitable is CZ120 compo engraving brass, with the addition of lead to aid free machining. Slices of normal extruded brass bar CZ121 can also be used; this can be parted off at the approximate thickness. It is also useful if a small wheel or brass pinion is to be cut, as it can be machined straight from the bar.

Normal rolled brass sheet CZ108 is not really suited for either wheels or plates, as it tends to be 'tough' and sticky when machining due to the work-hardening process in manufacture; it is therefore difficult to maintain a sharp edge after cutting. Problems are also experienced when crossing out the wheel (cutting out the spokes) with the piercing saw, as the blades snap like carrots!

If wheels for antique clocks are being cut, then yellow cast brass should be used to match the colour of the old clock brass. This was generally 70/30 – that is, 70 per cent copper and 30 per cent zinc. At this stage, however, this material will be disregarded until later, as a certain amount of preparation is necessary due to the fact that the material is quite soft in its 'as cast' state.

Engraving brass can be purchased in sheet form for the constructor to cut or trepan out himself, or it can be purchased as circular discs ready for use.

Select a suitable size of wheel blank, and drill and ream a hole in the centre. If the blank has been made from extruded brass, then this can be machined at the same time as the material is mounted in the three-jaw for parting-off. Now

Fig. 150 Drilling the blank.

Fig. 151 Parallel mandrels.

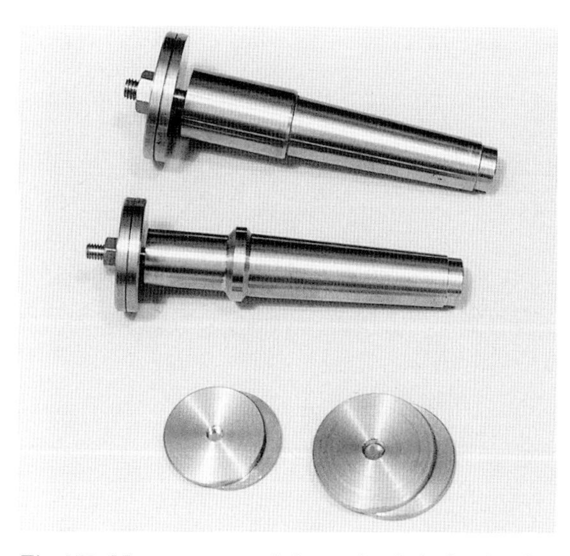

Fig. 152 Morse taper mandrels to suit a lathe headstock.

turn a mandrel from, say, 12mm (½in) dia. mild steel, down to 6mm (¼in) dia. and thread ¼in B.S.F. or M6 for part of the way. The suitable sizes of mandrel are 6mm (¼in) and 5mm (³⁄₁₆in) – the ³⁄₁₆in size can be threaded 2 B.A. It is sometimes necessary to use smaller thread sizes depending on the bore of the wheel required. If you have collets with your lathe, then the mandrels can be gripped accurately time and time again, whereas the three-jaw will not give the necessary repeat accuracy. In this case, therefore, it is better if all the wheel blanks are

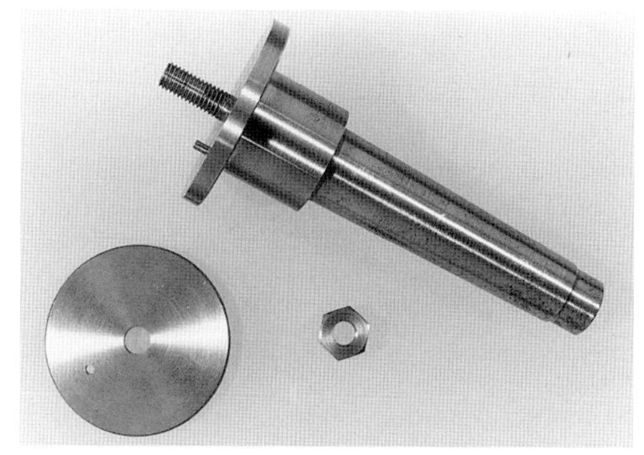

Fig. 153 Mandrel with a locating pin.

prepared first, then a suitable mandrel can be turned *in situ* in the three-jaw chuck. These take only a few minutes to turn down and thread, and this will ensure that everything is running true. Two types of mandrel are shown: the parallel types that are gripped in the lathe chuck, and the others that fit straight into the lathe spindle, in this case No. 2 Morse taper for the Myford lathe. A drawbar is used to ensure the arbor does not work loose.

Also included is a design with a solid backing disc and locating pin. This is sometimes beneficial when the wheel may have to be remounted, for example when replacing a missing calendar wheel where the module is difficult to determine. Note in the photograph there is a jig washer for positioning the hole in the wheel blank.

Mount the wheel blank onto its mandrel and turn to size. If the wheel to be cut is fairly thin, then use the backing washers to support the work whilst cutting is in progress. It is sometimes an advantage to make the backing washers the same size as the wheel being cut, then this will produce spare wheels which can often be useful on certain projects.

It is important that the wheel blank is secure, as the cutting forces can cause the locking nut to vibrate loose. Remove the lathe topslide, and mount the vertical slide on the lathe cross-slide. Note that a plug is fitted into the locating hole vacated by the topslide to prevent swarf from entering the slides, feed screw and nut.

Bolt the milling spindle on to the vertical slide as shown, and ensure that the spindle is square with the axis of the lathe. With a dial indicator, check the spindle axis is parallel with the lathe

Fig. 154 Wheel blank mounted on a mandrel.

Fig. 155 Checking the milling spindle with a straight-edge.

cross-slide. Mount the arbor into the bore of the milling spindle ensuring there are no dirt particles, as it is essential the arbor and cutter run true. All the cutter arbors are threaded to take a drawbar to ensure the arbor does not vibrate out when cutting takes place. Select the correct wheel cutter, fit to the arbor as shown with the washer or spacer, and tighten the locking nut.

Now the cutter requires to be centralized. Whichever method is used for holding the cutter, it is necessary that the cutter be absolutely central with the work; this is particularly noticeable when cutting pinions. There are various methods of centring the cutter.

Fig. 156 Checking the milling spindle with the dial indicator.

Fig. 157 Wheel cutter.

Fig. 158 Centring the wheel cutter with a fine needle.

Centring the Cutter

The first method is by holding a fine centre or needle centrally in the three-jaw chuck or collet. The cutter tip is adjusted until the position is exactly central – the final adjustment is made more accurately if an eyeglass is used. Note that the tool should be concentric, and gripped centrally in the chuck or collet.

The second method is to advance the rotating cutter down onto a piece of scrap brass very carefully until it just touches, then move the cutter laterally, to form a small ellipse. The cutter can then be accurately positioned centrally between the extremes of the mark; again, the use of the eyeglass will ensure greater accuracy.

A third method of centring is to cut a vee in a piece of brass held accurately in the lathe collet. The vee is cut as shown, then the work indexed through 180 degrees and another cut taken. This will ensure that the vee is exactly central with the lathe mandrel. To centre the wheel cutter, advance the cutting edge profile between the vee, observe with an eyeglass, and adjust until the cutter is central. This is possibly the most accurate visual method of the three described.

Fig. 159 Cutter-centring, forming an ellipse.

Fig. 160 First stage centring cutter, with vee.

Fig. 162 Third stage centring cutter with vee.

Fig. 161 Second stage centring cutter with vee.

Whilst the previous methods employ some form of guesswork, and a certain amount of skill is required, the next method, using a centring micrometer, will determine the actual position of the cutter beyond doubt. From the illustration it will be seen that the construction is quite straightforward, and the tool can be readily purchased. The shank can be either straight to fit in the collets of the headstock, or a morse taper to go into the bore of the lathe spindle.

Micrometer heads are readily available from most tool shops. The one chosen had a range of 0–0.50in and had a centring pin of 0.20in dia. If a metric micrometer is used, an equivalent range would be 0–12mm and a 5mm centring pin would be required. The micrometer spindle is

Fig. 163 Centring micrometer.

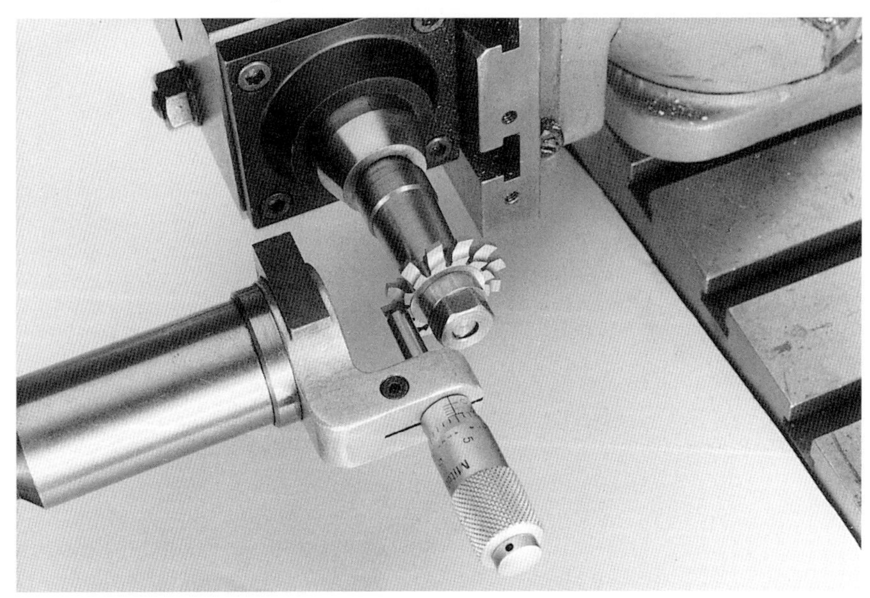

Fig. 164 Centring a wheel cutter.

zeroed on the 5mm (0.20in) dia. central dowel by setting the reading at 2.5mm (0.10in) – which is half the diameter – slackening off the clamping screw and carefully sliding the micrometer head down until the spindle end is in contact with the outside diameter of the dowel. Then tighten up the clamp screw. The micrometer spindle face, at its zero position, should then be exactly on the centreline of the lathe when the complete tool is located in the lathe headstock. Obviously it is necessary to set the tool in the exact vertical or horizontal position. To facilitate this, the body and arbor are fitted with a location spigot with friction device between to enable the micrometer to be rotated into any position. There are two flats at 90 degrees on the tool body to enable either of these to be set by the use of a toolmaker's square from the lathe bed. Measure the cutter as shown. For example, if the measurement is 0.140in, divide this by two, = 0.070in.

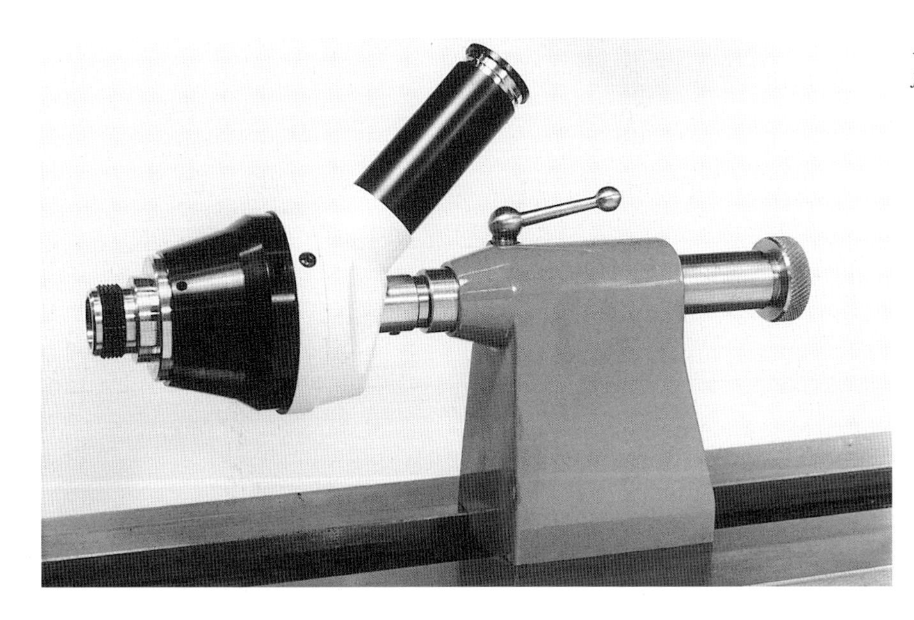

Fig. 165 Special tailstock for a centring microscope.

Fig. 166 Microscope, tailstock and milling spindle.

This, then, is the setting for the centring micrometer.

To centre the cutter, move the cross-slide to bring the side of the cutter into light contact with the micrometer spindle, face as shown, and lock the cross-slide – note the cutters are manufactured with the tooth form and the adjacent face width exactly equal. It is not suitable to use the ground face, as the form is not always central with this.

Where the alignment of the lathe tailstock can be relied upon, then it is possible to use a centring microscope.

Holding a Centring Microscope

The illustrated device for holding a centring microscope was constructed by the author to fit the Myford lathe, although similar arrangements could be manufactured to suit most lathes. To ensure that the bore for the microscope shank was true, it was machined actually located between the lathe-bed shears, and bored from the lathe headstock. The microscope holder can be located anywhere on the lathe bed, knowing that the microscope is exactly on the centreline. A fabricated construction could be made, but the holder shown is machined from a cast-iron

Fig. 167 Centring microscope aligning a wheel cutter.

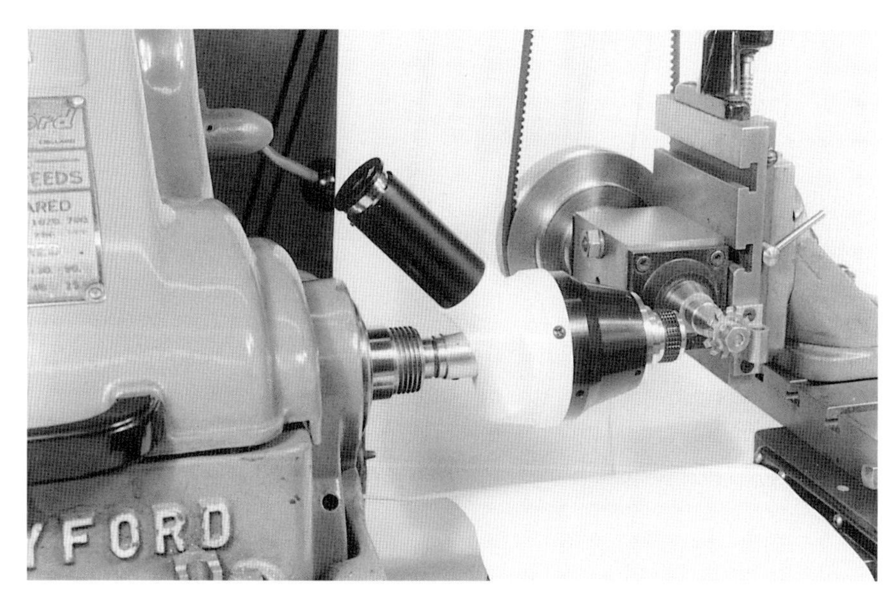

Fig. 168 Using the headstock to locate the centring microscope.

casting, a simple wooden pattern being made first.

It is important that the work is well illuminated, particularly when the microscope is used for setting up. It is also advantageous for checking the work to see how it is progressing. Alternatively, the microscope can be located in the headstock.

INDEXING

Now that the cutter-centring methods and work holding have been covered, we have to decide on the method of indexing to be used. Some of the methods were mentioned in a previous chapter. The simplest and quickest way is to use notched plates, but unfortunately a large number of plates are required if all wheel counts are to be covered. This method is ideal for pinion cutting when only six, seven, eight, ten and twelve divisions are normally required.

Using Division Plates

The second method is to use division plates. The ones shown are No. 1 and No. 2 plates, and each has fifteen rows of holes with counts from

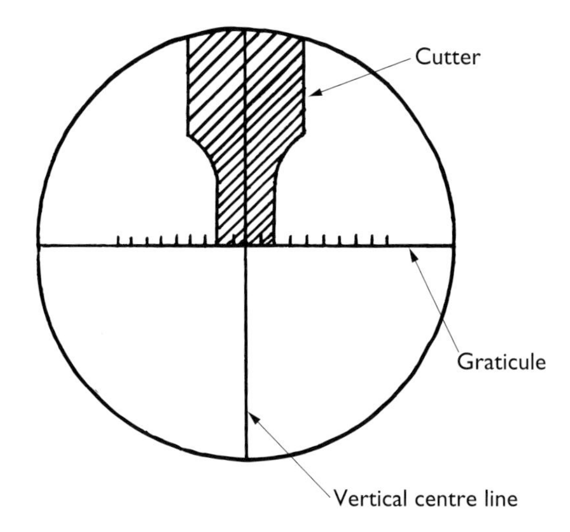

Fig. 169 View through the microscope, showing graticule and graduations, and cutter form.

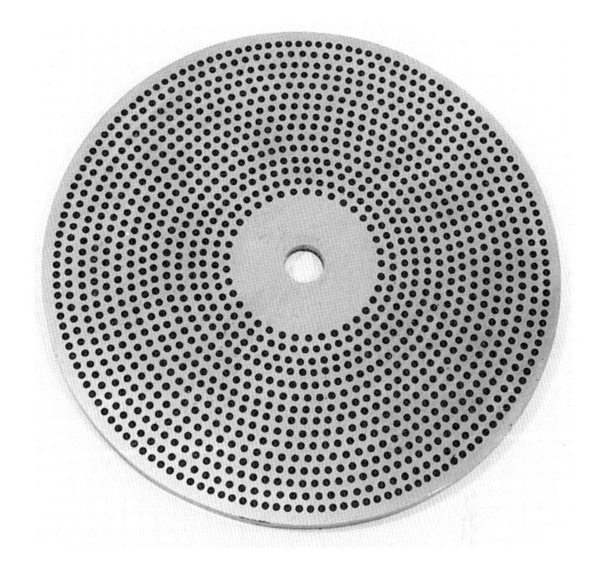

Fig. 171 Division plate.

Fig. 170 Notched plate indexing.

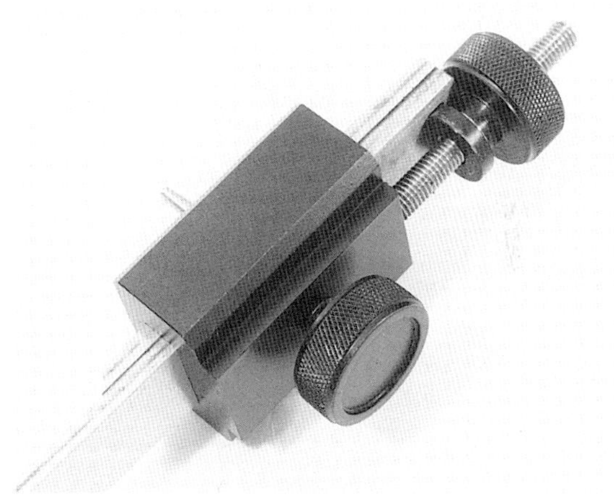

Fig. 172 Adjustable detent for division plate.

37–144; this covers thirty rows of divisions and all the in-between divisible counts – quite a range, but this will not cover some of the less common wheel counts. Note the adjustable detent: this is useful if an antique clock wheel is being replaced and it is necessary to widen the tooth spaces to match the original exactly. All teeth on the wheel are cut, and then the index on the detent is adjusted the appropriate amount, and the wheel can be re-cut by just taking a shaving off on one side of the tooth.

The dividing head will accommodate a vast range of divisions, but extra care is required when using it as mistakes are more easily made. It is normally supplied with at least three division plates. Dividing heads generally have either sixty- or forty-tooth worm wheels, the sixty-tooth model being generally of more use to the clockmaker; the Myford dividing head and the headstock dividing attachment previously referred to both have sixty-tooth worm wheels. Basically, the unit is used in the following manner: if a sixty-tooth wheel is to be cut, then the crank which rotates the worm is turned one complete revolution; this moves the work $1/60$, which is obviously one tooth. If 120 divisions are required, then the crank will be turned a $1/2$ revolution. The rule is to divide the number required into the number of

Fig. 173 Sector arms and division plate on a worm and wheel dividing head.

teeth in the worm wheel. Therefore, if an eighty-tooth wheel is required to be cut = $^{60}/_{80}$

Taking factors	60	=	3 × 20	=	3
Out of	80		4 × 20		4

We are then required to revolve the crank ¾ of a turn. Assuming a division plate that has a row of divisions that can be divided by four, for example forty-eight, we need to keep the same ratio of ¾.

Then $$\frac{3}{4} \times \frac{12}{12} = \frac{36}{48}$$

The sector arms would be set at thirty-six holes. Always count the holes that are being encompassed with the sector arms, plus one hole – this is the one with the detent pin located in it.

The angular movement of the crank would thus equal $^{1}/_{80}$, or one division of an eighty-tooth wheel. Likewise if the number of divisions required is less than the number of worm-wheel teeth, say thirty, then $^{60}/_{30}$ would give two complete turns of the crank.

This is basically how the dividing head is used. For more detailed information, refer to the *Machinery Handbook* or *Dividing & Graduating* by George H. Thomas, published by Argus Books 1983. The latter book is excellent and also covers the use and construction of the headstock dividing attachment shown here.

Another advantage with this unit is the addition of a second worm and wheel, called the micro attachment, which enables differential indexing to be carried out and makes it possible to cover the division of all prime numbers. Angular dividing is also possible.

Now that the necessary equipment has been described, the first steps in wheel cutting can commence. If this is the first attempt, it may be advisable to use the following on a wheel of determined size as an example.

Fig. 174 Wheel-cutting set-up, showing headstock dividing attachment.

FIRST STEPS IN WHEEL CUTTING

Assume a wheel with a 0.8 module is required with sixty teeth. From the previous formula No. 1:

$$OD = [N + 2.76]M$$

$$OD = [60 + 2.76]0.8$$

$$= 50.21\text{mm } (1.977\text{in})$$

Mount the blank on a suitable arbor and turn to 50.2mm (2in). With a black marking pen or marking out blue, cover the top edge of the disc; this will enable the setting of the depth of cut to be ascertained more readily. Remove from the lathe and replace with the centring micrometer. To centre the 0.8 module cutter, the flank measurement is 3.68mm (0.145in), divide by two, and this equals the reading required on the micrometer. Position the cutter gently against the micrometer anvil, checking carefully that no movement takes place, then lock the cross-slide. Remove the centring device from the lathe headstock and replace with the wheel blank to be cut.

Ensure that all backlash is taken out of the worm and wheel if this is the method you are using for indexing. Note that if the headstock dividing unit is used, practically all backlash can be eliminated by adjustment. If the direct index method is used, select the correct hole circle and engage the detent pin in the first hole, set the speed at approximately 4,000rpm, and carefully advance the cutter so that it just touches the blank. Check the theoretical tooth depth from the table, and then move the cutter in further to nearly the full depth. Take a cut, slowly traversing the saddle. A steady hand is essential, taking care not to force the cutter through the work; if the cutter is advanced too quickly, an uneven finish will result. Check that everything is all right, then index to the next tooth. When this has been cut, deepen the cutter slightly and take another cut, and then check the land on the tooth tip. Keep indexing and taking small cuts until there is only a slight land at the tip of the wheel tooth. This will show up more significantly with the black marking previously placed on the wheel-blank rim.

Ensure that all the necessary slides are locked, then proceed to index and cut until the wheel is complete. Make sure that a finishing cut is taken through the first two or three teeth that had not been taken to full tooth depth. This is most satisfying, and well worth the effort of making and setting up the equipment.

When this type of work is commenced, total concentration is necessary, as it is quite easy to miscount and then at the end you have half a tooth that you do not know what to do with.

Sometimes, when cutting fairly deep teeth, it is better to take two cuts, go right round the wheel once, and then take a finishing cut of about 0.127mm–0.250mm(0.005in–0.010in).

Fig. 175 Cutting the wheel.

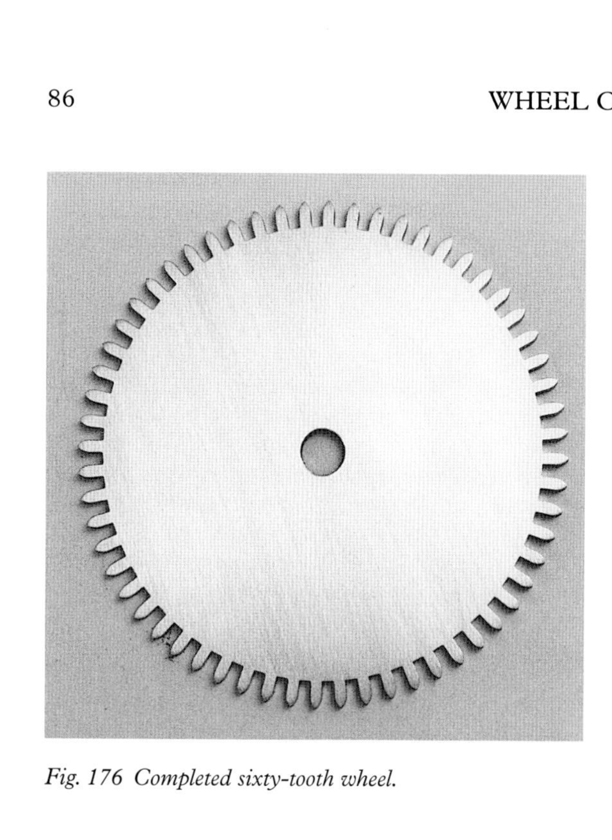

Fig. 176 Completed sixty-tooth wheel.

Whilst holding a wheel on a mandrel turned in the lathe will give extremely good results as far as concentricity is concerned, a more accurate method is to mount the finished wheel in a boxwood chuck as shown. Hold the wood disc in a three-jaw chuck, and bore out to give a slight fit and press the wheel home; the wheel bore can then be opened out slightly to suit its collet. The same wheel blank is shown being gripped in a precision six-jaw chuck; this will give exactly the same results. These chucks are expensive to purchase but will give a lifetime of service, and save many hours on set-up when holding discs, wheels, tubes and so on.

Fig. 177 Holding the finished wheel in a boxwood chuck.

Fig. 178 Gripping the wheel in a precision, six-jaw chuck.

Apart from the wheel cutting previously covered, there are a number of differing types of tooth form in clock movements.

CUTTING RECOIL ESCAPE WHEELS

Here again, commercial cutters are available, but home-made fly cutters can also be used. The diagram opposite shows the design of tooth of a typical recoil or anchor escapement wheel, usually found in longcase, bracket and skeleton clocks. It is without doubt the most prolific escapement ever used. It was invented, or evolved, some time between the mid to latter part of the seventeenth century, and is still just as popular today, more than 300 years later!

Fig. 179 Commercial recoil escape-wheel cutter.

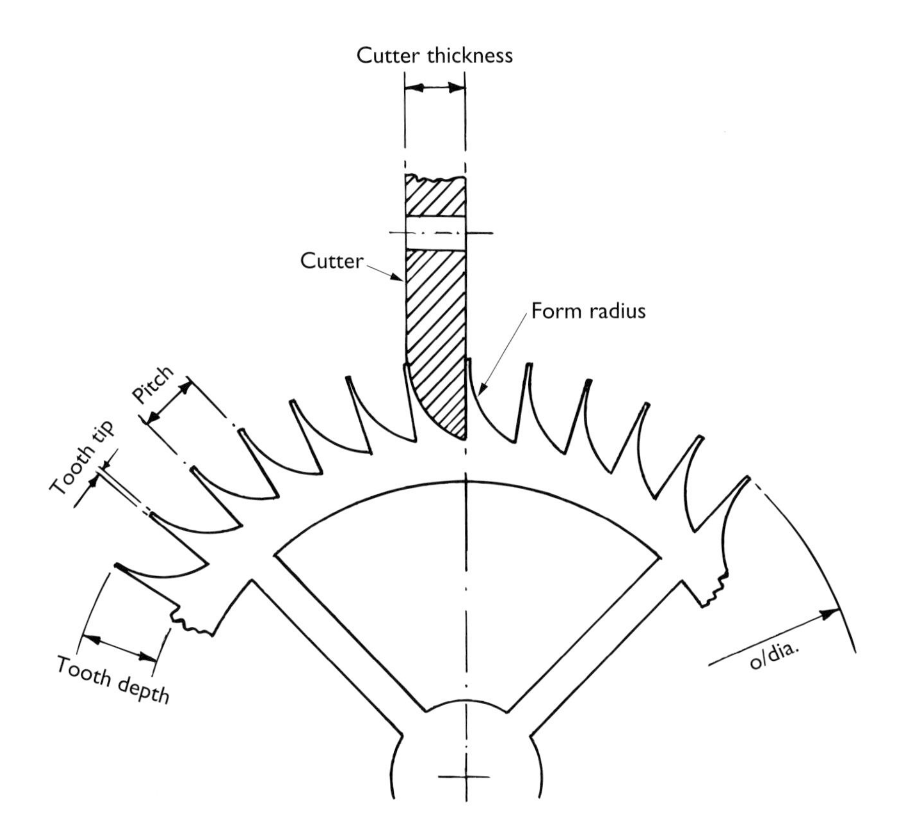

Fig. 180 Anchor- or recoil-escapement wheel and cutter.

As the individual design of escapements is a subject on its own, only the basic requirements will be covered when cutting this type of wheel.

The equipment used is exactly the same as for normal wheel cutting (as previously described), the only difference being the cutter. Whether a commercial cutter or a home-made fly cutter is used, the procedure is the same. Obviously a commercial cutter will give slightly better results as it is a multi-tooth cutter, and certainly when cutting thin escape wheels it will ensure that at least one cutting edge will always be in contact with the work at any one time, whereas a single-point cutter will be cutting intermittently. However, quite satisfactory results can be obtained by this latter method. If a replacement escape wheel is to be cut for an old clock, then it is possible a commercial cutter may not match the contour of the existing teeth exactly. If this is so, then it is quite an easy task to file and polish a piece of silver steel to the form, then harden and temper.

Note that when using fly cutters, speeds in excess of 4,000rpm are beneficial. Recoil escape wheels are covered by a range of seven cutters (*see* table).

Fig. 181 Various fly cutters.

The way to select a suitable cutter is as follows: multiply the diameter of the proposed escape wheel × π, and divide by the number of teeth to be cut. In this case assume an escape-wheel diameter of 35mm and the number of teeth to be thirty. Then:

$$\text{pitch} = \frac{35 \times \pi}{30} = 3.66\text{mm}$$

Recoil Escape-Wheel Cutters – Dimensions in mm

Size	Cutter thickness	Radius R.	Cutter dia.	Pitch less tooth thickness	Cutter bore
1E	1.64	3.28	26mm	1.31	7mm
2E	2.05	4.10	26mm	1.64	7mm
3E	2.56	5.12	26mm	2.05	7mm
4E	3.20	6.40	26mm	2.50	7mm
5E	4.00	8.00	26mm	3.20	7mm
6E	4.50	10.00	26mm	4.00	7mm
7E	5.80	12.25	26mm	5.00	7mm

The nearest cutter in the range is number 6E, at 4mm pitch. This will allow for the slight flat or land at the tip of the tooth. If a wider or narrower tooth tip is required, then the cutter can be fed into the wheel accordingly.

Now the cutter size is established, the escape-wheel blank is mounted on its mandrel. Ensure that it is well supported by back-up washers: this is necessary when cutting takes place, to prevent the teeth bending under the cutting load. Sometimes it is beneficial if two wheels are cut together, as this gives added rigidity.

Turn the outside diameter of the blank true to 35mm (1.377in). As the depth of cut will be much deeper than when cutting normal wheels, care is required when feeding in. Two cuts can be taken, but one is normally satisfactory if care is taken when making a pass; feed through slowly and evenly. A cutter speed of 4,000rpm will ensure that a good finish is produced.

Fig. 183 Completed escape wheel.

Fig. 182 Cutting a recoil escape wheel.

Ensure the blank is gripped tightly, as all the pressure of the cut is on one side of the tooth. This is particularly important when using fly cutters, as these create more vibration.

CUTTING DEAD-BEAT ESCAPE WHEELS

These are cut using a similar method to cutting recoil escape wheels. The dead-beat or Graham escapement is generally found in precision clocks, turret clocks and high-grade regulators. It was first used successfully by George Graham *c.* 1715. Unlike the anchor escapement previously mentioned, there is no recoil; the escape wheel remains stationary during the completion of the arc of vibration of the pendulum. The escape-wheel teeth are more fragile than on the anchor or recoil escapement, therefore more susceptible to damage.

Replacing the escape wheel is sometimes necessary, and Thornton's supply a range of cutters suitable for wheels, all having thirty teeth but different diameters. Whilst not all escape wheels have thirty teeth, the cutters will cover a fair range; they are also designed for an undercut of 6 degrees, which is necessary to give clearance so that only the tip of the tooth is resting on the pallet. On some escape-wheel teeth, the undercut is 10 degrees, but it is generally never greater, as this would make the teeth very weak.

To calculate the cutter size for a wheel having thirty teeth and an o/dia. = 45mm (1.772in):

Fig. 184 Commercial dead-beat cutter.

Fig. 185 Cutting a dead-beat escape wheel.

$$\text{Pitch} = \frac{\pi D}{N} = \frac{\pi \times 45}{30} = 4.71\text{mm } (0.186\text{in})$$

From the table (*see* page 91), the nearest is cutter 'A'.

The undercut is produced by offsetting the cutter from the centreline the calculated amount, as follows:

Calculation for 45mm diameter escape wheel:

Offset x = radius × tan 6 degrees

(From tangent table, tan 6 degrees = 0.1051)

x = 22.50mm × 0.1051

Offset x = 2.36mm (0.093in)

The centring micrometer device previously described can be used to advantage here. Determine the offset, and then set the micrometer thimble to the appropriate reading and adjust the cutter until it just comes in contact with the micrometer spindle face. Lock all slides, and cutting can commence. As a large amount of metal is being removed, a slow feed and a speed of 3,000 to 4,000rpm is required or the resultant wheel teeth will be damaged. Two views are shown of the dead-beat wheel in the process of being cut.

As previously mentioned, if a suitable commercial cutter cannot be matched to the required replacement wheel from either the dead-beat or recoil range, then a fly cutter will have to be made (*see* Chapter 8).

Fig. 186 Supporting the wheel blank with a backing washer.

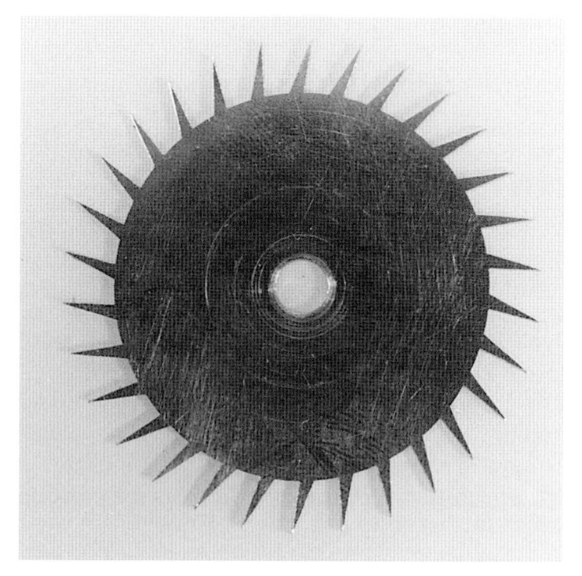

Fig. 187 The completed dead-beat escape wheel.

Fig. 188 Dead-beat escape wheel and cutter.

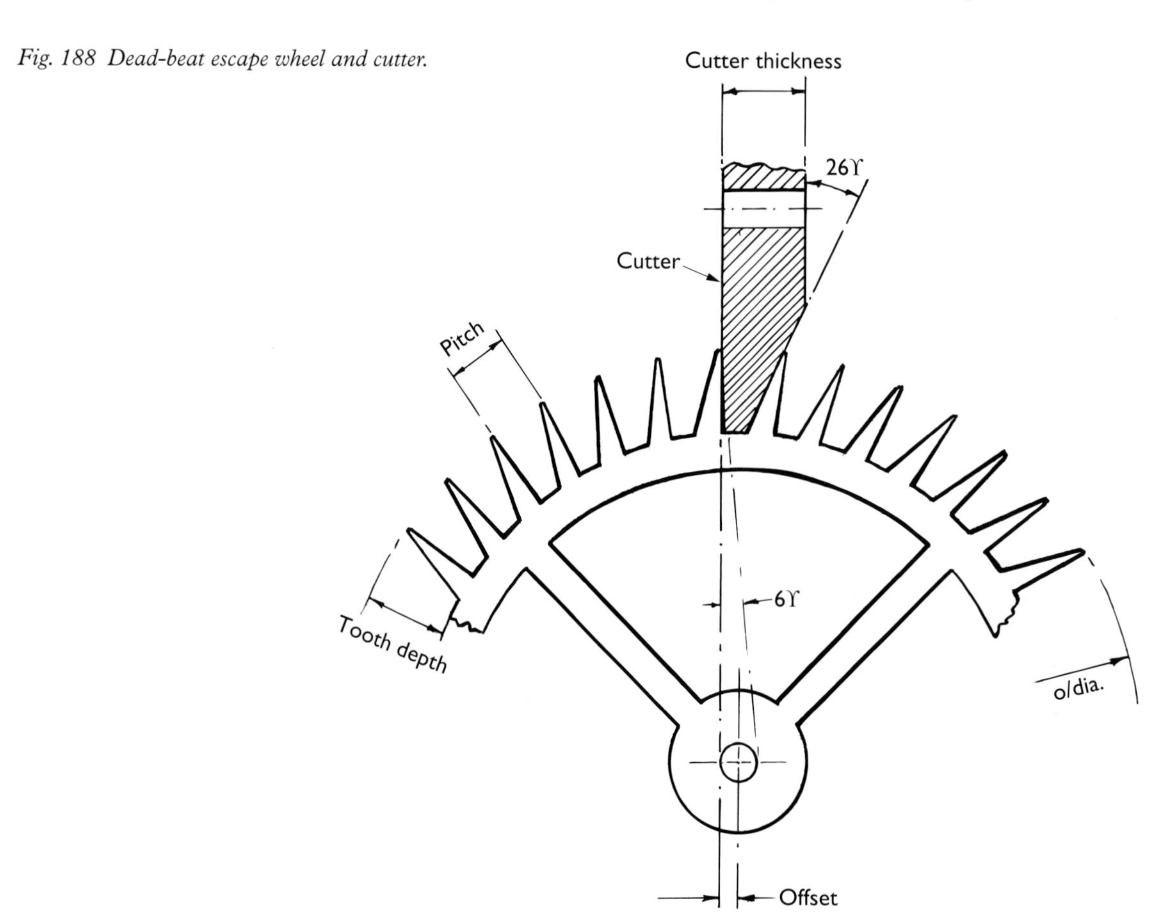

Dead-Beat Escape-Wheel Cutters – Dimensions in mm

Cutter dimensions		Escape-wheel dimensions				
Cutter ref.	Cutter thickness	Escape wheel dia.	Pitch	Depth	Cutter offset	No. of teeth
A	5.0	44.4	4.64	5.0	2.30	30
B	4.1	36.0	3.75	4.1	1.88	30
C	3.1	27.4	2.88	3.1	1.42	30
D	2.2	19.0	1.98	2.2	1.00	30

Dead-Beat Escape-Wheel Cutters – Dimensions in inches

Cutter dimensions		Escape-wheel dimensions				
Cutter ref.	Cutter thickness	Escape wheel dia.	Pitch	Depth	Cutter offset	No. of teeth
A	0.197	1.75	0.183	0.197	0.091	30
B	0.161	1.42	0.148	0.161	0.074	30
C	0.122	1.08	0.113	0.122	0.056	30
D	0.087	0.75	0.078	0.087	0.039	30

Dead-beat escape-wheel cutters have a bore of 7mm and o/dia. of 30mm.

CUTTING CONTRATE WHEELS

The illustrations show first, the back of a typical eighteenth-century bracket clock with verge escapement, and also the actual escapement, where the contrate wheel is clearly visible. The crown wheel is also shown, and the cutting of this will be described later.

Quite often it is necessary to replace the contrate wheel if it is worn or damaged. To calculate the module or tooth size, the outside diameter of a contrate wheel is equal to the pitch diameter of a wheel from a standard wheel train. This would normally be required if a calculated module was being used, but it is more likely that the original tooth form would have to be matched with the nearest standard cutter available, or a special cutter made to match the original tooth profile and the outside diameter of the blank already determined.

In the case of a missing contrate wheel, to obtain the module it would be necessary to measure the matching escape-wheel pinion and

Fig. 189 Verge bracket clock movement.

Fig. 190 Verge escapement showing contrate wheel and crown wheel.

calculate the module. A typical pinion would be six-leaf 4.4mm o/dia, then using formula 9:

$$M = \frac{OD}{n + 1.71} \quad M = \frac{4.4}{6 + 1.71}$$

$$M = 0.57$$

A typical contrate-wheel tooth count is seventy-two. Applying formula 12:

pitch dia. = M × N

PD = 0.575 × 72

PD = 41.4mm

Note: This is the finished outside diameter of the blank before the teeth are cut.

Machining Procedure

The series of photographs below shows the machining procedure. The wheel blank is prepared first, and mounted on a suitable arbor. Follow the procedure outlined previously for

Fig. 191 Machining the contrate-wheel blank.

Fig. 192 Parting off the blank.

Fig. 193 The contrate-wheel blank on its arbor.

Fig. 194 Cutting the contrate wheel.

Fig. 195 Close up of the cutter and blank.

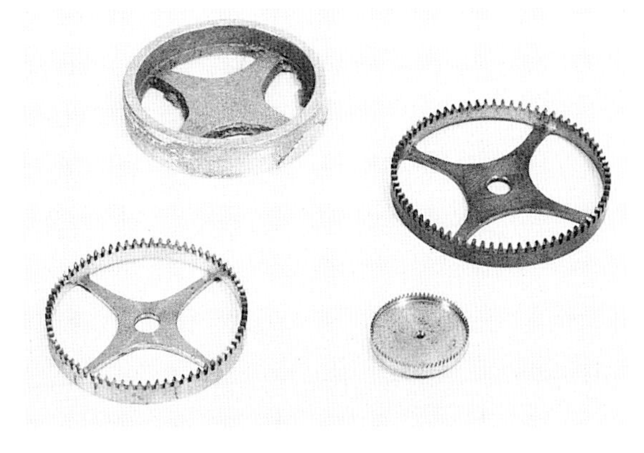

Fig. 196 The finished contrate wheel and wheel blanks.

cutting wheels, but in this instance it is necessary to use the vertical slide to traverse the cutter through the work. If a cutter frame is being used, then the cutter is operating in the horizontal plane, and would be traversed through the work by the lathe cross-slide. Centre it as before, and obtain the correct depth for cutting the tooth; the theoretical depth is obtained from the table shown in the Appendix. Lock up all slides before commencing wheel cutting. Note the counter-sunk screw holding the wheel blank in place, which gives clearance for the wheel cutter. Alternatively, the wheel can be cut on the round bar whilst still in the lathe chuck. Two finished wheels are illustrated, and a small carriage clock contrate cut, ready for crossing out; also shown is a cast yellow-brass wheel blank with crossings already formed. Early clock wheels were made in this manner.

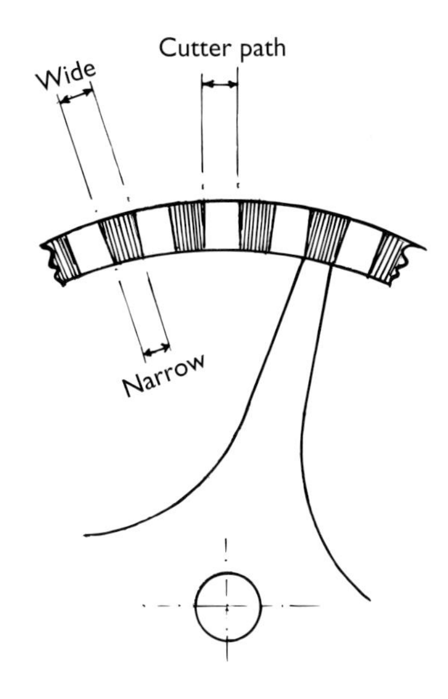

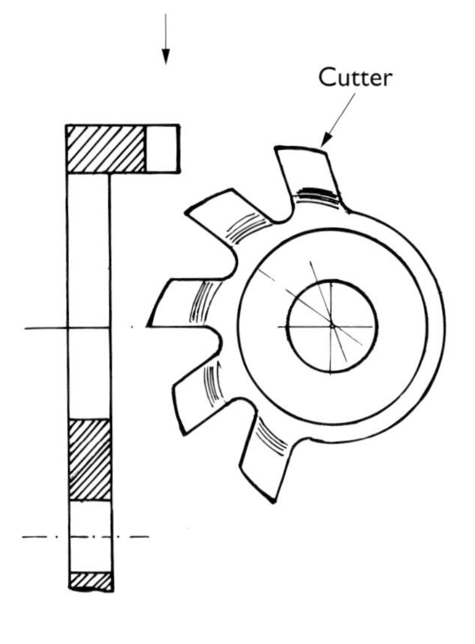

Fig. 197 Enlarged view of the tooth form – cut by gashing.

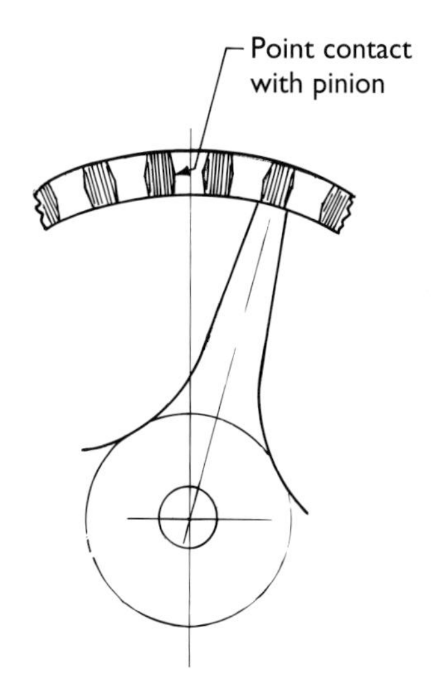

Fig. 198 Enlarged view of tooth form – machined by the hobbing method.

In the enlarged view of the tooth form (*see* page 93), it can be seen that the gashing process will leave a tooth form with a narrower section at the inside, and the tooth will drive the matching pinion on its outer diameter. If the contrate tooth form had been generated on a gear-hobbing machine, then the tooth form would be as illustrated above, and in this case the pitch diameter would be located mid-way along the radial width of the tooth and a smoother drive would ensue.

Antique clock contrate wheels often have the rim thinned by chamfering internally. This is rarely seen on modern clocks.

Contrate wheels are also used on carriage clocks, both French and English, to convert the direction of the drive to a vertical plane, enabling the contrate wheel to mesh with the escape-wheel pinion. The modules of these are much smaller, therefore finer and more accurate work is required when cutting.

CUTTING CROWN WHEELS

Referring to the photographs from the previous section on cutting contrate wheels, it will be seen that the verge escapement crown wheel is similar in design. The cross-section of the two types of wheel is almost identical, the main difference being the shape of their respective teeth. A similar wheel-cutting procedure is therefore used.

Verge crown wheels almost always have an odd number of teeth to enable the verge staff to lie directly across the centreline. Some continental verge escapements have an even number of teeth, when the verge staff lies at an angle – not as neat an arrangement as the English-style verge.

A typical tooth count would be twenty-seven, twenty-nine, thirty-one. This type of wheel is often used for the alarm work on lantern and bracket clocks. Crown wheels are more likely to be replaced than contrate wheels due to wear on the escapement decreasing the depth of engagement of the pallets to the crown-wheel teeth, and the wheel going on a ‘run’ and so removing the tips of the teeth. This also happens if someone inadvertently commences to remove the verge staff whilst power is still on the train.

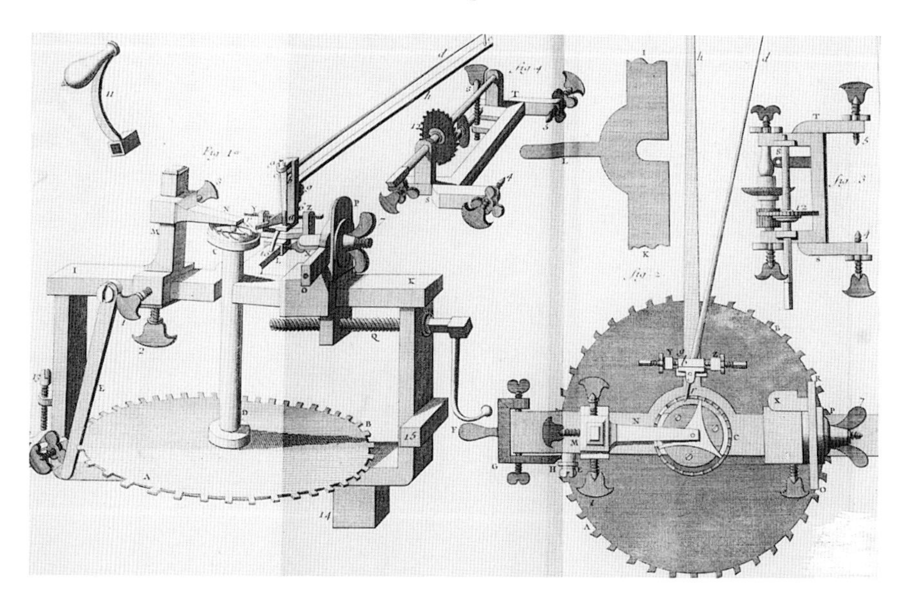

Fig. 199 Early machine from Antoine Thiout, showing crown-wheel cutting.

Fig. 200 Commercial crown-wheel cutter and wheels already cut.

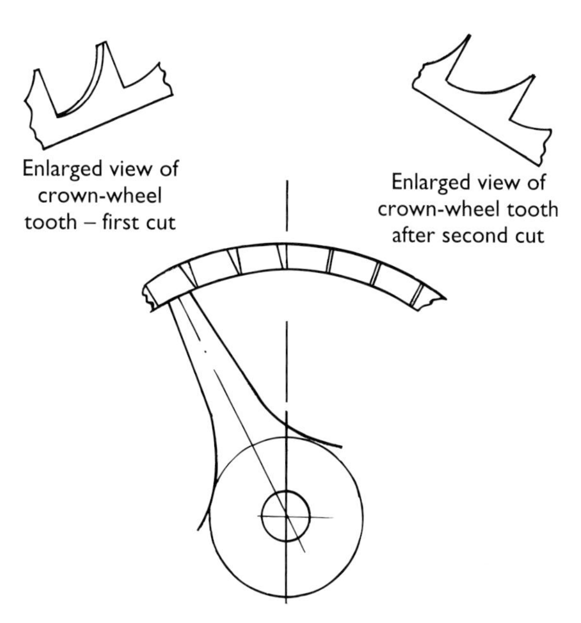

Fig. 202 Enlarged view of crown-wheel teeth.

Fig. 201 Cutting a crown-wheel for a verge escapement.

Fig. 203 Wheel tooth backed off.

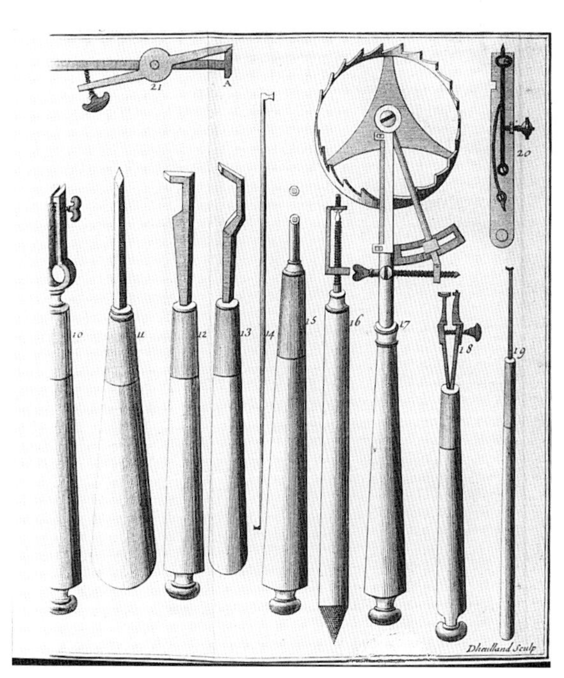

Fig. 204 Tool for checking the pitch of a crown-wheel tooth; engraving from Antoine Thiout, 1741.

Verge crown-wheel cutters are now available commercially. These are similar to recoil escapement cutters but have a longer radius, sizes 1E–4E inclusive. As mentioned previously, with escape-wheel cutters, it is sometimes necessary to make a special cutter to match a tooth form that falls outside the standard range.

A crown wheel is shown (page 95) in the process of being cut. Note that the teeth faces are cut vertical for English work, although on continental work the teeth are quite often undercut. On English work, if commercial cutters are used, it is necessary to cut upwards from the inside, due to the direction the verge teeth are facing. Some continental clocks, for example French Comtoise clocks, have teeth facing in the opposite direction, as do verge watches.

If the cutter is set vertically through the wheel centre, then a tooth is produced where its outer thickness is greater than the inner one. Whilst this does not affect the working of the escapement, the tooth would normally be backed off to make it more aesthetically pleasing. This can be carried out either by the use of needle files, or by re-setting the cutter. The teeth tips should finish with a slight land and follow a line parallel with the vertical face that passes through the centre of the wheel. When verge crown wheels for watches are being cut, the clearance is essential for the pallets to operate successfully.

If the vertical slide is set over, it is necessary to determine the cutting angle. The Peak optical scale is ideal for this purpose. Magnifications of 7× and 10× are available, with various scales; the angular scale is used for measuring the tooth already cut, and a typical angle would be 15 degrees. When cutting, ensure the cutter does not touch the face of the adjoining tooth. With this method, it is possible to obtain a nice parallel land.

With the next method, a stop should be fitted to the vertical slide, as shown. Only one cut is taken, but great care is necessary in setting the depth stop. The cutter should be set so that it does not traverse right through the work, only partially, the cutter radius forming the parallel land. Using this method, cutting has to take place from the top, therefore a fly cutter was produced with the opposite form.

Particularly with fly cutters, keep your hands and loose clothing well away when they are revolving. Even when static, a fly cutter is dangerous; it will easily gash your hand. This also

Fig. 205 Milling spindle set over to give relief to the tooth.

Fig. 206 Peak optical measuring scale.

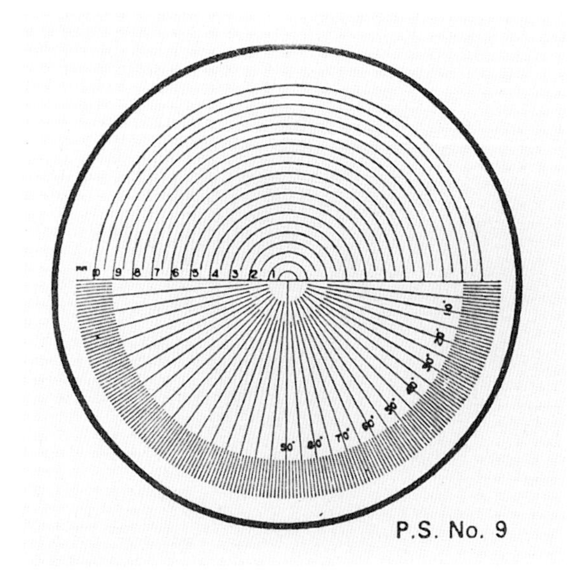

Fig. 207 Peak protractor scale.

Fig. 208 Cutting the crown wheel with a depth stop.

Fig. 209 Fly cutter and arbor.

Fig. 210 The Aciera F1 set-up for wheel cutting.

Fig. 211 Close-up of the cutter and blank.

applies when using multi-tooth cutters, though to a lesser degree.

Whilst the set-up for cutting crown wheels has been shown on the lathe, this can also be carried out on the milling machine. An Aciera F1 precision milling machine is shown, set up with the horizontal spindle in position. Another view gives details of the actual cutting taking place.

GOING BARRELS

Fig. 212 Going barrel and short-form cutter.

Short-form commercial cutters are available for cutting French and German going barrels. The dedendum is reduced from 1.57 x M to 1.07 x M. If the barrel and tube are one piece, then with a standard cutter there is a danger of cutting into the barrel outside diameter. Obviously this would not occur with a two-piece barrel, as the two components are soldered together and can be separated for machining. Short-form teeth are also stronger; in some instances these have radiused roots, which also add to the strength. This sort of short-form cutter has been found useful for many types of American clock; sizes would range from 0.40M to 0.65M. A typical French going barrel and a short-form cutter of corresponding size are shown; note the wheel-cutting marks on the barrel tube, possibly a replacement.

WINDING AND RATCHET WHEELS

The cutting of these follows the same procedure as before, it is just the form that is different. Here again commercial cutters are available to make the job easier. Clocks manufactured in Germany around 1900 and later, have winding wheels and some train wheels with round bottoms: these

Fig. 213 German mass-produced movement showing winding wheels.

Fig. 214 Straight-tooth commercial ratchet cutter.

Fig. 215 Typical French mantel clock.

Fig. 216 Striking movement from a French clock showing winding ratchet wheel and click.

were produced for tooth strength. This type of wheel was also produced in England on later clocks, but the teeth were cut by hobbing or gear-generating methods. As production methods improved, some of the wheels were stamped out, but it is often necessary to cut a wheel to match a damaged component, particularly winding wheels.

Straight-tooth winding ratchet wheels often have to be replaced. Two cutters are available, 60 degrees and 70 degrees. Most French clocks have winding ratchet wheels with radius teeth, and again, these can be cut quite easily. A range of four commercial cutters is available, with radii ranging from 4–13.5mm.

CUTTING WHEELS FOR WATCHES

In the last twenty years there has been a resurgence in watchmaking and watch restoring. Cutting small wheels for pocket watches is not difficult, though a small precision lathe is required to carry out the work. The Schaublin 70 precision lathe shown on page 100 is ideal, but any similar lathe would be suitable. The lathe can be fitted with either the standard wheel-cutting attachment as shown, or direct indexing plates. A commercial centring microscope is shown, that is very useful for accurate work: as can be seen, the

Fig. 217 Schaublin 70 precision bench lathe.

Fig. 218 Set-up showing dividing head and milling spindle on the Schaublin 70.

milling spindle is quite robust, having large bearings, always an excellent feature of a well designed spindle. The set-up shows a small wheel being cut. Follow the same procedure as described for cutting clock wheels. This lathe is also quite capable of producing small steel pinions, and the cutting of these will be shown later.

Early tool catalogues have shown attachments for cutting wheels on watchmakers' lathes since the late 1900s, although it would be difficult to cut pinions on this type of machine due to the lack of rigidity; a more heavily constructed machine such as the Schaublin 70 just described would be more suitable. The Bergeon 8mm watchmaker's lathe is shown on page 102. It has a well constructed vertical slide and milling spindle, and a large division plate for direct indexing. Greater care is required when centring the cutter; a miniature centring micrometer is being used here.

Fig. 219 Isoma centring microscope.

Fig. 220 Cutting a small wheel on the Schaublin 70.

Sometimes special arbors have to be machined to suit the small bores of the watch wheels. The illustrations show a pocket-watch wheel being cut, using a 0.15 module cutter. To true the centre hole, the wheel blank can be mounted on a wax chuck with shellac. It is preferable to make a brass chuck with a recess to locate the outside diameter of the wheel, which will ensure concentricity. Use shellac to hold the blank in place, and bore out the centre hole in the lathe with a small boring tool. Remove the wheel from the chuck by soaking in methylated spirits.

The illustration shows a special mandrel for holding watch wheels with very small bores. The locating portion of the arbor is plain with no thread; as threading is difficult on such small diameters, locking pressure is applied by the use of the tailstock holding a hardened steel ball against the hollow spindle, as shown in the drawing. Four wheel blanks can be successfully held at any one time by this method.

Commercial wheel and pinion cutters for watches are available in a limited range, sizes as follows: wheels 0.15 and 0.18 Module, and matching pinion cutters for each size in six, seven, eight, ten, twelve and sixteen leaves. When

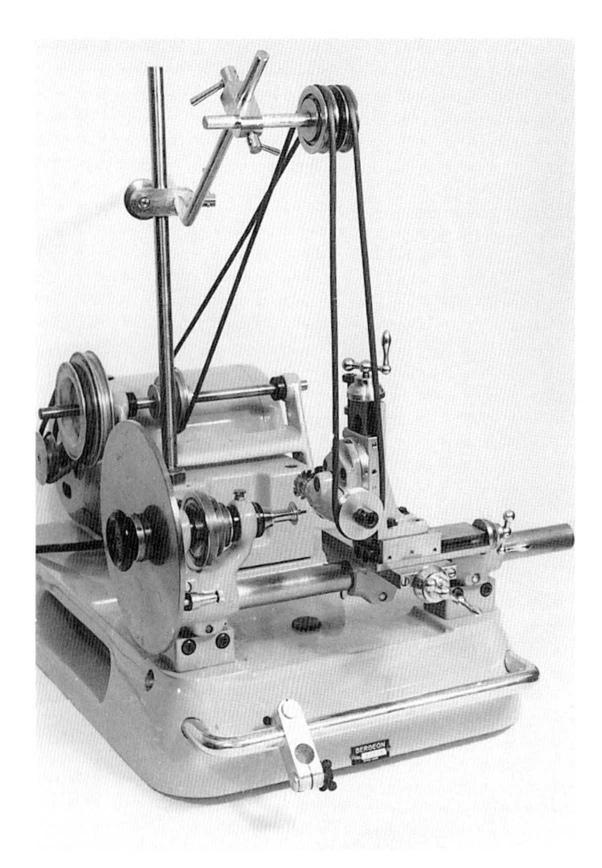

Fig. 221 Bergeon 8mm watchmakers lathe set-up for wheel cutting.

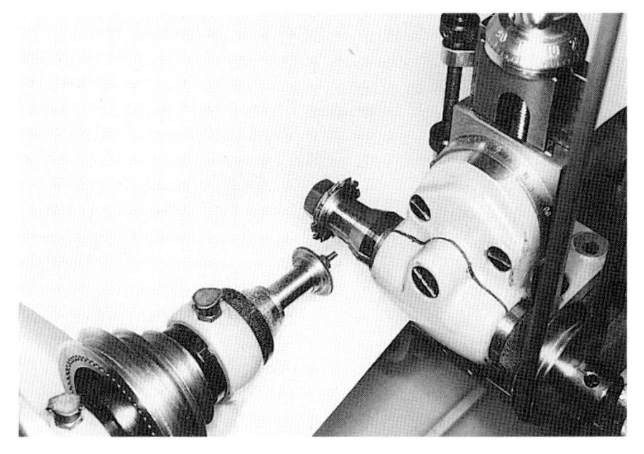

Fig. 223 Milling spindle fitted to the Bergeon 8mm lathe.

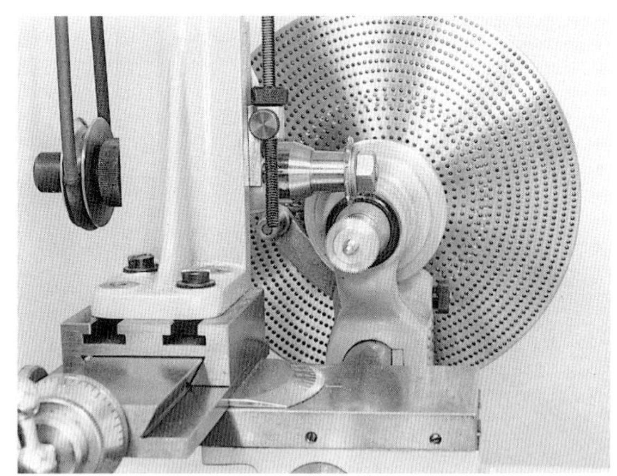

Fig. 224 Cutting a pocket-watch wheel.

Fig. 222 Miniature centring micrometer.

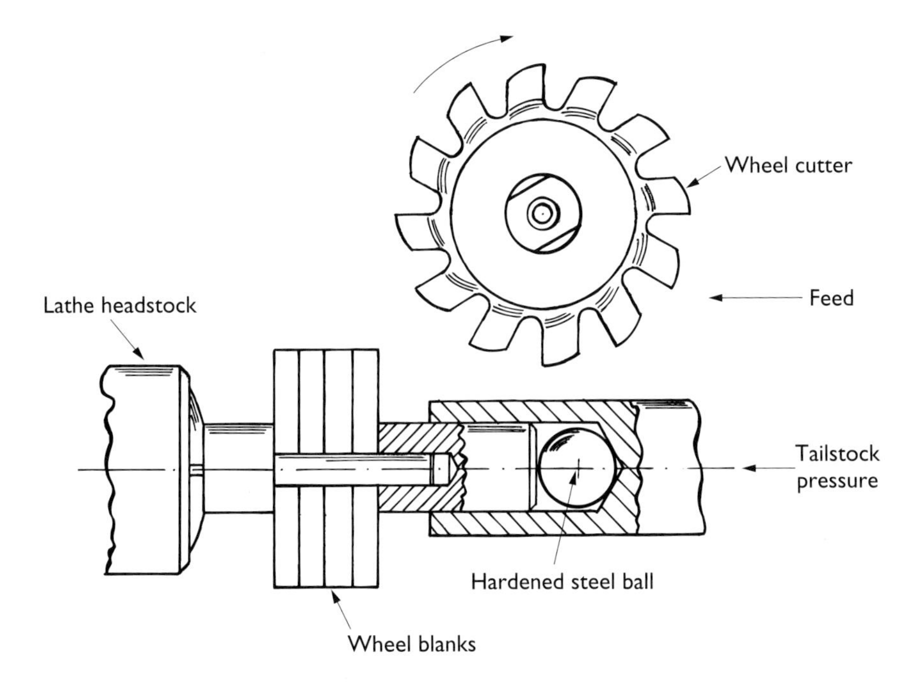

Fig. 225 Special mandrel for holding small watch wheels.

a small wheel has been cut, the problem is then the accurate crossing out required for the wheel to have the correct proportions. This will be covered in Chapter 9.

The sizing of wheels and pinions is more important on watch trains. However, for replacing a wheel in an English lever or verge pocket watch, it is not a difficult job to match up with the nearest cutter, then check by calculation. If finer work is to be encountered where the wheels and pinions have to be replaced from precision watches, then it is more important that the modules are carefully matched. The theory of gearing is covered in *Watchmaking* by G. Daniels, and *Gears for Small Mechanisms* by W. O. Davis. Using calculations from these two sources would probably entail requiring cutters that are not commercially available, and the only way to progress would be to make one's own. Fly cutters are suitable for cutting brass wheels, but with the profile being small, some form of optical projector would be required.

7 PINION CUTTING

Having now covered the cutting of brass wheels, the slightly more difficult process of cutting steel pinions is addressed. The basic procedure is the same, namely that the blank is indexed and the cutter is fed through the work, but this is a much slower process; the cutter speed is now considerably reduced and a cutting lubricant is used.

STEELS FOR PINIONS

Until recently, the most commonly used steel for pinions was a free-cutting silver steel; its reference was KEA108, and it was manufactured by Sanderson Kayser in Sheffield. As anyone who has machined normal silver steel will know, it is tough, and it is not easy to obtain a good finish. KE108 had an addition of 0.2 per cent selenium, which gave it its free machining properties. This is now considered to be a dangerous substance, however, and can no longer be used in steel making.

Alternative steels for pinions are EN24t, EN8DM and silver steel.

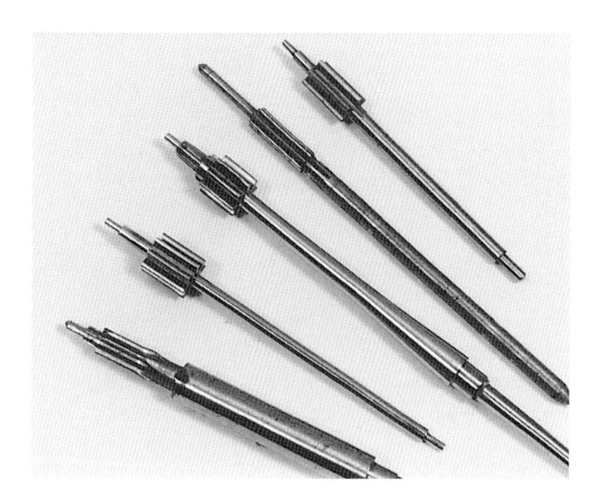

Fig. 226 Selection of cut pinions.

EN24t

EN24t (BS Spec. 817 M40) is a steel supplied already hardened and tempered. Whilst it is tough, it will machine quite well, and a good finish can be obtained. It is a nickel chromium molybdenum steel with a carbon content of 0.36–0.44 per cent. The advantage is that once machined and polished, no further heat treatment is required. Sizes available are from 6mm (¼in) dia. upwards, and equivalent metric sizes, and the condition of finish is bright drawn.

EN8DM

A steel that is more easily machined than EN24t is EN8DM (BS Spec. 212 A42), a medium carbon steel. Whilst it can only be hardened by flame or induction-hardening methods, it will give reasonable results for long-term wear. The steel has a carbon content of the range 0.40–0.45 per cent, but with a higher sulphur content than normal EN8, which gives improved machinability. Sizes available are basically as the EN24t range.

Silver Steel

Finally there is silver steel, a high-carbon tool steel. This varies considerably in machinability, but once finished and polished, will give excellent long-term wear results. Carbon content 'C' is 1.10–1.20. It is supplied in the annealed condition, and is available in centreless ground lengths in sizes from 3mm to 40mm (⅛in–1¼in) dia. Smaller sizes than 3mm are also available in the drawn and sometimes ground finish; this makes the material extremely suitable for arbors as well as pinions. Hardening and tempering can quite easily be carried out in the home workshop without any specialist furnace or heat-treatment facilities, although more controlled results can be achieved by the use of a small muffle or pot furnace, which can be either made or purchased. With some form of temperature control, improved heat-treatment results can be achieved.

Recommended Uses

As a general guide, for English clocks EN24t would be recommended, although EN8DM is easier to machine. Clock pinions manufactured some thirty years ago using this material show no sign of wear.

For French clocks where the pinions and arbors are usually hardened and tempered, silver steel is recommended. There is a high grade silver steel on the market (*see* List of Suppliers) that is manufactured to closer specification than normal, and machines quite well. Whilst silver steel in its un-heat-treated state would be quite satisfactory for French clock pinions as replacements and would give long-term wear characteristics, the small-diameter arbors of around 1mm are not strong enough unless heat-treated.

CUTTING THE BLANK

The illustration shows the pinion blank ready for cutting in the lathe; as can be seen, the set-up is identical to the wheel-cutting method used previously. The cutter is run considerably slower when cutting carbon steel, however, and a positive drive by a miniature vee-belt is obtained from the countershaft via the $\frac{1}{8}$hp motor. Referring back to the calculations for the pinion in Chapter 5, it was determined that the outside diameter was 7.77mm (.306in) and an eight-leaf 0.8 module cutter was to be used. The table in the Appendix gives all dimensions required for tooth depth and root diameter; the latter information is particularly useful for machining pinion heads to fit existing arbors. From the table we can see that the calculated full tooth depth for this pinion is 2.2mm (.087in).

Select a piece of material 9mm ($\frac{3}{8}$) diameter. Make sure that it is longer than the total length of the arbor, to allow for centring the end; this is required to support the work, but will later be faced off. Mount the material in the three-jaw or collet, centre drill the end, and support the work with a male centre in the lathe tailstock. Note that there are other ways of supporting the work, and these will be covered later. The material can then be machined to the required dimensions.

Now that the blank is ready for cutting, mount the vertical slide, together with the milling spindle, on to the lathe cross-slide. The dividing head, or division plate, can also be mounted to the lathe mandrel. The illustration shows a notched division plate mounted on the rear of the lathe spindle; this is a rapid method of indexing when cutting pinions. Before cutting commences, there are a number of important points to be observed.

Cutting Speed

The surface cutting speed is important, because if it is run too fast there is a danger of ruining the cutting edges. The recommended speeds are as follows:

Fig. 227 Pinion blank ready for cutting.

Fig. 228 Notched index plate fitted to lathe mandrel.

Cutter o/diameter	14mm	=	430/500rpm
"	20mm	=	300/350rpm
"	24mm	=	250/290rpm
"	26mm	=	230/270rpm

In this instance, therefore, with the cutter 0.8 module with an outside diameter of 24mm, the spindle speed should be around 250rpm, and the cutter should have a copious supply of cutting oil to lubricate both it and the work. The lubricant can be applied with a brush if no pump is fitted. If the set-up is rigid, then it is often possible to take quite heavy cuts, three cuts being sufficient to achieve full depth. It is better to feel your way, however, and not push the cutting procedure to its maximum. A light finishing cut of 0.13mm (0.005in) would give a good finish, although this would possibly be reduced on a pinion having a module of 0.3 or less. Ensure that driving belts are tightened sufficiently to give a positive drive. Sometimes it is preferable if the bulk of the material is removed with a high-speed steel slitting saw to three-quarters of the depth of tooth, as this will often save both time, and wear and tear on the pinion cutters.

It is always an advantage if all the leaves are roughed out, leaving a small amount of material for the finishing cut. Then with the final cut adjustment locked up, each tooth is indexed in turn and the cutting is carried out. This will ensure that all leaves are equal in depth.

Whilst it was noted that the theoretical depth of tooth was 2.2mm (0.087in), as with wheel cutting, it is better to feed down with successive cuts until the addendum curves just meet. A slow and even feed is necessary, and this can be accomplished by traversing the saddle of the lathe, using the leadscrew handwheel.

Ensuring an Even Feed

A method of traversing the lathe leadscrew automatically is shown. This is a design made by the author: it is quite simple, and will give an extremely smooth finish to the pinion leaves. A vee pulley is fitted to the lathe leadscrew and is driven by a miniature vee belt via a small,

Fig. 229 Slow-motion drive to leadscrew.

fractional horsepower DC motor with Thyristor control with reduction gearbox. The drive to the leadscrew is infinitely variable and can be adjusted to suit the work. A description of this attachment was published in *Model Engineer* November 1987, under the heading 'Driving the Saddle' by J. Malcom Wild.

In operation, the pinion is roughed out by using the leadscrew handwheel or saddle handle, and then the leadscrew half nuts are engaged and the drive set in operation. A light finishing cut is set, and as soon as the cut is complete, the leadscrew nuts are disengaged and the saddle wound back for the next cut. This arrangement can also be used for a fine feed for turning if no gearbox is fitted to the lathe.

Ensuring Rigidity

Before cutting takes place, ensure that the cutter spindle is square with the work, and that the cross-slide and vertical slide are locked firmly in position. This is most important in pinion cutting, as there is far more cutting resistance experienced than when cutting brass wheels, and rigidity is the key to a good finish. Note the chamfer at the front of the pinion outside diameter on the completed pinion; this assists the cutter to 'lead in' when cutting takes place. It is beneficial if the cutter arbor is threaded left hand, because if a right-hand thread is used, there is a tendency for the cutting action to unscrew the cutter locking nut, particularly if heavy cuts are taken.

Cutting Lubricants

The importance of cutting lubricants has been mentioned, and various types are suitable. Multi-spec cutting and tapping fluid is a neat cutting oil which can be applied using either a small pump or a brush; this will help to give a good finish to the work. There are also pastes that can be applied easily with a brush, for example Rocol RTD metal-cutting compound. This has a consistency like treacle, and has extreme high pressure cutting agents as additives, which prolongs the tool life and will cling to both tool and work-piece. Trefolex cutting compound is also very good: it is a well known product for various machining operations, and again, this can be applied with a brush.

A continuous stream of neat cutting oil is obviously the best solution when machining, as it keeps the work cool and washes away the swarf. This arrangement is not always possible in the small workshop, however application by brush is quite satisfactory where continuous production is not being carried out.

Centring the cutter

When wheel cutting was described, the importance of centring the cutter was referred to, and when cutting pinions this is doubly important, as any off-centre cutting will show up dramatically due to the depth and low number of teeth on a small diameter. If the micrometer centring tool or centring microscope is used, no problem should be encountered, but if using any of the

Fig. 230 Completed pinion.

Fig. 231 Profile of a pinion cut off-centre.

other methods, great care is required or the teeth or leaves will lean.

If no accurate method of centring is available, a trial run should be carried out on a piece of brass until the cutter is exactly on centreline. Check the tooth form with an eyeglass; it will be quite apparent if it is incorrect, as only a small amount of offset will show up with a tooth form that leans and is not central.

A sample module 1.0 six-leaf pinion deliberately cut with the cutter offset 0.127mm (0.005in) is shown above. The profile clearly shows how such a small amount of offset can affect the shape of the leaf or tooth.

CUTTING SMALL PINIONS

Pinions for French clocks are quite small, often around 0.25 module, and have arbors 1.25mm diameter (0.049in) and smaller. These are more difficult to machine than pinions for English clocks which usually have a module around 0.8, as we have seen previously, and are, of course, substantially more rigid.

When machining slender arbors, a travelling steady or support is required, otherwise the arbor will bend or break due to its length-to-diameter ratio. A method of overcoming this is to use the small diameter turning tool. This accessory is based on the toolboxes used on Swiss automatics for slender turning. The illustrations show the tool in operation reducing a length of silver steel of 3mm (⅛in) diameter in one pass, to 1.25mm (0.050in) dia. Note the top rake on the cutting tool.

When using the tool it is necessary to prepare a bronze bush with a reamed hole to match the outside diameter of the material being turned. Once the tool has been used a few times, a good range of bushes will have been built up, so that most standard bar sizes will be covered.

Fig. 232 Slender turning tool.

Fig. 233 Turning a slender arbor.

Fig. 234 Close-up of the slender turning tool in action.

Fig. 235 Turning tool showing cutting rake.

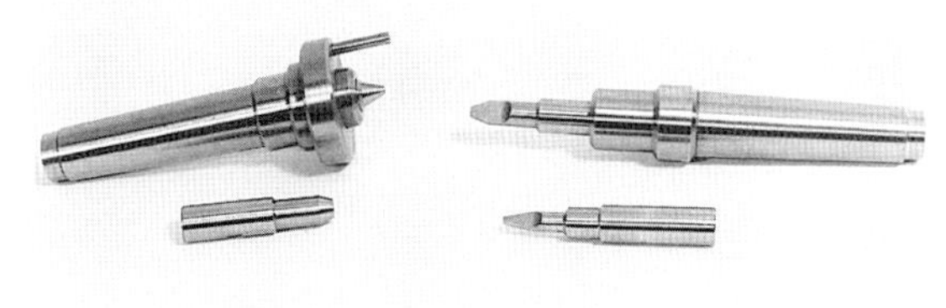

Fig. 236 Driving plate and special centre for pinion cutting.

Hold the 3mm (⅛in) diameter material in the collet or chuck, and machine a 60-degree male cone to act as a centre. Turn the material down to the required diameter for the arbor. The cone centre will then be true, and can be supported by the tailstock centre whilst the pinion is being cut and the arbor is finally sized and polished.

When turning takes place, oil the bush and work, and lubricate the edge of the cutting tool. Set the cutting tool to the required diameter and cutting can then commence, either by hand feed or a slow automatic traverse.

Fig. 237 Jack supporting a small pinion.

Accessories to Support the Work

It is an advantage to make various accessories to assist in cutting small pinions. Figure 236 shows various centres, and a driving plate for holding the pinion whilst it is being cut and also for finishing the arbor and pivots. The smaller centre, at lower left, has a female cone with internal grooves cut to enable the work to be gripped without using a carrier.

Figure 237 shows the set-up and a completed pinion of M0.25. Note the supporting jack to prevent the work from flexing. This single support jack is quite satisfactory, but if pinions with an odd number of leaves are cut, each time the pinion blank is indexed the jack has to be released and then reset.

Fig. 238 Double support jack for small pinions.

The pinion supported in Figure 238 shows a jack with two supporting vees; each vee supports the work on either side of the portion of pinion being cut. With this method it is possible to index the work without any adjustment to the jacking screws.

Another method of supporting the work if the arbor is slender is shown, mounted in the lathe tailstock. Standard 8mm watchmaker's lathe collets are used. As can be seen, the work can be held very close to the action of cutting, and it is supported true so as to give complete concentricity.

The basic components of the tool are shown in the illustration. The main body is machined from a No. 2 Morse taper blank; the inner sleeve can be either brass or bronze, machined to accept standard 8mm collets. A lock nut holds the collet in position, and enables it to be gripped onto the work. This is necessary because, as with most small bench lathes, there is no drawbar hole through the tailstock barrel. The locking nut is operated by a small tommy bar through a radial

Fig. 239 View of jack showing detail of construction.

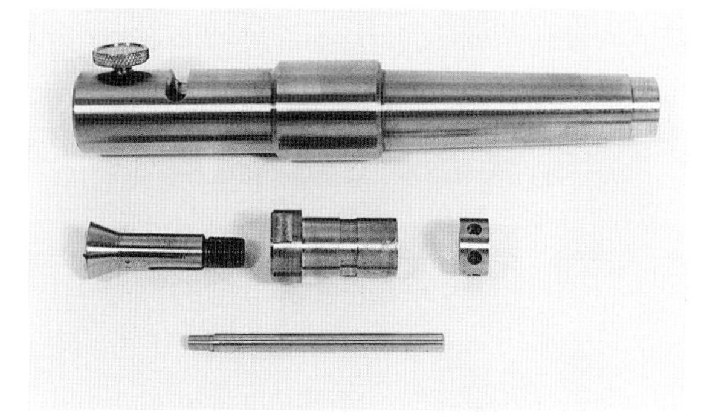

Fig. 241 Components of the tailstock support.

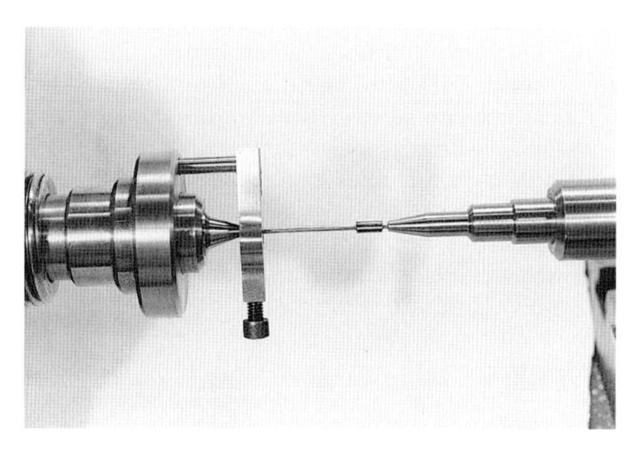

Fig. 242 Holding the work between female centres.

slot in the main body. A knurled locking screw is provided to lock the bronze sleeve when pinion cutting takes place. By using standard 8mm collets, it enables the tool to accommodate most small arbors.

A dual purpose for the tool is as a revolving support. If the bronze sleeve is lubricated and the knurled locking screw released, additional turning operations can be carried out to the work whilst the bush revolves in its support of the main body.

Fig. 240 Tailstock support to take 8mm watchmaker's lathe collets.

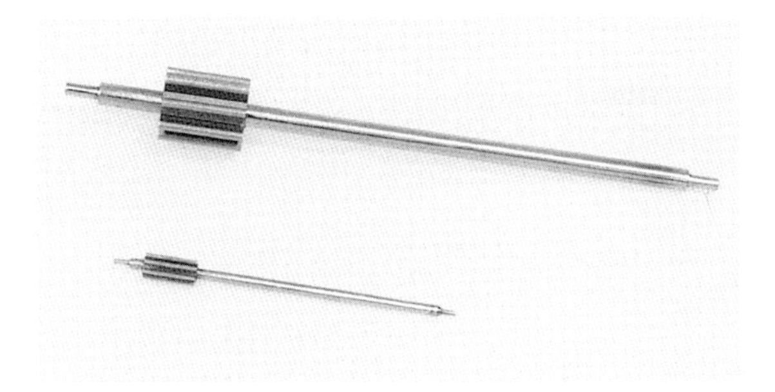

Fig. 243 Comparison of 0.25 module and 0.8 module pinion.

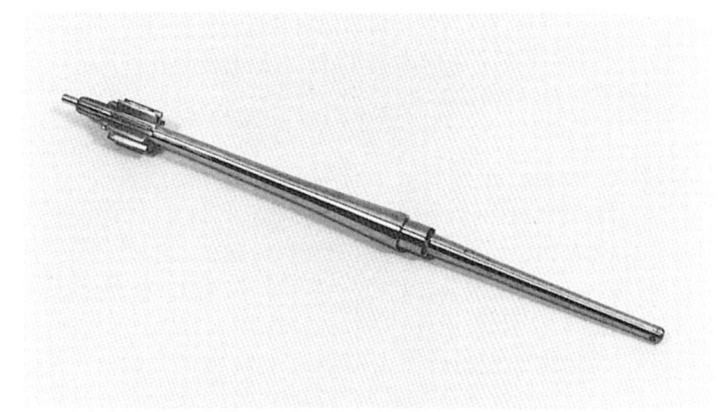

Fig. 244 Completed centre pinion and arbor.

A set-up is shown using the driving plate and the special centres previously referred to. A small French clock pinion and arbor is held between two female centres to enable the arbor to be polished. As can be seen from a number of the previous photographs, even fine work can be carried out on a 3½in bench lathe, although obviously it would be considerably easier on a good instrument or clockmaker's lathe.

When calculating modules for French clocks, it is often necessary to make an adjustment as, when using the standard cutters available today, the dedendum is excessive and will cut into the arbor. It is sometimes better, therefore, if a final cut is applied using a cutter with 0.05 module less on a scrap piece of metal to check the pinion leaf depth before machining the complete pinion. This can often save a considerable amount of time.

VARIOUS OPERATIONS IN PRODUCING SOLID PINIONS

Centre pinions and arbors are not easy to machine due to their length and because they are very slender. The following drawings show a method that works quite well, using the tool shown previously.

The various steps in the machining operation are shown in the illustration. The material is held in the lathe collet and supported by a lathe centre or, alternatively, the small-diameter turning tool can be used. The diameter for the cannon pinion is turned, then the shoulder for the friction spring. Using the special tool previously described, select a collet to suit the minute-hand arbor and lock the collet up so that it grips the work securely. Lubricate the bronze bush. This will act as a revolving centre, thus supporting the work and ensuring it is concentric. Set the lathe topslide over to the required angle and the taper

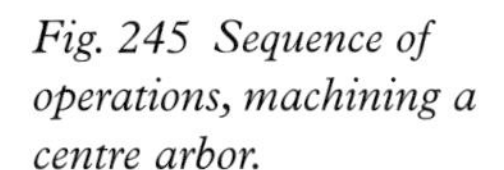

Fig. 245 Sequence of operations, machining a centre arbor.

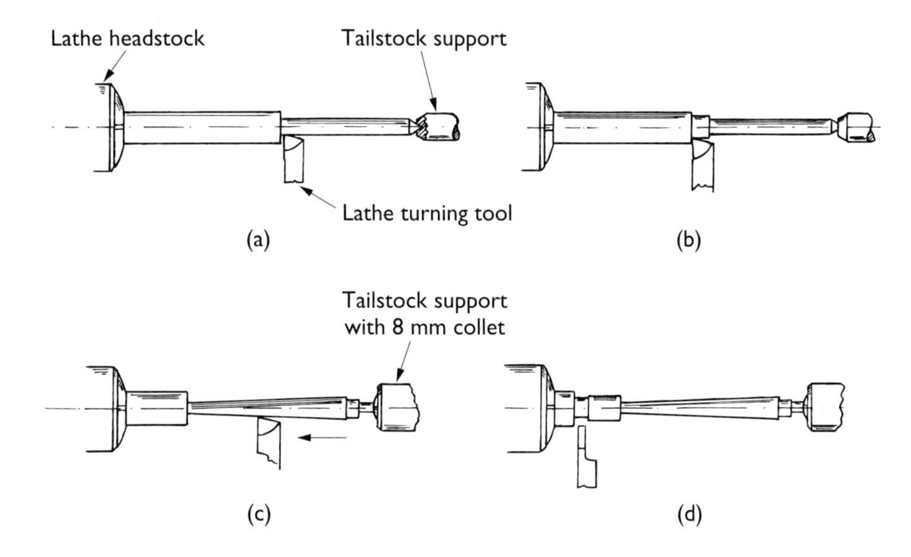

can be machined. The pinion can then be cut as before, and the remainder of the machining carried out.

Replacement Pinion for an Old Clock

When making a replacement pinion for an old clock, the late Philip Thornton recommended that when cutting pinions of six, seven and eight leaf between 0.2 and 0.5 module, use a cutter 0.05 module smaller than that recommended for the wheel, but the outside diameter of the pinion should be calculated using the corresponding wheel module, and for pinions between 0.5 and 1.0 module the module is reduced 0.1. This may sound a little complicated, but the following example should help to clarify matters.

Assume that a pinion of eight leaves is required to match with a wheel of 0.45 module. Then:

Outside diameter of pinion

$$= (8 + 1.71) \times 0.45$$

Outside diameter $= 4.37$mm

Outside diameter $= 0.172$in

When cutting the pinion, however, select a 0.4 module cutter. This, being smaller, will give a slightly shallower and thicker tooth form which can be an advantage, particularly when restoration work is being undertaken, as this form is closer to the pinions cut by the old clockmakers. As noted, when cutting pinion heads the case can arise where the tooth depth can leave a very weak core. If the above method is used, then it can help to overcome this problem. The replacement of the pinion head only onto an existing arbor is an advantage, as the original arbor can be preserved.

General Principles

The following applies as a general guide. For all new work – that is, when making a clock from scratch – use solid pinions, and work out the sizes of both wheels and pinions, size for size, as described previously when general formulae were covered. When carrying out restoration work on antique clocks, and sometimes when six-leaf pinions are used, use Philip Thornton's method.

The late Claude Reeve constructed most of his clocks with separate pinion heads onto plain parallel arbors, they being either chemically bonded together or soft soldered. If the joint is machined to a nice location fit and the items thoroughly degreased, then Loctite 601 will make a perfect bond. To remove the pinion head, all that is required is to heat gently until the two parts can be separated. If a greater shear strength is required, Loctite 638 can be employed.

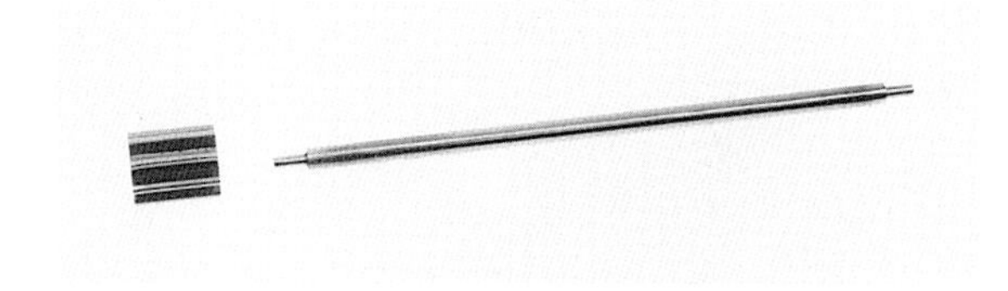

Fig. 246 Pinion head and arbor.

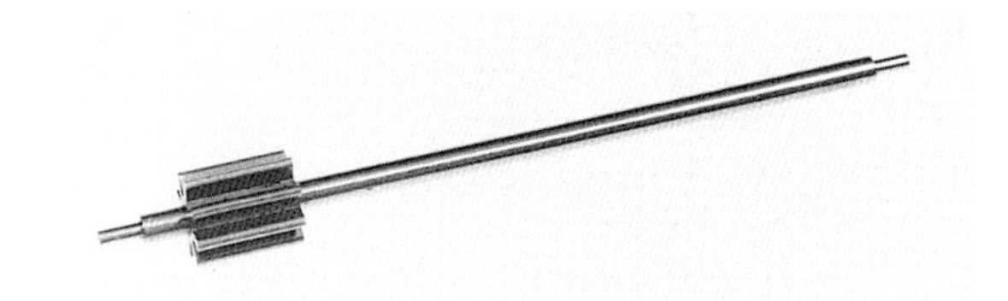

Fig. 247 Two parts assembled.

Holding the Work Between Centres

Gripping the work in the collet is often sufficiently accurate for general work. When absolute concentricity is required, however, it is essential that the work is held between centres: this can be either male or female centres. If female centres are used on the workpiece, then conventional methods can be used: the work can be centre-drilled with a standard high-speed centre drill and the work mounted between the lathe centres with a lathe carrier attached to the work. Additional material has to be added to the length of the arbor so that the female centre can be machined off once the work is complete.

Male centres on the work are more suitable, as the centre can be incorporated into the arbor and pivot when final finishing takes place. The ones shown are for heavier work. Special female centres are made to accommodate the arbors. Two lathe centres are shown that have been adapted for holding work with male cones; these are soft No. 2 Morse taper arbors machined to accept small, hardened silver-steel female cones.

These can be seen in use when finishing the pivots of a sixteen-leaf pinion for a month-going longcase clock. In the illustration showing a brass motion work pinion being cut note that no tail-stock support is necessary. Whilst most clock-

Fig. 248 No. 2 Morse centres modified to take male work centres.

Fig. 249 Sixteen-leaf pinion machined between centres.

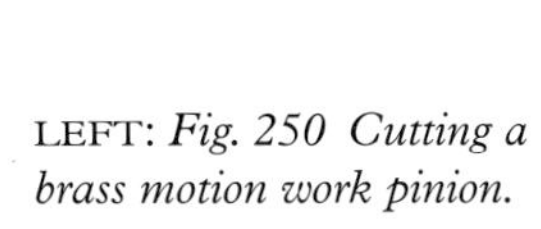

LEFT: *Fig. 250 Cutting a brass motion work pinion.*

Fig. 251 A small Centec milling machine set up for cutting pinions.

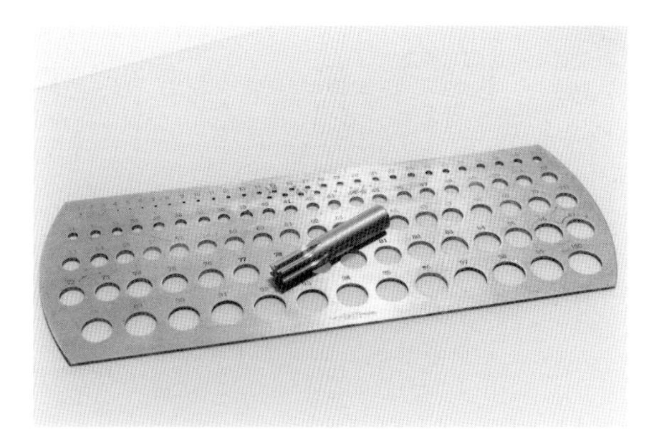

Fig. 252 Drill gauge for measuring pinions with an odd number of leaves.

makers will cut wheels and pinions on the lathe as this will be the only major machine tool in the workshop, some will prefer to cut pinions on the mill if they have one available. A Centec milling machine is shown set up with the vertical head and dividing head cutting an eight-leaf pinion. Where pinions with an odd number of leaves have to be measured on the outside diameter, a drill gauge is used. Select the correct hole that the pinion will just enter, and note the reading on the gauge.

FINISHING PINIONS

Whilst the finish obtained straight from the cutter should be fairly smooth, this is not normally satisfactory for a perfect action between the wheel and pinion. The higher the finish, obviously the less friction is encountered, and this in turn means less motive force required when driving the wheel train. The leaves therefore have to be individually polished. This can be carried out either by hand methods, or on the lathe with a set-up similar to that for pinion cutting.

Hand Polishing

If the hand-polishing method is to be used, select a length of pegwood or wood from an old emery stick, and sharpen the end to a chisel section to fit roughly between the leaves of the pinion. Prepare a small quantity of oilstone dust or emery powder of medium grit by mixing it well with a spot of oil. Dip the pointed pegwood into the abrasive and begin polishing the pinion leaves with a backward and forward motion, with the pinion resting against the bench. As the work continues, the wood will take up the shape of the teeth spaces and will also absorb the abrasives more readily. Continue until all the teeth have been polished evenly.

When cleaning the component between the differing polishing grades, it is important to remove all traces of the previous polish before attempting to use a finer grade. To clean, use Rodico or pith, and then degrease thoroughly. Check that all scratches have been removed, and a good finish imparted. This method is rather laborious, but works well with a little patience.

Fig. 253 Polishing the pinion leaves by hand.

Once again, clean off the abrasive and finish off with Solvol Autosol. This is a metal polish readily available from motor spares shops; it cuts well and imparts an excellent finish. A more traditional method is to use diamantine for the final finish, but this requires practice.

Using Diamantine

To prepare diamantine, use a small piece of glass or a clean dish. Place a small amount of diamantine on the clean receptacle, make a small hollow or indentation, and apply a drop of oil. Mix well with a glass rod – if a steel rod is used it will tend to darken the paste. It is important that no dust particles are allowed to enter the mix, so ensure that the glass dish and mixer are scrupulously clean. Mix thoroughly until the oil is completely absorbed by the powder and the mixture forms a stiff paste.

To apply the paste, use either a zinc, tin or soft steel polisher in a metal strip of suitable proportions. With practice the diamantine will impart an excellent polish to the work. This is termed 'black polish': it is a deep polish that reflects the light in a certain way; it works particularly well on hardened steel.

Polishing on the Lathe

A quicker method is as follows. Machine two boxwood or fruitwood (pear or apple) circular discs of similar diameter, thickness and bore to suit a standard threaded mandrel. Whilst the pinion cutter is still mounted on its arbor, select a piece of bright drawn mild steel 3 × 12 × 50mm (⅛ × ½ × approximately 2in); trim the ends, and then clamp between the three jaws of the three-jaw chuck. Now mill the form of the cutter into the end of the mild steel, remove from the three-jaw, and file a 10–15 degree front rake angle onto the steel as you would for a normal turning tool. Having done this, remove the previous cutter and replace it with one of the boxwood circular blanks. Put the mild steel form cutter back into the three-jaw, set the blank revolving, and feed into the tool; this will transfer the shape of the cutter onto the boxwood blank. All this takes only a matter of minutes, and will produce a lap that is the exact profile of the pinion leaves.

As before with the pegwood, charge the boxwood lap with oilstone dust, and feed the revolving lap into contact with the pinion. Traverse along as for pinion cutting; the speed is not important – somewhere around 1,000rpm.

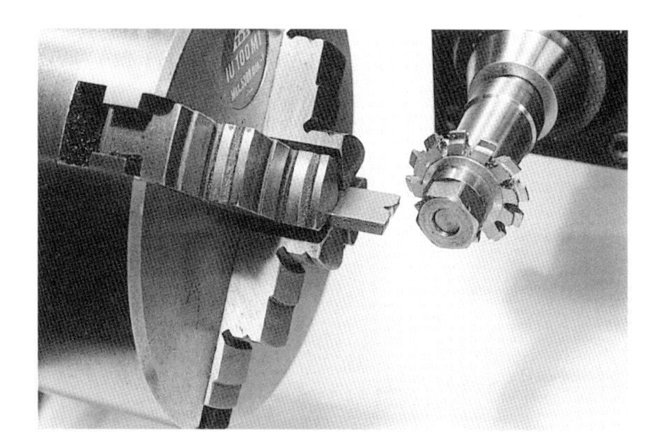

Fig. 254 Milling the tooth form.

Fig. 255 Transferring the tooth form onto the circular wood lap.

Check the finish and remove the pinion from the lathe; clean it thoroughly so that no trace of abrasive remains. Replace in the lathe, and now charge the second boxwood lap with Solvol Autosol. Diamond paste can also be used successfully, but in place of the wood lap use one made from hard plastic such as Delrin. When the final polish is complete, check the finish – which should now be free from scratches. Clean off as before, and the pinion arbor is ready for placing on one side for the final turning operation to be carried out as required.

Special machines were manufactured for the production polishing of pinion leaves. Two views of a Swiss machine made by Hauser are shown. The large wood-polishing disc can be seen clearly; this is mounted at a slight angle to the work, and when the revolving wood disc charged with polish is lowered onto the work, it will cut a helical groove into the wood and this, in turn, will rotate the pinion between the two steel discs.

Fig. 256 The circular lap polishing the pinion leaves.

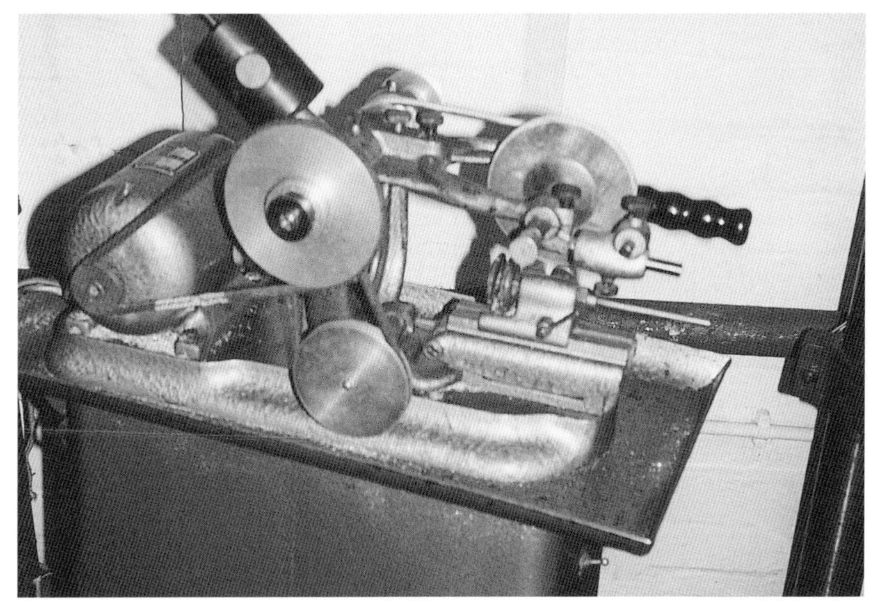

Fig. 257 The Hauser commercial pinion-leaf polisher.

Polishing will take place when the wood wheel is traversed backwards and forwards. This method is quite rapid.

Polishing Pinion Faces

It may be that the faces of the pinion require to be polished. It is advisable to leave the pinion just a fraction longer for this process, as when polishing the leaves there is a tendency for the lap to slightly radius the corners of the leaves as it passes over the ends. This can either be trimmed back by turning, or when the final polishing takes place.

Make two polishers as shown in the drawing, one in brass and the other in either zinc or boxwood. Set the pinion with its arbor running true in the lathe, making sure it is well supported. Charge the brass polisher with oilstone dust, and hold the polisher against the revolving work. Ensure that the hole in the polisher is at least 25

Fig. 258 Close-up of the work-head on the Hauser.

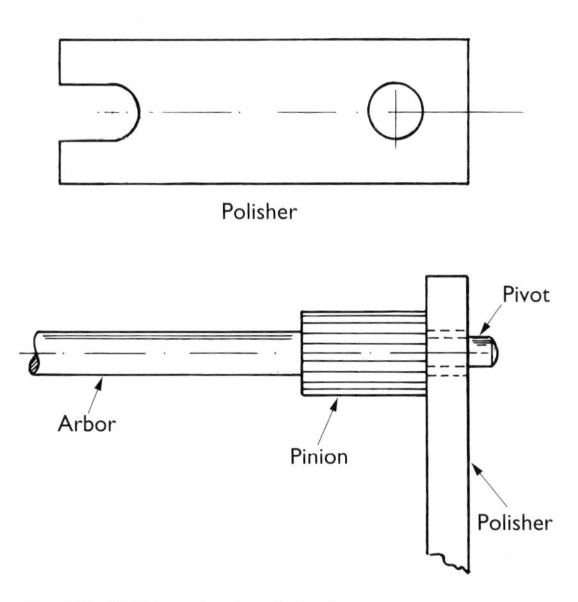

Fig. 259 Polishers for the pinion faces.

GRIMSHAW, BAXTER, & J. J. ELLIOTT Ltd.,

PINION FACING TOOLS.

D.R.G.M. No 57504

No. 49.—Set of 24 pinion facing tools, copper and iron, fitting into double-ended handle with swivelling arrangement.

Price, in case 5/-

Fig. 260 Set of pinion facing polishers from the Grimshaw, Baxter & J. J. Elliott catalogue c. *1920.*

per cent greater than the arbor; then when the pinion is rotating, the polisher can be moved in a backward and forward motion, ensuring it is kept completely flat and square against the work. When a good finish has been imparted, clean off thoroughly. Now mix a small quantity of diamantine with a spot of clock oil until it forms a thick paste. *Do not* inadvertently mix in with it any dirt or dust, as this will scratch the work and it will be impossible to get anything like the required finish. Now charge the boxwood polisher with the diamantine, and polish as before until the finish is of the required standard. This method, if used carefully, should give a really good finish to the edges of the pinion leaves.

Another method that is guaranteed to keep the pinion faces flat and polished is by using a circular carbide lap mounted on the milling spindle. A proprietary tool for pivot polishing is available. This also has a carbide lap to carry out the polishing on pivots, but can be used to equally good effect on polishing pinion faces. This tool will be covered in more detail later (*see* Chapter 10).

HEAT TREATMENT OF PINIONS

On high grade work, for example on regulator clocks and French clocks, it is usually necessary to heat-treat the pinion and arbor for hardness. The heat-treatment procedure is as follows: normal silver steel is used. To prevent scaling, it is necessary to cover the complete arbor with soap, then bind with soft iron wire and rub soap on the wire, making sure that all the grooves are completely covered. Heat evenly to cherry red, then quench in a 10 per cent solution of brine water, lowering the arbor down vertically into the water to minimize distortion. Remove the wire and check for hardness with the use of a fine cut file. If not satisfactory, the whole process has to be repeated – but before doing that it is necessary to anneal the steel to reduce the chance of stresses being present, and possible cracks in the material after re-hardening takes place. All that is required is to heat the steel to cherry red and cool slowly.

When hardening steels, the use of plain cold tap water to quench them should be avoided because of the risk of cracking. Steam bubbles are produced on the surface of the work as the item is immersed, and tiny pockets of water flash off into steam vapour. These steam bubbles insulate the local area from the water and form soft spots, and in an intrinsically hardened workpiece, a crack can therefore be produced due to stress at this point.

Fig. 261 Pinion and arbor wrapped in iron wire, ready for hardening.

The use of a 10 per cent solution of brine-water mix as a quenchant greatly reduces any risk of cracking, as brine does not form steam bubbles easily. The strength of the brine solution should, however, be checked regularly with a hydrometer, since any dilution of the mix will again bring with it a risk of cracking.

Now the arbor with its pinion is at the correct hardness, it is necessary to temper it down, as in its present state it is far too brittle. If the following method is used, it will ensure the material is heated gradually and evenly, and this is necessary with a small item of changing section: if the pinion with its arbor were heated over a normal flame it would be very difficult to

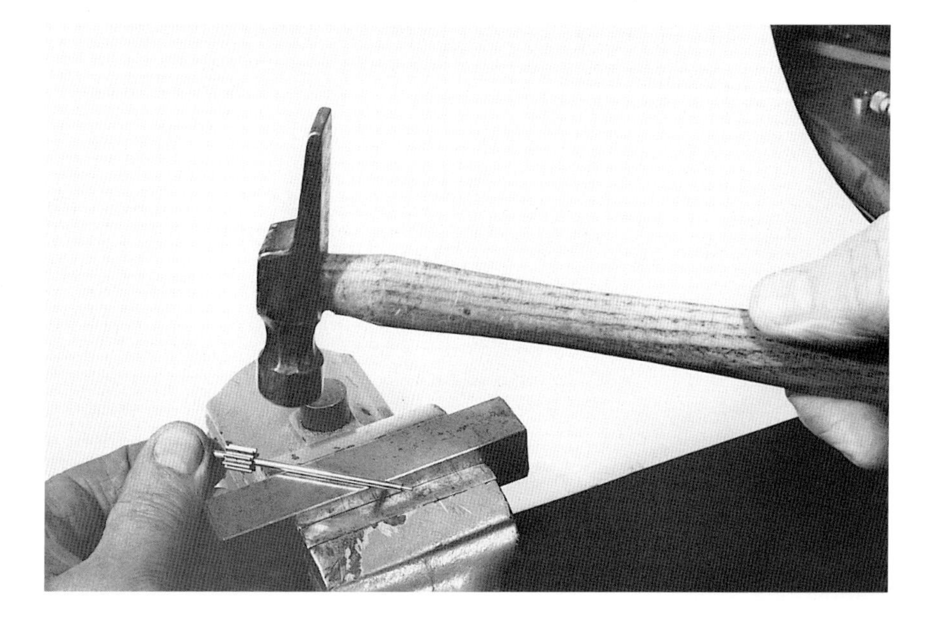

Fig. 262 Removing distortion of the arbor by peening.

heat it evenly; consequently the tempering would not be consistent.

Polish off any scale, if present, then fill a small tin with fine sand; embed the arbor and pinion into the sand with just a small portion of the pinion showing. Now heat the tin evenly from beneath and watch for the colour change taking place on the work. A light straw colour will be apparent at first, this will then change to a dark straw, after which it will turn a lovely bluish-purple. This is when the quenching takes place: remove from the flame and quench vertically in brine, as before, and the completed item can either be left purple or polished bright. Alternatively, leave the items to air-cool, but ensure there is no colour change.

It is important that the tempering process is carried out carefully. If the pivots are left too hard, then it is quite likely that they will suffer breakage when being assembled between the plates of the movement. If pinion heads only are to be hardened, then harden and temper as before.

Distortion can, of course, take place and arbors can be straightened by peening. To do this, use a smooth block to act as a small anvil, also a hammer with a polished face. Care is required in straightening the arbor; only gentle taps are required, and these should be on the side to be stretched. Rotate the work whilst carrying this out.

Once the pinion has been hardened and tempered, the final polishing can take place. Methods described previously can be used, but now the metal is hard, diamond paste can be applied. A hard plastic, for example Delrin, can be used. Make a separate lap for each grade of diamond paste. Other useful materials are brass, zinc and boxwood, all of which can be used to good effect.

This is only a brief description of hardening and tempering, but the method described will give satisfactory results. There are, of course, books covering the procedure in more detail; for example, *Hardening, Tempering and Heat Treatment* by Tubal Cain.

LANTERN PINIONS

Lantern pinions were produced on many American and German clocks for cheapness of manufacture; they were described in Camus as early as 1735. The pinion teeth are short lengths of hardened and polished steel wire trapped between two sheaves of brass that are mounted onto the clock arbor. This is an efficient method of transmission when the contact for engagement is on the exact pitch-circle diameter of the mating wheel; the illustration shows clearly the tooth form of the wheel engaging with the pin wire.

Various views are shown of a machine for producing lantern pinions, believed to be of

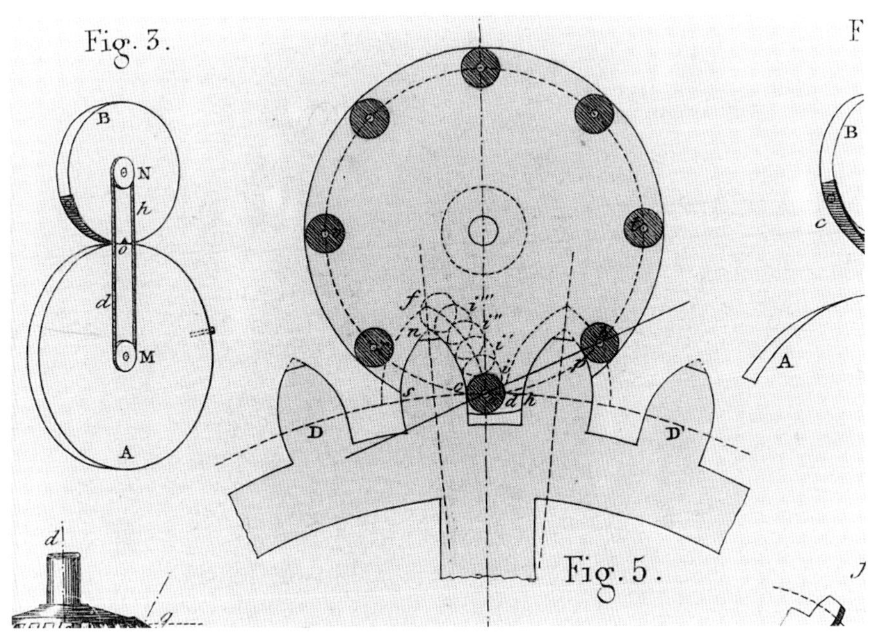

Fig. 263 Lantern pinion meshing with the mating wheel.

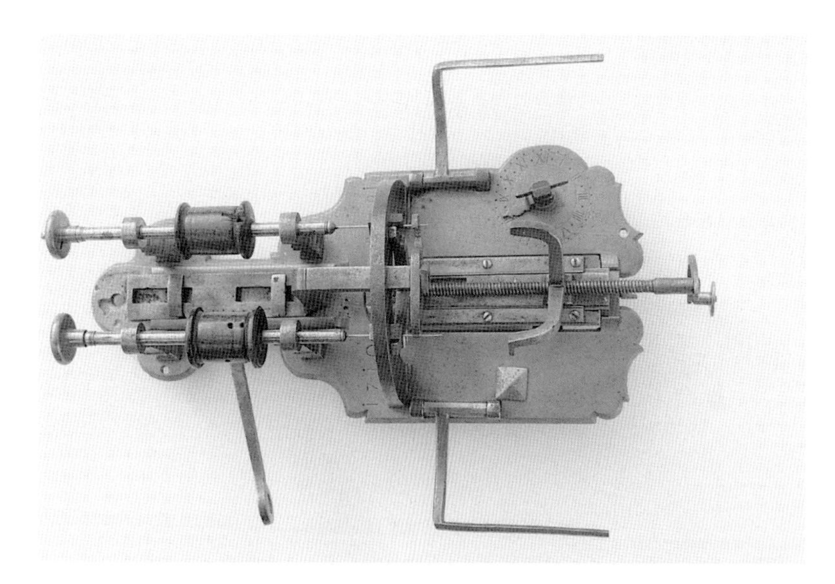

Fig. 264 Early machine for drilling lantern-pinion bobbins.

German origin. It is engraved 1710–BA91, and the engraved numerals and letters can be interpreted in at least two ways: first, the date is 1710, and BA91 refers to the maker or such; and second, the date is 1791, and I (J) O B A could be the initials of the maker or owner. The only other known, documented example of such a tool is in the Deutsches Uhrenmuseum in Furtwangen, Black Forest, Germany, which is described in an old catalogue of the museum published in 1925: A. Kistner, *Die historische Uhrensammlung Furtwangen*. This latter tool seems to be of a later date, and is not so decoratively finished.

Drilling Lantern Pinions

Two lantern pinions of the same dimensions – that is, of the same pitch circle – can be drilled at the same time. The discs or shrouds to be drilled are mounted permanently on the arbors. Dividing discs with the same number of divisions as holes to be drilled are fixed firmly to the arbors, but so that they can be removed later (it is suggested in Kistner that normal gear wheels were used for the purpose). The two loose arms at the side are engaged in the divisions as indexes. The arbors are fixed between female centres by means of the long thread and crank so they can just rotate without wobble.

The long drills on the ends of the two sliding spindles are permanently attached, and cannot readily be interchanged with drills of varying sizes as the holes they pass through in order to keep them centred and in line are also a permanent fixture. (One hole has a diameter of 1.2mm, the other 1.0mm; considering the age of the machine, they were probably originally intended to be the same.) The second holes they pass through are in a sliding bracket, which can be brought right up to the first set of discs by means of the lever. (The same principle is used in the self-centring pivot-drilling attachments of standard watchmakers' lathes.) This ensures that the drill is centred at the exact position required, and the tendency of the drill to bend and distort is also reduced. The bending of the drill is the

Fig. 265 Close-up of machine, showing the date.

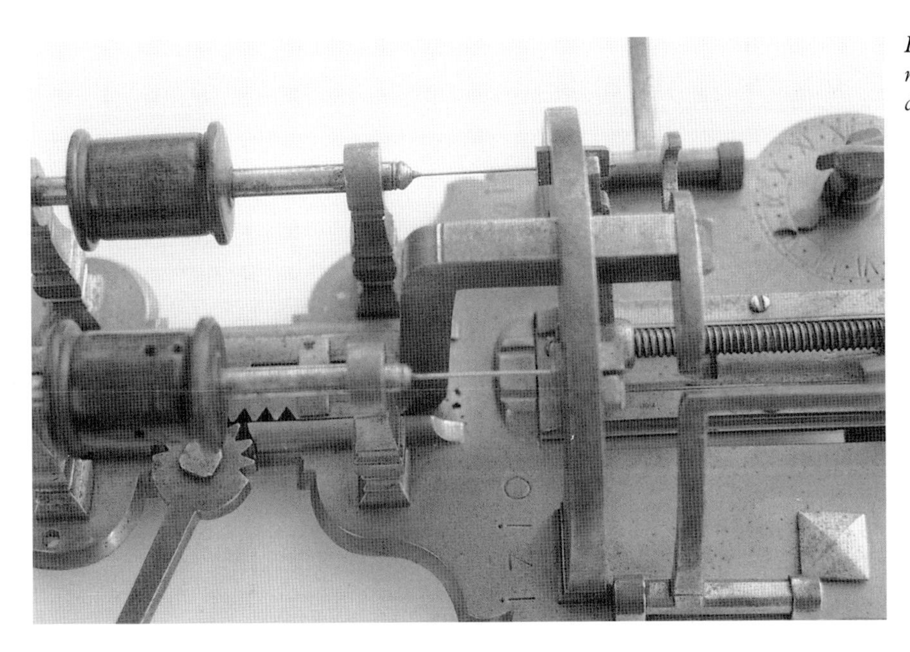

Fig. 266 Lantern bobbin machine showing two drilling heads.

biggest disadvantage to making lantern pinions, as the trundles or pins would not be parallel.

The size of the drills (1.0–1.2mm) indicates the minimum size possible of the pins of the lantern pinions. Once the holes had been drilled parallel, they could subsequently be opened out off the machine by hand with larger drills for pinions with larger trundles or pins.

The diameter or pitch circle of the lantern pinion to be drilled is set by turning the winged knob and indicator on – what looks like – a twelve-hour dial. The twelve numbers and thirty-six divisions are most likely arbitrary, and serve a purely comparative purpose. The smallest pitch circle that can be drilled is about 5.5mm, and the largest about 25mm. The knob and indicator are fixed to a screw that protrudes under the frame onto a steel block attached to the frame in which the long thread and crank are mounted. This frame can thus be tilted so that the distance

Fig. 267 Close-up showing actual drilling.

between the female centres and the drills can be altered. The drilling spindles have wooden bobbins onto which a driving belt may have run, although the 1.2mm diameter drills can be easily made to cut by turning manually.

The Use of Lantern Pinions

Whilst the use of lantern pinions was much maligned, they were employed successfully in many types of clock. The illustrations show first, a lantern pinion from a late eighteenth-century German longcase clock, and then a wheel and two pinions from German clocks made in the late nineteenth century, when clocks were starting to be produced in large numbers in the Black Forest. A most unusual application is the use of a lantern pinion in a ship's chronometer, which proves this type of gearing was successful in the higher grade of timepiece.

Fig. 268 Lantern pinion from a late eighteenth-century German longcase clock.

Fig. 269 Nineteenth-century lantern pinions from mass-produced Black Forest clocks.

Fig. 270 The use of lantern pinions in a ship's chronometer.

Lantern pinions are suitable for going and striking trains, but only when the wheel drives the pinion. A data sheet from W. O. Davis *Gears for Small Mechanisms* gives proportions for both pinions and wheels with corresponding pin diameters (*see* Appendix).

Machining a Lantern Pinion

To machine a lantern pinion, first make the bobbin. A piece of brass bar of suitable diameter is mounted in the three-jaw chuck or collet, and machined to the dimensions required. Sufficient material should be left projecting to enable the work to be parted off when machining is complete. Accurately centre-drill, and then drill and ream the centre hole for the arbor. As Jobber-type twist drills are sometimes difficult to control in sizing a hole, and because there is a tendency for them to wander if not correctly centred, 'D' bits will give a more constant hole size, these can be purchased, but are easily made from silver steel.

Set the milling and drilling spindle up with a small drill chuck as shown, and select a Eureka-type drill that matches the lantern pinion wire. If the drill is held with a short proportion of the flute projecting, it will be quite easy to drill the holes for the pins accurately. Always carry out a test-drilling first on some scrap material, as drills do not always cut to size.

If only twist drills are available, then it will be necessary to centre each hole first before commencing drilling. Sometimes drills will not be readily available in exactly the diameter of the wire required, in which case use a drill slightly

Fig. 271 Modern mass-produced movement with lantern pinions.

Fig. 272 Machining the lantern bobbin.

Fig. 274 Drilling the bobbin for the wires.

Fig. 273 Centre drilling.

smaller in diameter than the wire and broach to size.

The wires are made from blue pivot steel, which is in a hardened and tempered state: it will machine, but with difficulty. A simple jig is shown, to locate the wire to enable the length to be determined. It is possible to cut the pivot steel with a thin abrasive disc mounted in a hand-held electric tool, such as the Mini-Craft; this will cut a square edge and an accurate length.

Alternatively, good quality side-snip pliers can be used, but this will leave a rough edge that will require squaring up on the edge of fine grinding wheels. A third way is to nick the pivot steel on the bench grinder; it will snap quite easily, but again, the edges will have to be cleaned up.

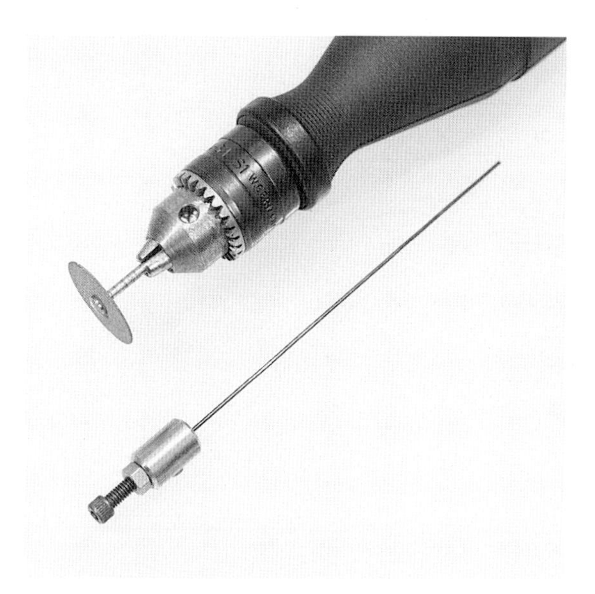

Fig. 275 Jig for cutting the wires to length.

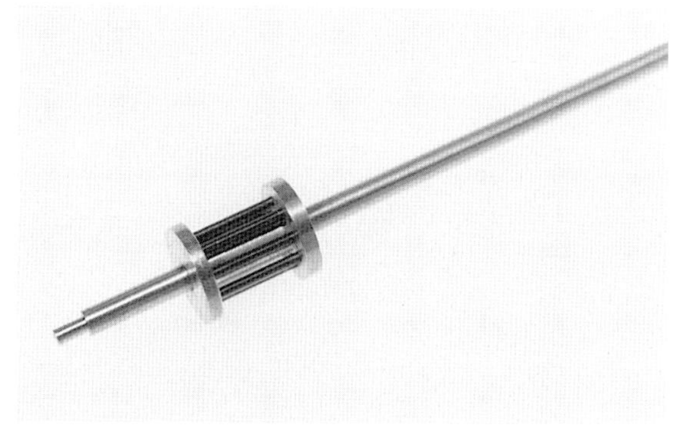

Fig. 277 Completed lantern pinion mounted on its arbor.

Ensure the pins are a nice tight fit; if there are any loose ones, these can be fixed with Loctite 601 or 638, or soft soldered. Trim back to finish level and smooth with the face of the bobbin. Mount the completed lantern pinion in the lathe, and stone the pins back flush, then finish with carborundum paper. The pinion is now ready to be mounted on its arbor.

We have already seen that early machines for producing lantern pinions were produced by Morat and others. Two views of a twentieth-century machine constructed by Boley are shown here, using what appears to be a standard watchmaker's lathe headstock.

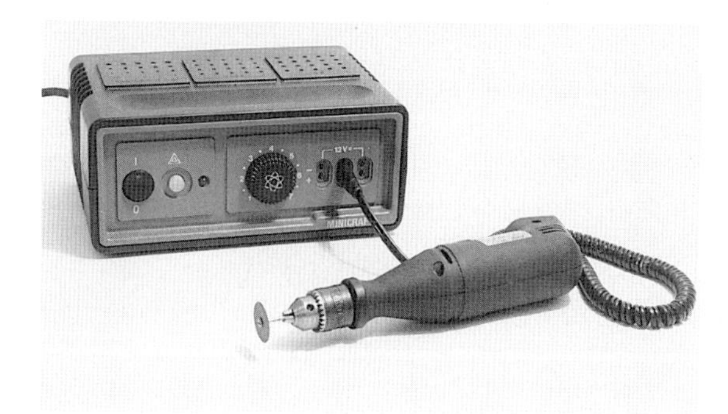

ABOVE: *Fig. 276 Minicraft drill with cut-off abrasive disc.*

Fig. 278 Special machine by Boley for drilling lantern pinions.

Fig. 279 The Boley headstock.

CUTTING PINIONS FOR WATCHES

The set-up for cutting clock pinions has been covered in detail, and small pinions for watches are cut in a similar manner. Whichever machine is used, rigidity is essential. The following set-ups are shown using the Schaublin 70 precision lathe. It has an excellent milling spindle that is extremely solid in its construction, having large bearings. The spindle takes 12mm collets, the same as the lathe headstock.

Unfortunately the range of watch pinion cutters currently available is limited, although Thorntons intend to extend the range down to 0.1 module in the near future. It is possible to make one's own, although access to an optical projector is essential. The centring of the cutter is very important, and the microscope shown previously is necessary to obtain the accuracy required.

On the lathe shown, the drive for the milling spindle via the overhead jockey pulleys is directly from the lathe motor, which is 1,400rpm and too fast for pinion cutting. The Schaublin 102 has a geared drive fitted to the milling spindle to give the necessary reduction to the cutter spindle. With modern electronics, it is now possible to obtain speed control on a three-phase motor quite easily by use of an inverter.

The speed of DC motors can also be successfully controlled by electronic means. However, if the motor is AC single phase, as the one used on the lathe shown, then some form of mechanical speed reduction is necessary. As can be seen, the

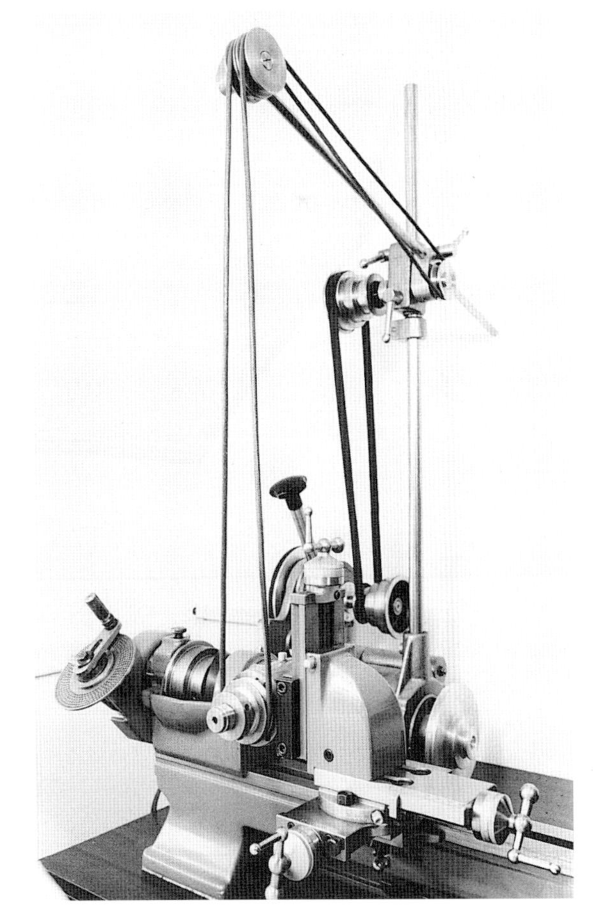

Fig. 280 General view of set-up on Schaublin 70 for cutting pinions.

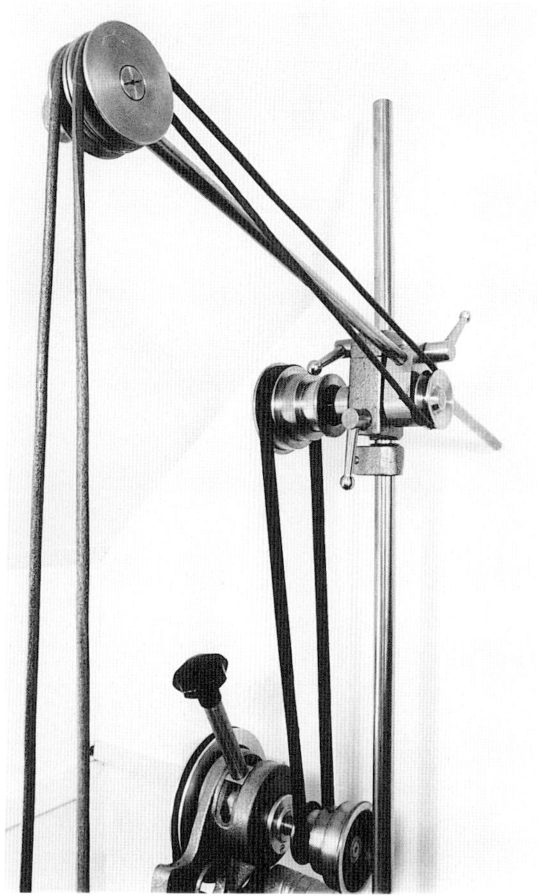

Fig. 281 Modification to Schaublin 70 overhead drive.

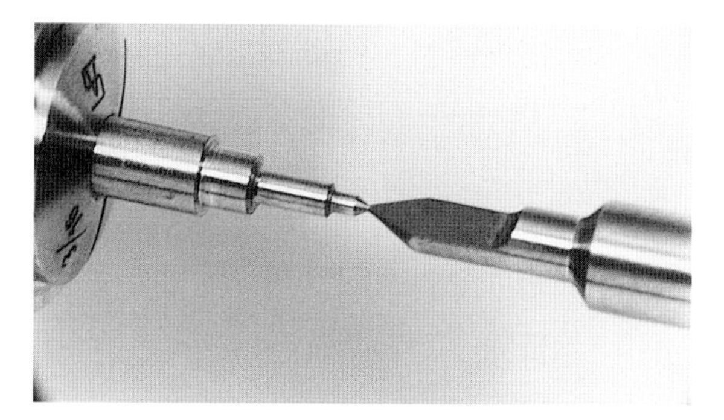

Fig. 282 Watch pinion blank, machined ready for cutting.

standard Schaublin drive has been modified so that the speed reduction is achieved by driving the milling spindle through the lathe countershaft. The speed required for a watch pinion cutter is approximately 400–500rpm.

Use high-grade silver steel for the material on watch pinions; this can then be hardened and tempered. After the pinion has been cut, the final finishing can then be carried out. It is essential that the material is heat-treated to provide strength when finishing the small pivots; if left in its soft condition it is likely to bend with the force of the cutting action. On the other hand, if it is too hard, it will not be possible to machine it. Grip the material in a collet and turn to the required dimensions, as in the example shown; note the blank is left with plenty of surplus material near the collet to avoid flexing of the pinion whilst it is being cut.

Fig. 283 Cutting a pinion.

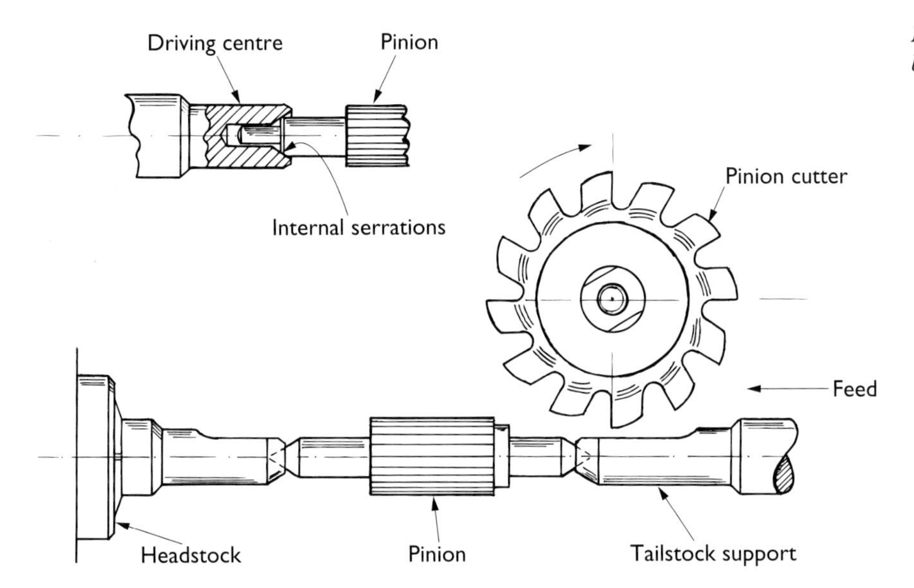

Fig. 284 Pinion supported between female centres.

On very short pinions, it is possible to cut without a tailstock or other type of support.

The set-up shows a ten-leaf M0.18 centre pinion being cut for an English lever pocket watch. Note the extended support centre. When the pinion leaves have been cut, the remainder of the arbor can then be re-finished. Sometimes it is preferable to machine the pinion between centres, which will ensure complete accuracy. Male centres are turned on the pinion blank, and a special female cone with serrated teeth is required.

The illustration above shows how pinions were held for production runs on Swiss automatic pinion-cutting machines, for example Wyssbrod and Safag. Female centres can be made by first machining a cone using a small centre drill, then make a male punch with a matching angle of 60 degrees, mill serrations on the taper, harden and temper. The female centre and the hardened male are then driven together, being sure to keep them in alignment: this will transfer serrations to the female centre, then harden and temper. Another method that has been used successfully is to groove the female cone with a small lathe tool ground to a sharp point. Set the topslide over to match the cone, and form the grooves with the sharp tool, traversing with the lathe topslide

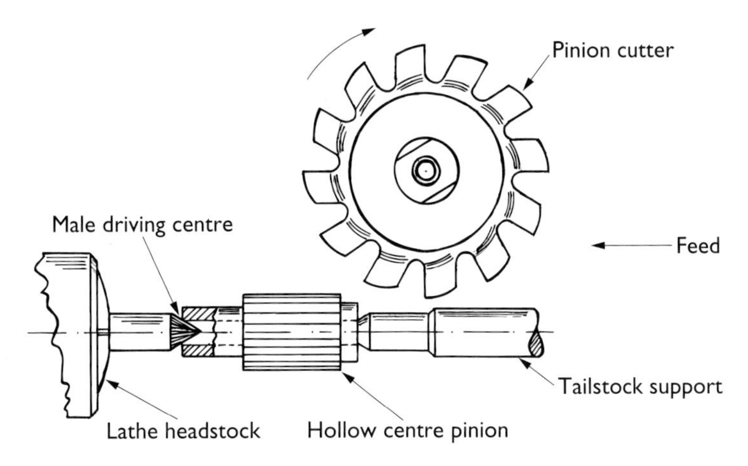

Fig. 285 Male centres being used for a hollow pinion.

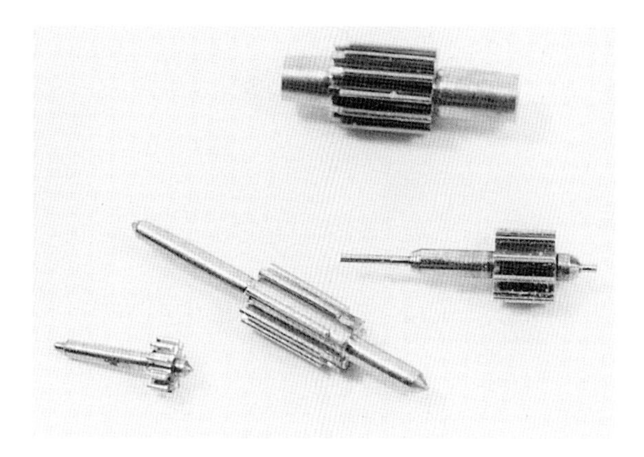

Fig. 286 Various completed pinions.

handwheel. Some form of indexing will be required to divide the centre equally.

Hollow pinions can be held with a male cone centre with serrations or flats, as shown; note the tailstock centre is machined away locally to allow clearance for the pinion cutter.

When cutting takes place, apply only light cuts. The depth of cut and feed will depend on the material being used and the rigidity of the machine. Use a lubricant to improve the surface finish, and to keep the work and the cutters cool. Generally three or four cuts are required to obtain a full depth of cut – it is better to rough out, then take the final finishing cut on all pinion leaves at one setting. Take care when approaching full depth that the addendum curves from each side of the pinion leaf match up equally.

The lever feed was shown in the chapter on wheel cutting, and for cutting wheels this is a rapid method of feeding when cutting brass. When steel pinions are being milled, apply the cut with the topslide feed screw; this will ensure an even feed and a good finish to the work.

Whilst the methods shown cover basic watch pinion cutting, it is often necessary to make special centres and accessories to complete the job accurately.

8 WHEEL AND PINION CUTTERS

Wheel and pinion cutters are readily available; as previously shown in Chapter 3, they are manufactured to Swiss Standard NIHS, except that the teeth are flat-bottomed. Constructors may wish to make their own cutters, and various methods will be covered, including single-tooth fly cutters and multi-tooth form-relieved types.

There are thirty-eight operations in manufacturing a commercial wheel or pinion cutter, from basic turned blank to completed cutter. Extremely accurate carbide form tools are required; these are form-ground on a Studer pantograph profile grinder. In the roughing-out process, ten cutters are produced before regrinding of the cutting face is required; but for the finishing process, only one cutter is produced between sharpenings. The form tool is mounted on a Safag semi-automatic form-relieving machine, with the appropriate cutter blank. A continuous flow of cutting oil is used to ensure a first-class finish, as this form will eventually produce the completed cut wheel or pinion. Faces of the cutter blank are ground, as is the cutting edge. The bore is honed. A general view of the cutter manufacturing shop shows various automatic machines producing cutters.

Sets of cutters are also available, as shown, by Bergeon, in both round- and flat-bottom types.

If you should be fortunate enough to obtain old cutters, many of these were produced with fine teeth and it is not possible to resharpen them, but they will give good service if treated well. Escape-wheel cutters are still produced in fine tooth form. The cutters are produced on Swiss automatic indexing machines, with cams

Fig. 287 Commercial form-relieved, constant profile wheel cutter.

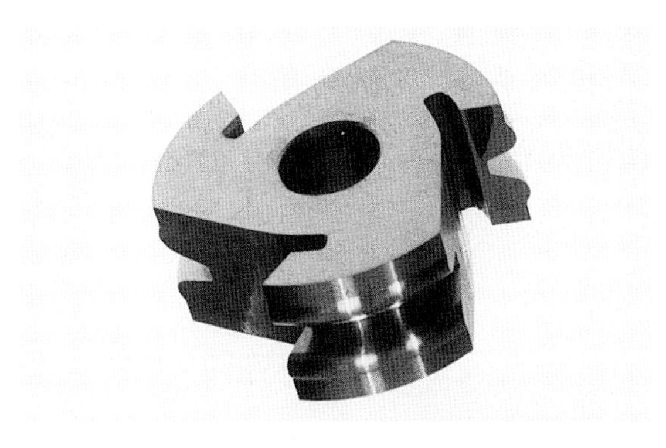

Fig. 288 Carbide form tool for producing pinion cutter blanks.

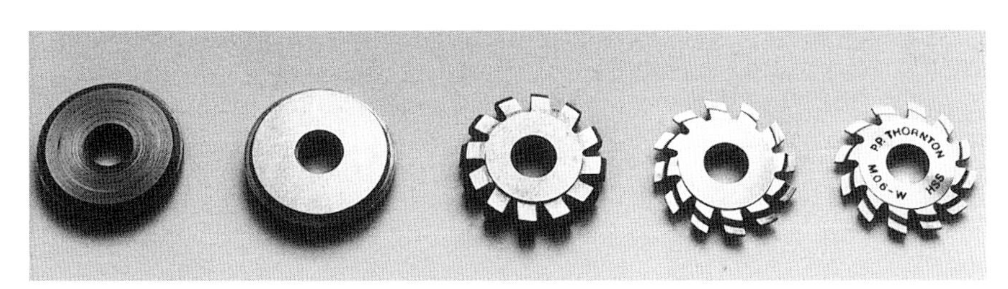

Fig. 289 Various stages of cutter manufacture.

Fig. 290 Safag automatic form-relieving machine.

that produce the required radius, which the rotating cutter follows. The cutter is similar to a large dental burr.

The making of this type of cutter is shown in *The Watchmaker's Handbook* by Saunier, *c.* 1870, plate XIV. In this instance a template is used to guide the revolving cutter. The template-to-cutter ratio is 10:1, and reduction is by a pivoted lever. A similar adaptation was recently constructed by a retired doctor, Peter Clark of Southwold; his attachment was beautifully designed and made to fit the Aciera F1 milling machine. A master template is used; working from a gimballed mounted reducing lever, this controls the cutter spindle. Ratios of 1:4 and 1:10 were used, and cutters as small as 0.5mm diameter can be accommodated.

Another type of cutter produced by this method is used with the topping or rounding-up machine, as previously described in the chapter on wheel-cutting machines.

Fig. 291 Close-up view of a carbide form tool at work on a cutter blank.

Fig. 292 General view of a cutter manufacturing workshop.

ABOVE: *Fig. 293 Boxed set of Bergeon cutters.*

ABOVE RIGHT: *Fig. 294 Selection of early cutters.*

Fig. 295 Machining fine-tooth escape-wheel cutters.

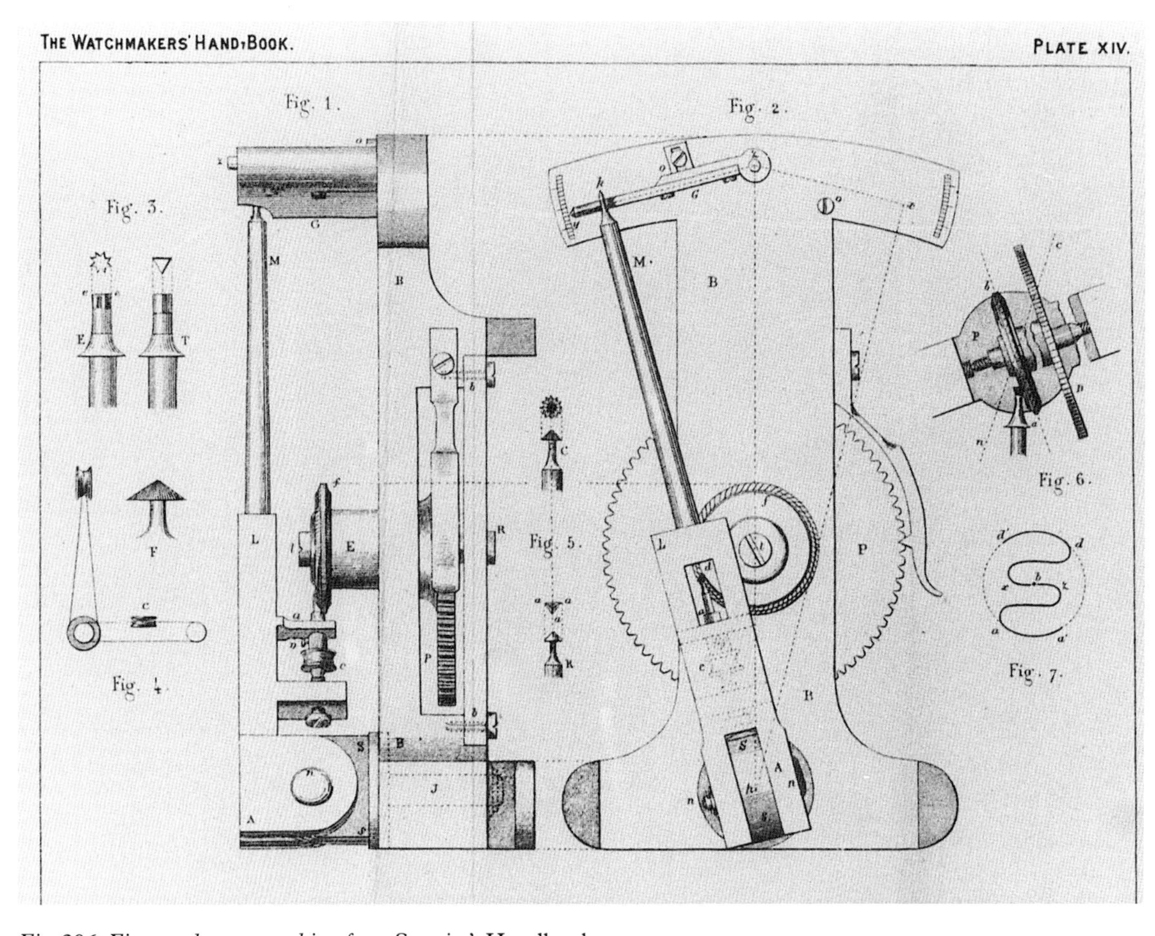

Fig. 296 Fine-tooth cutter making from Saunier's Handbook.

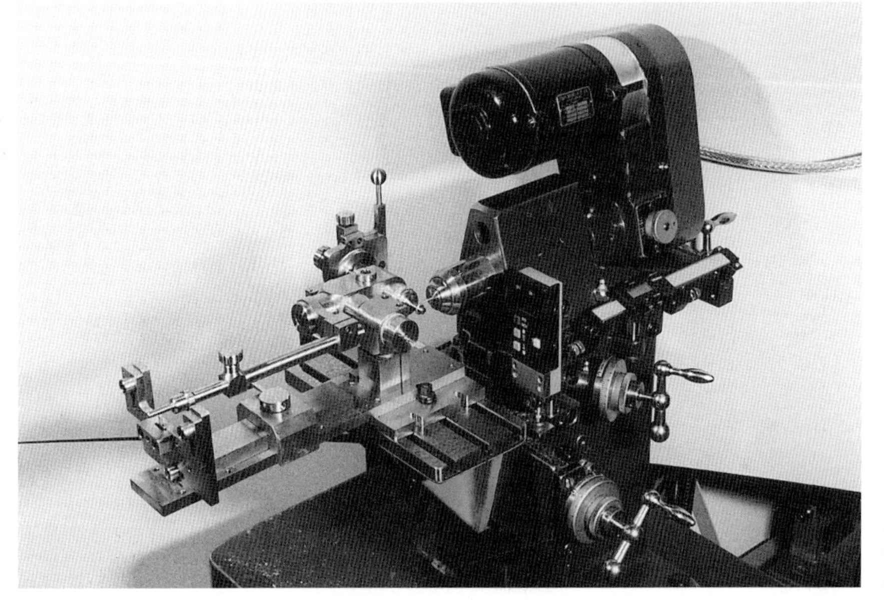

Fig. 297 Aciera F1 milling machine with fine-tooth cutter attachment.

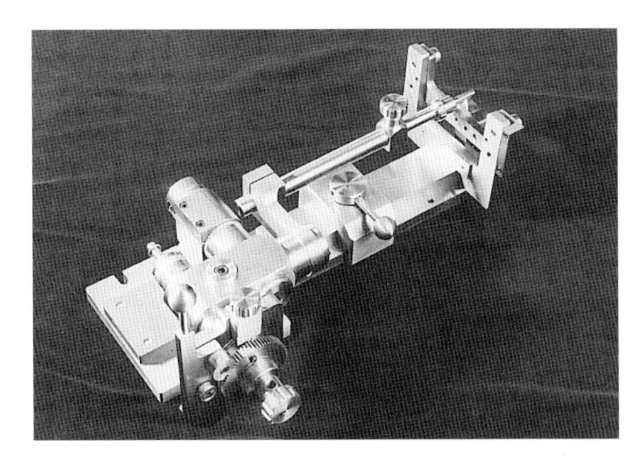

Fig. 298 Fine-tooth cutter attachment.

Fig. 299 Commercial fine-tooth cutter.

CARE OF CUTTERS

Modern multi-tooth form-relieved cutters are relatively expensive, as so many operations are required in their manufacture. They are made from a high hardness type of high-speed steel such as AISI M42 (HSS-E), giving a hardness of 68–69 Rockwell C. This good quality steel gives a maximum life to the cutting edge, and therefore preserves the accuracy of the profile. They will not, however, stand abuse, and lack of rigidity at any point may result in slight chipping or even tooth breakage. The simple guidelines given should be followed in order to give a long life to the cutter.

CUTTERS MUST BE SHARP

Keep the cutter teeth very sharp. Carry out frequent inspections of each of the twelve teeth with a ×2 magnification eyeglass, and a ×12 if below module 0.5. Blunt teeth produce an increased operating temperature at the cutting edge, and this undesirable state demands greater rigidity from the machine.

Keep all cutters stored separately to ensure that teeth do not get damaged accidentally when they come into contact with one another.

Fine-toothed cutters with sixty or so teeth should give a long and useful service. They cannot, however, be re-sharpened. If rough cutting or overheating is experienced, examine the teeth carefully with an eyeglass: if they are found to be blunt, the cutter must be discarded and replaced without delay.

GRINDING WHEELS

For sharpening form-relieved cutters, green-grit wheels work moderately well when sharpening one or two cutters, but CBN (Borazon) wheels far outperform any other grinding wheel due to their efficient stock removal and cool cutting characteristics. Use Mobil Centrex-1102 or paraffin as a coolant for both types of grinding wheel.

RIGIDITY

When cutting wheels and pinions, lack of rigidity, either in the machine or work holding, is a frequent cause of poor work-surface finish, or cutter failure. Poor work-surface finish on pinion leaves involves further polishing, and extra polishing can cause loss of profile accuracy and shape.

Using a dial gauge equipped with a magnetic base, check the following carefully:

a) cutter arbor end-shake should not exceed 0.01mm (0.0004in);
b) machine slides – to improve rigidity, lock any not in use;
c) work holding – check arbors and headstock for both end- and side-shake;
d) work support must always be of the very best for pinion cutting.

GENERAL POINTS

Constant profile cutters are sharpened without loss of form. Wheel cutters, particularly those with square bottoms, have a reduced cutting efficiency on the flank as compared with pinion cutters. This is due to their flank angle of only 2 degrees per side, which offers minimum clearance for the cutter flank as it cuts through the work. The tip and addendum radii of these cutters do not present any such problem: their clearance is similar to that of a pinion cutter.

Constructors who cut their own wheels may have had experience of the brassing-up problem on wheel cutter flanks. Therefore there is a constant need for vigilance as to the state of the cutting edges of all the teeth. A sharp wheel cutter will have much less tendency to brass up. With the aid of an eyeglass and a good light, rotate the tooth cutting edge (at the extreme periphery) around its axis in order to catch the slight reflection of the light which is indicative of an edge that needs sharpening.

Hold between thumb and finger, and manually index around until all the teeth have been examined in this way. It has been said that if you can see the cutting edge of a tool it is not sharp.

Wheel cutters these days have a better finish on their flanks, therefore brassing-up is less of a problem. Removal of brass impacted onto the flanks of wheel cutters can be achieved by soaking them in nitric acid for a few minutes. Make sure the cutter is scrupulously clean and free of oil and brass chips before immersion in the acid. Rinse under a running cold-water tap (never hot water as this will stain the cutter) and dry immediately. Inspect with an eyeglass to see if all the brass is removed, and if not, repeat the process.

Goggles and protective clothing should be worn when handling this acid. Such treatment is best done before the cutter is sharpened.[16]

SHARPENING CUTTERS

An early machine by Rehe for sharpening multi-tooth cutters is reproduced from *Rees Cyclopaedia*, 1819. It is unlikely that the small clockmaker's workshop will have facilities for sharpening cutters, although it is possible to set up the lathe for grinding with a suitable spindle. This is not recommended, however, as grinding particles that are produced will inevitably deposit themselves under the lathe slide-ways, and these are difficult to remove.

A machine that is most suitable for sharpening all types of cutter is the Quorn tool and cutter grinder. This was produced from a kit of castings and materials originally described in *Model Engineer* and subsequently produced in book form, giving constructional and fully operational instructions on how to use the machine to sharpen all types of cutters. The title of the book is *The Quorn Universal Tool & Cutter Grinder* (Professor D. H. Chaddock), and it was published in 1984. Many of these machines have been constructed, and if you are fortunate enough to own one, then wheel- and pinion-cutter sharpening is relatively straightforward. The position of the grinding wheel relative to the cutter tooth is shown in the illustration.

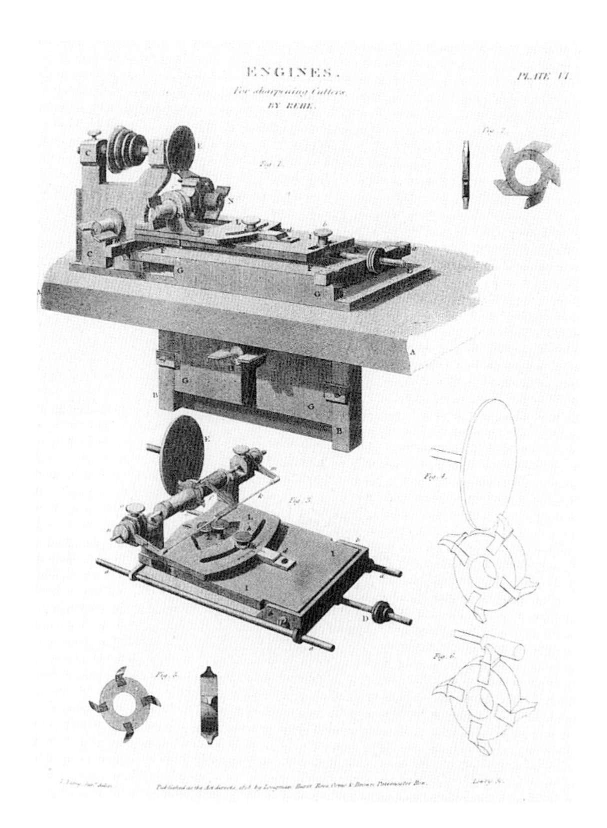

Fig. 300 Early machine by Rehe for sharpening cutters.

Cutters can be returned to the manufacturer for sharpening, or to a local workshop that specializes in tool and cutter grinding.

EARLY TWENTIETH-CENTURY CUTTER MANUFACTURERS

There were a number of cutter manufacturers in business at the end of the nineteenth and into the twentieth century. Various cutters are shown in Grimshaw, Baxter & J. J. Elliott Ltd's early tool catalogue; an excellent range was available. Most

Fig. 301 The Quorn tool and cutter grinder.

Fig. 302 Sharpening a wheel cutter using a Borazon grinding wheel.

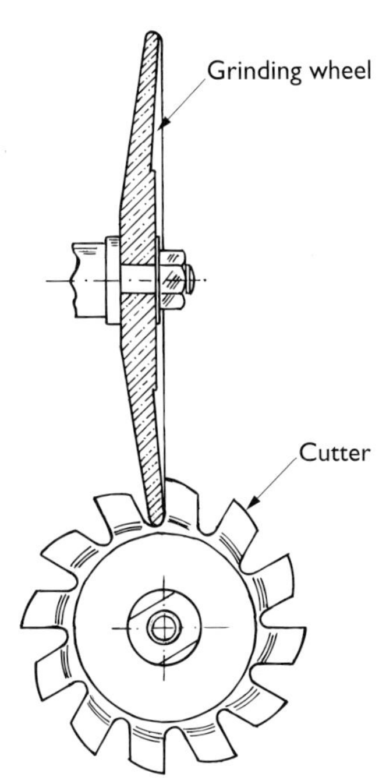

Fig. 303 Position of cutter and grinding wheel when sharpening.

of these were produced by companies such as L. Bretton, Koepfer and Carpano, and many of them are still around today. Although manufactured in carbon steel, they are still capable of producing good work and can often deal with a job with an odd size module.

The only problem with many of these early cutters is that the sizing of the module is often difficult due to manufacturers having their own system of identification. Below is information from an early Carpano catalogue that solves this problem for wheel cutters. The information has been translated to enable the tables to be understood.

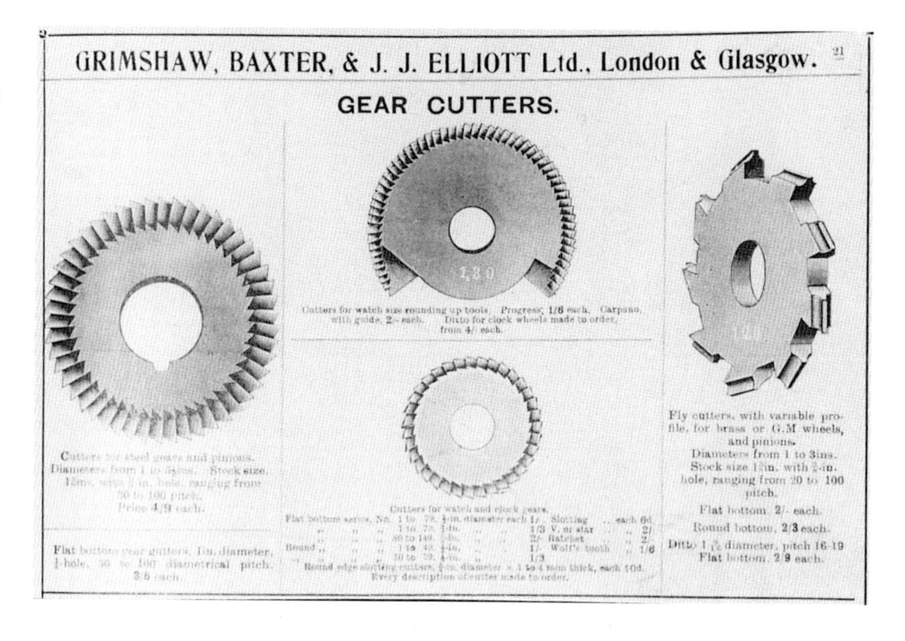

Fig. 304 Cutters available in the late nineteenth century and early twentieth century, as evidenced in a tool catalogue.

Fig. 305 Carpano catalogue.

CHARACTERISTICS OF 'L CARPANO' MILLING CUTTERS (FRAISES)

Several factories have a numerical series of size numbers similar to our original 'Carpano' system. In order to avoid every kind of error we advise our customers to use the modular notation as far as possible when submitting their orders.

Carpano Cutter Data

$$P = \text{Module} \times 3.1416$$

$$H = h + h'$$

$$H = E \times 1.10$$

$$E = \frac{P}{2}$$

$$E' = \frac{P}{2}$$

MODULES

For greater precision, the modules are given in increments of 0.0025 up to module 0.2, and thereafter by 0.0050 up to module 1.

DIAMETRAL PITCH

This gives a figure that is the number of teeth on a wheel divided by the diameter in inches. The

FRAISES A PROFIL CONSTANT « L. CARPANO »
POUR TAILLAGE DE ROUES ET PIGNONS
EN LAITON

Ces Fraises sont employées en Pendulerie et pour la Fabrication de Roues ou de gros Pignons en Laiton pour Phonographes, Compteurs, etc...

Elles conservent exactement le même Profil jusqu'à usure complète et, par conséquent, durent plus longtemps que les Fraises affûtables à denture dégagée sur les côtés. Mais elles ne permettent pas une production aussi intense.

Diamètre 25 m/m Alésage 7 ou 8 m/m	du Module 0,15 au Module 0,80 N° Carpano 8 au N° 120	**N° de série**........ E. 1.
Diamètre 30 m/m Alésage 10 m/m	du Module 0,20 au Module 100 N° Carpano 17 au N° 163	**N° de série**........ E. 2.
Diamètre 35 m/m Alésage 10 à 12 m/m	du Module 0,25 au Module 100 N° Carpano 26 au N° 163	**N° de série**........ E. 3.
Diamètre 40 m/m Alésage 12 m/m	du Module 0,25 au Module 100 N° Carpano 26 au N° 163	**N° de série**........ E. 4.
Diamètre 45 m/m Alésage 16 m/m	du Module 0,30 au Module 100 N° Carpano 35 au N° 163	**N° de série**........ E. 5.
Diamètre 50 m/m Alésage 16 m/m	du Module 0,30 au Module 100 N° Carpano 35 au N° 163	**N° de série**........ E. 6.

Pour les Commandes, indiquer :

1° Le Diamètre des Fraises.
2° Le Module **ou** le N° Carpano **ou** le Diamètre total du Mobile à tailler.
3° Le Nombre de dents du Mobile à tailler et de celui qui engrène avec lui.
4° Le Profil adopté : *Voir Note sur le* « CHOIX ET EMPLOI DES FRAISES CARPANO ».

DIMENSIONS ET PROFILS SPECIAUX : **PRIX SUR DEMANDE**

11

Fig. 306 Carpano pinion cutters.

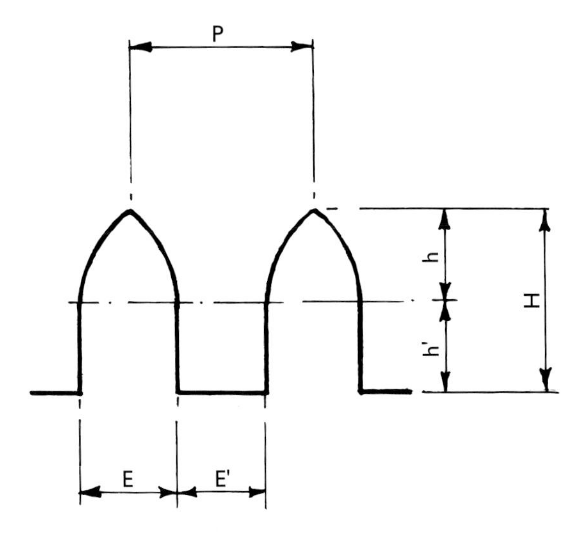

Fig. 307 Tooth-form data from the Carpano catalogue.

English and Americans use this system for their gear calculations. We are of the opinion that it would sometimes be useful for our customers to understand the value of the diametral pitch in relation to the module and Carpano numbers.

IMPORTANT NOTE

The values and numbers of corresponding cutters that follow in the tables below are theoretical. It is well known that a new cutter cuts a space that is oversize. It is good practice, therefore, to choose a cutter that is much smaller than the next smaller size indicated in the tables for a particular D P or module. Experience shows that the deviation from theoretical size is approximately as follows:

- Two numbers less than cutter size for No. 0 and smaller
- Three numbers less than size for No. 0–No. 20
- Four numbers less than size for No. 20 and above

Modules commence at 0.04 going right through to module 1; the smaller sizes covering watch wheel cutters. To use the tables, assume a M0.65 wheel cutter is required. This is a No. 94 in the Carpano table.

Then the addendum h	=	E × 1.10
Therefore h	=	1.021 × 1.10
	=	1.123mm
Total tooth height	=	1.123 × 2
	=	2.25mm

For the complete Carpano table, see Appendix IV.

Carpano are still in business in France, but no longer produce cutters. They now manufacture electric motors for garage doors, and electric windows for the automotive industry.

Another manufacturer of wheel and pinion cutters was Smith's Industries of Cheltenham. They produced a catalogue in the 1970s showing gashing cutters and hobs. Two tables are reproduced here, as the information may be useful for anyone who purchased this make of cutter; as can be seen, the system used was the Berners Swiss Standard. Unfortunately Smith's coding system was unique to the company. Most of these cutters had a 'K' reference number, but all the

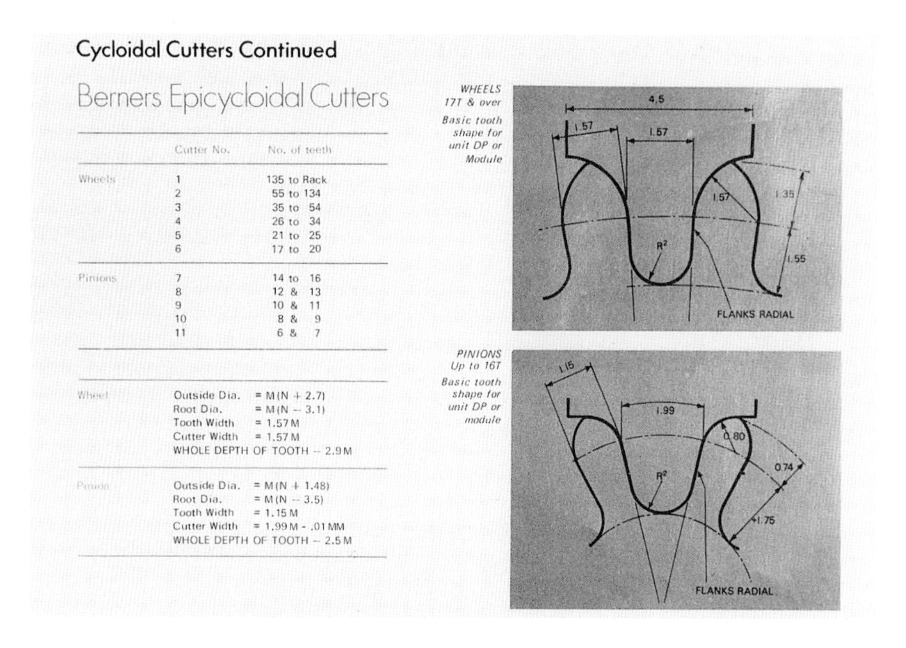

Cycloidal Cutters Continued

Berners Epicycloidal Cutters

	Cutter No.	No. of teeth
Wheels	1	135 to Rack
	2	55 to 134
	3	35 to 54
	4	26 to 34
	5	21 to 25
	6	17 to 20
Pinions	7	14 to 16
	8	12 & 13
	9	10 & 11
	10	8 & 9
	11	6 & 7

Wheel	Outside Dia.	= M(N + 2.7)
	Root Dia.	= M(N – 3.1)
	Tooth Width	= 1.57 M
	Cutter Width	= 1.57 M
	WHOLE DEPTH OF TOOTH – 2.9 M	
Pinion	Outside Dia.	= M(N + 1.48)
	Root Dia.	= M(N – 3.5)
	Tooth Width	= 1.15 M
	Cutter Width	= 1.99 M - .01 MM
	WHOLE DEPTH OF TOOTH – 2.5 M	

Fig. 308 Berners epicycloidal cutters – manufactured by Smith's Industries.

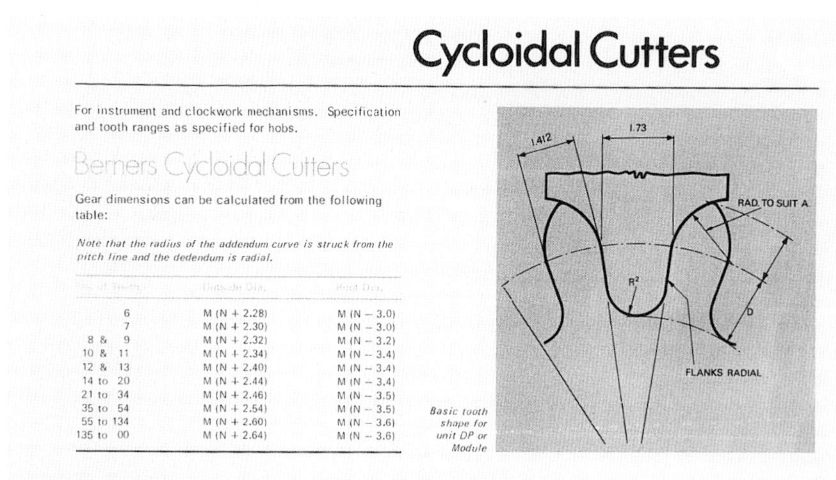

Cycloidal Cutters

For instrument and clockwork mechanisms. Specification and tooth ranges as specified for hobs.

Berners Cycloidal Cutters

Gear dimensions can be calculated from the following table:

Note that the radius of the addendum curve is struck from the pitch line and the dedendum is radial.

No. of Teeth	Outside Dia.	Root Dia.
6	M (N + 2.28)	M (N – 3.0)
7	M (N + 2.30)	M (N – 3.0)
8 & 9	M (N + 2.32)	M (N – 3.2)
10 & 11	M (N + 2.34)	M (N – 3.4)
12 & 13	M (N + 2.40)	M (N – 3.4)
14 to 20	M (N + 2.44)	M (N – 3.4)
21 to 34	M (N + 2.46)	M (N – 3.5)
35 to 54	M (N + 2.54)	M (N – 3.5)
55 to 134	M (N + 2.60)	M (N – 3.6)
135 to 00	M (N + 2.64)	M (N – 3.6)

Fig. 309 Berners cycloidal cutters – manufactured by Smith's Industries.

company records concerning cutters have been destroyed. All manufacturing of clocks, watches and cutters ceased in the late 1970s.

MAKING CUTTERS

It is possible to manufacture your own cutters; all formulae and dimensions have been made available. Various methods will be covered for both multi-tooth and single-tooth fly cutters, the latter being most easily produced in the home workshop.

The dimensions shown are the ones used by commercial cutter manufacturers, and are based on the Swiss Standards previously discussed. The range covered is from 0.2–1.5 module, which should cover sizes most likely to be encountered.

For wheels, an addendum factor of 2.76 is used. This is based on a wheel-to-pinion ratio of 7½ to 1 – that is, a wheel of forty-five teeth driving a pinion of six leaves. This is a good average, and one that is likely to be found in the wheel train of a typical longcase clock and many other similar types of clock.

WHEEL CUTTERS

Dealing with wheel-cutter forms first, the formulae used are as follows:

For sizes 0.2–0.45 and 1.1–1.5 module

Addendum 'A'
'A' = 1.38 × module

Dedendum
'D' = 1.57 × module

Tooth height
'H' = 2.95 × module

Addendum radius of tooth
'R' = 1.93 × module

Tooth thickness
'T' = 1.57 × module

Tooth space
'S' = 1.57 × module

Angle of cutter flank
= 2 degrees

Root dia.
= (N–3.14) M

For sizes 0.5–1.0 module

Dedendum
'D' = 2 × module

Tooth height
'H' = 3.38 × module

Root dia.
'R/DM' = (N–4) M

The remaining sizes are as above. When using this data, refer to Figure 310.

As was seen in the chapter on wheel and pinion theory, when calculating the outside diameter of a wheel, O/Dia = (N + 2.76) M, 2.76 being the addendum factor used by P. P. Thornton, cutter manufacturers, adopted to match closer to old English clocks. This could vary for different manufacturers.

$$\text{Circular pitch CP} = T + S$$

$$\text{or} \quad CP = \frac{\text{Pitch dia.} \times \pi}{\text{No. of teeth}}$$

$$= \frac{PD \times \pi}{N}$$

A complete table of wheel-cutter dimensions has been calculated from the preceding formulae to help where cutting wheel teeth for full tooth depth; other dimensions will assist if anyone should require to manufacture their own cutters. See the chart in the Appendix.

PINION CUTTERS

Pinion leaf calculations are slightly more complicated. Here again a chart has been worked out covering all the standard leaf modules likely to be

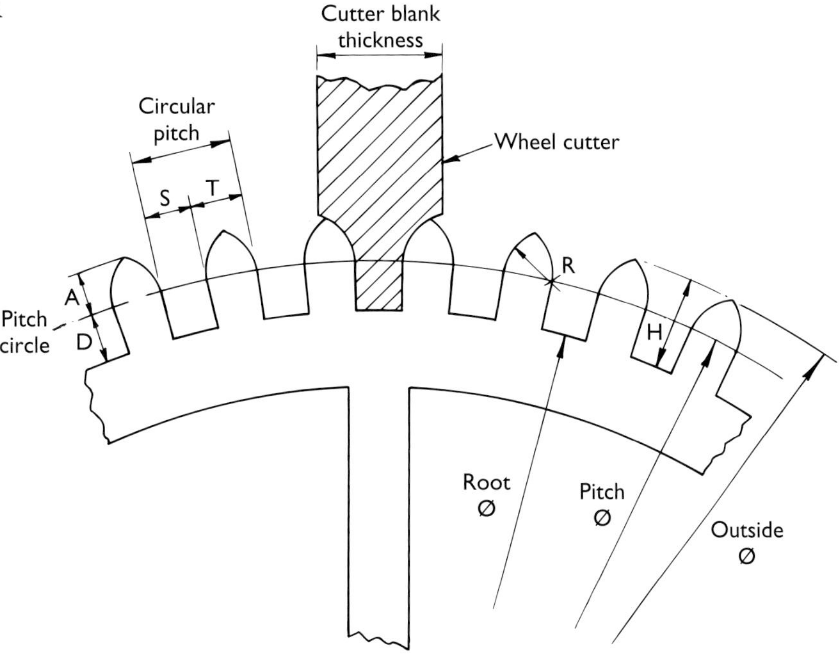

Fig. 310 Wheel and cutter data.

Wheel and cutter data (see table for dimensions)

Fig. 311 Pinion leaf profiles.

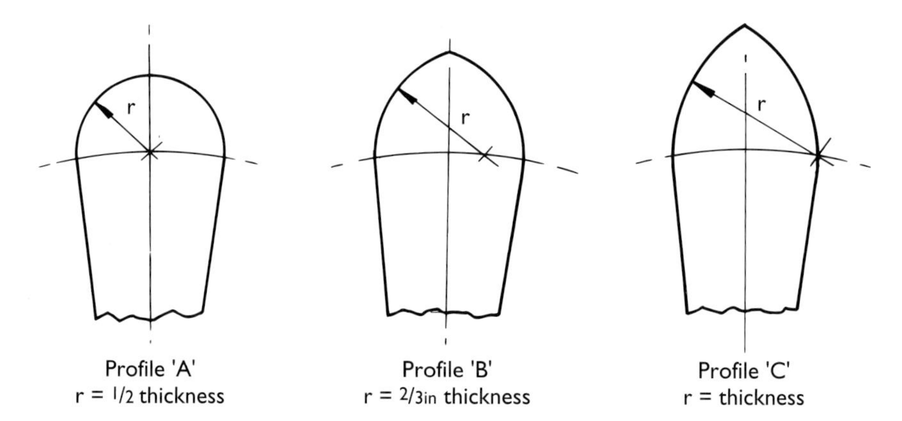

encountered for pinions six, seven, eight, ten, twelve and sixteen. Three different pinion profiles are shown in the drawing, 'A', 'B' and 'C':

Profile 'C' is used for pinions six, seven and eight leaf; this is a full ogive form.

Profile 'B', ⅓ ogive, is used for pinions ten, twelve and sixteen.

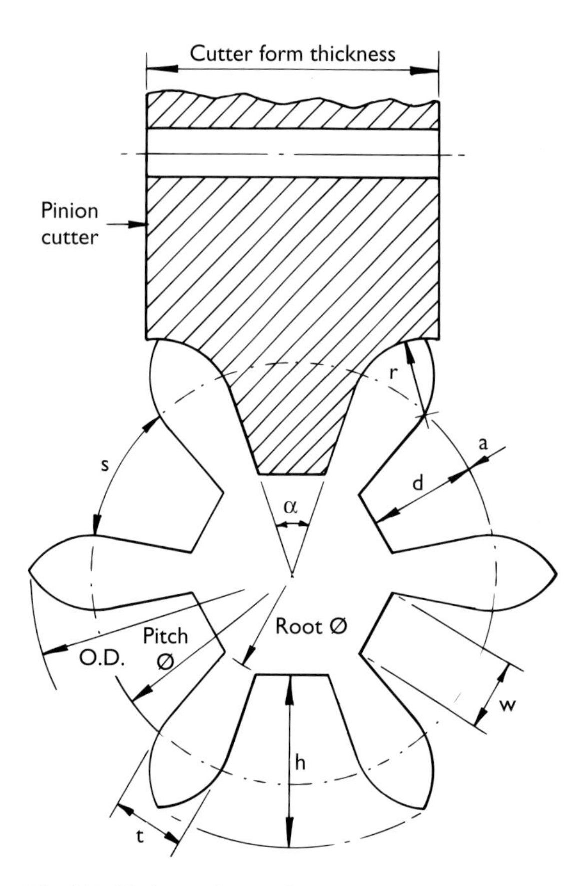

Fig. 312 Pinion and cutter data.

Profile 'A' would be used on very high count trains, and can be disregarded in this instance. Profile 'B' would cover a high train count quite adequately. The profile radius is determined by the number of leaves in the pinion.

Pinions of six, seven and eight leaf, where the drive starts before the line of centres, require to have a profile high enough, and with sufficient curve, to give a smooth transmission of power when the wheel commences to drive the pinion.

The formulae for calculating the pinion tooth form are as follows:

For six, seven and eight leaf, full ogive form Profile 'C'

Addendum
'a' = 0.855 × module

Dedendum
'd' = 1.58 × module for 6 leaf
1.85 × module for 7 leaf
1.90 × module for 8 leaf

Leaf or tooth thickness
't' = 1.05 × module for 6, 7 and 8 leaf

Addendum radius of tooth
'r' = 1.05 × module

Full tooth depth
'h' = 2.435 × module for 6 leaf
2.705 × module for 7 leaf
2.775 × module for 8 leaf

Tooth-to-pitch ratio
= 1/3

For ten, twelve and sixteen leaf 1/3 ogive Profile 'B'

Addendum
'a' = 0.805 × module

Dedendum
'd' = 2.05 × module for 10 leaf
2.10 × module for 12 and 16 leaf

Leaf or tooth thickness
't' = 1.25 × module

Addendum radius of tooth
'r' = 0.82 × module

Full tooth depth
'h' = 2.855 × module for 10 leaf
2.905 × module for 12 and 16 leaf

Tooth-to-pitch ratio = 2/5

When using this data, refer to Fig. 312.

Theoretically, wheels have teeth and spaces that are equal, although old clocks have spaces much wider: teeth proportions varied dramatically, each clockmaker having his own standard. With pinions of six, seven and eight, however, the leaves or teeth are ⅓ of the pitch and the space is ⅔ of the pitch, the dimensions being measured along the pitch circle.

For pinions of ten, twelve and sixteen leaves, the thickness of tooth is ⅖ and the space is ⅗. The following example should clarify this:

Example:

Six-leaf pinion: divide $\frac{360}{6} = 60$ degrees

This angle, then, is the space occupied by a pitch, or one tooth and one space: this has to be split up into the ratio of ⅓ and ⅔.

$$\text{leaf} = \frac{60}{3} = 20 \text{ degrees}$$

$$\text{space} = \frac{60 \times 2}{3} = 40 \text{ degrees}$$

Applying the same method for seven- and eight-leaf pinions, the following is obtained:

Seven-leaf tooth
= 17 degrees space = 34 degrees

Eight-leaf tooth
= 15 degrees space = 30 degrees

This is to the nearest degree, which is as near as can be measured unless an optical projector is used for checking. The tooth spaces then represent angle α in the table, and this will be the same for any size pinion of one count.

As with the wheel-cutter dimensions, a table has been calculated to cover all pinion sizes; this is shown in the Appendix.

Now that all the pinion dimensions have been calculated, there is sufficient information for cutters to be manufactured.

FLY CUTTERS

Fly cutters for cutting wheels are quite useful, and can be made in the workshop. These are not suitable for cutting steel pinions.

Fig. 313 A completed fly cutter for cutting wheels.

The first method for making fly cutters is quite straightforward, and cutters can be made fairly quickly. Use 5mm (3⁄16in) dia. silver steel, and turn the blank to the appropriate diameter with a left-hand knife-tool ground to the correct radius as shown. This can be determined by the use of radius gauges, or by drilling a hole of the correct size in a piece of thin sheet and then cutting away the surplus to leave the required radius.

The tool requires stoning on all cutting edges so that it will impart a good finish to the cutter blank. After this, polish with fine crocus or wet-and-dry paper to give the final polish. Next, file or mill a flat to the centreline, then relieve the tip to approximately 10 degrees. Check that a good finish has been obtained on all cutting edges before heat-treating. Heat the cutter to cherry red, and plunge in brine as described previously. Dry off and hold in the lathe chuck or drilling machine for polishing.

Fig. 314 Sequence of operations in making simple fly cutters.

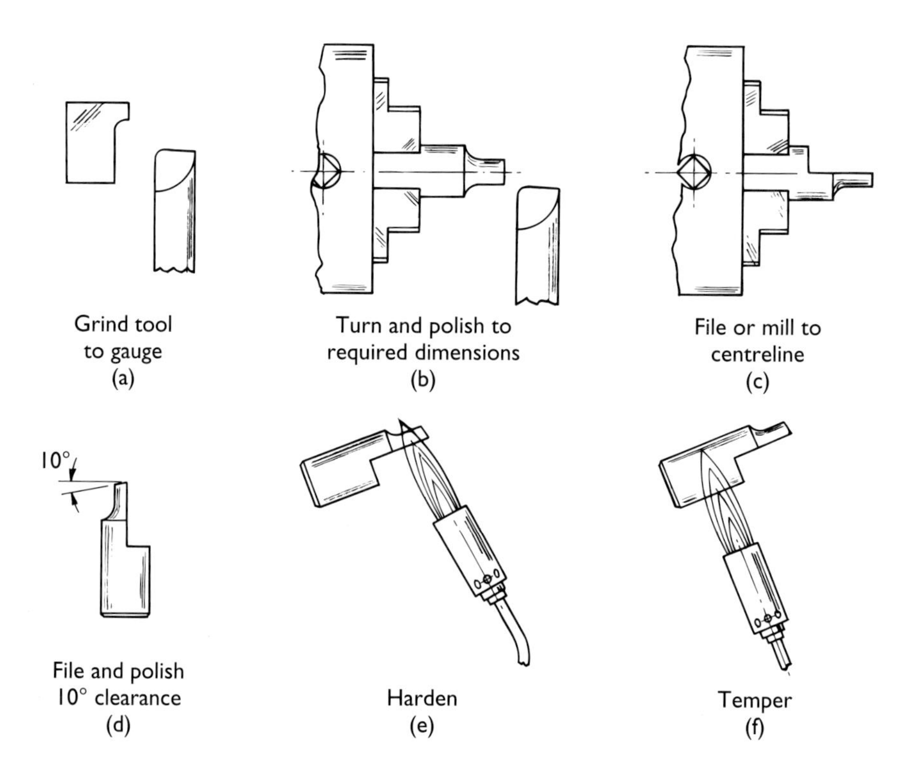

The cutter now requires tempering back from the very brittle state it is in. Heat the shank – not the tip – in a low flame, very gently, and watch for the colour to change, first to a light straw then to a deep straw. As soon as the colour changes on the shank it will run very quickly towards the tip, this being a smaller section. At this stage, plunge in brine solution as previously described.

The cutter now only requires stoning with an Arkansas stone to give a really keen edge. Be careful not to let the colour pass through the deep straw to a blue, otherwise the necessary hardness will be lost and the whole heat-treatment process will have to be repeated. These cutters work very well on brass for normal size clock wheels. However, small cutters break easily and are unsuitable for smaller modules.

Fig. 315 Early-style fly cutters.

The advantage of the next method is that the fly cutters are form-relieved, constant profile and can therefore be resharpened quite a few times and are much stronger; whereas the previous type cannot, because once you have ground the cutting face below the centreline, the dimensional form is lost. Use 5mm (3⁄16in) square-section silver-steel instead of round. A piece of mild steel 38mm (1½in) dia. is cross-drilled as shown, and at 11mm (7⁄16in) below the centreline. The hole is required to be of sufficient diameter to accept the across-corner dimension of the cutter blank. A hole is then drilled and tapped 2BA for a locking grubscrew. Mount the bar in the three-jaw chuck. It will be necessary to grind two tools similar to the tool used in the previous method, but in this case it will require both right-hand and left-hand.

As this is intermittent cutting, care is required when feeding in. Remove the cutter and file, or grind a slight top rake to the cutting face. Harden and temper as before. Now make a small cutter

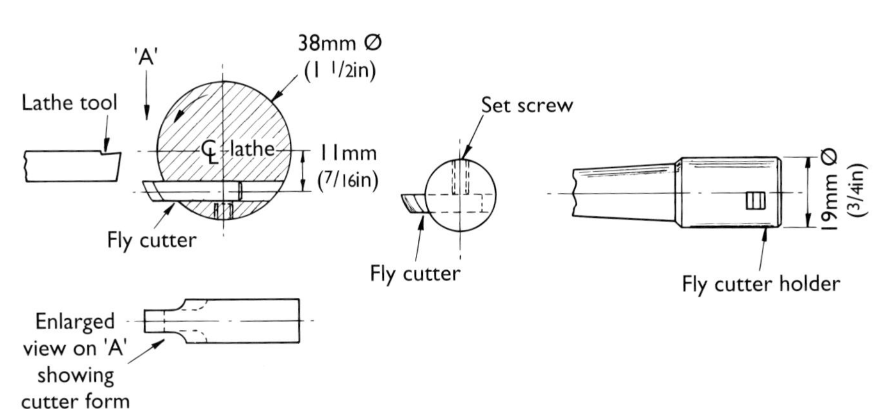

Fig. 316 Making form-relieved fly cutters.

Fig. 317 Modern fly cutter – manufactured from a circular blank.

Fig. 318 Lapping tool showing the fly cutter ready for sharpening.

bar to the dimensions shown, with a shank either taper or parallel, depending on which type of milling spindle is being used; if a taper shank is employed, always thread and use a drawbar. The cross-hole for the cutter is drilled so that the top cutting edge of the form-relieved cutter is on the centreline; then, when cutting takes place, this will give the necessary clearance.

SHARPENING FLY CUTTERS

We have previously mentioned the necessity of obtaining a good finish to the cutter edge, as this will determine the finish applied to the work. Normally an Arkansas stone is used for the final polish after the use of various grades of carborundum; however, it is not easy to obtain a true, flat, smooth finish to the cutting edge with hand-held, unguided polishing stones, and with this in mind, the lapping tool was developed. A fly cutter is shown being held ready for sharpening. Laps are circular and operate in two

Fig. 319 Using the lapping tool.

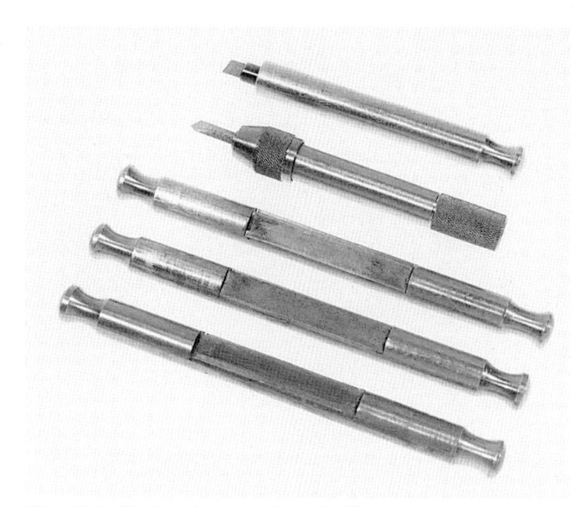

Fig. 320 Various laps and tool holders.

guides; a flat is machined to accommodate diamond shim. Two grades are used, 600 and 1,200, the latter being approximately 9 microns, whilst an extremely good finish is imparted.

The final finishing process is applied with a copper lap using 6-micron diamond lapping paste; this will give a mirror finish. The fixture is held in the vice and the tool to be sharpened can be rotated through 360 degrees, as also can the lap. Many different tool angles and configurations can therefore be accommodated. The lapping tool proved to be so useful that a facility for holding gravers was made; a few strokes from each lap produces an excellent finish to the graver cutting face.

Fig. 321 A graver ready for lapping.

MULTI-TOOTH CUTTERS

Multi-tooth, constant profile, form-relieved wheel and pinion cutters can be made in the home workshop with simple jigs and lathe attachments.

Cutters with twelve teeth similar to commercial cutters can be manufactured with care, and will give excellent results. Various form-relieving methods have been described over the years in the *Model Engineer* magazine; an excellent article by Duplex[17] describes the complete methods of making cutters using a simple form-relieving attachment – this ran for several issues and was very descriptive. However, a most ingenious device was described in 1987. This was called the 'Eureka' continuous form-relieving attachment, and was designed by Ivan Law and the late Professor D. H. Chaddock.[18] A complete set of working drawings is available to make up the tool.

Fig. 322 Eureka continuous form-relieving attachment.

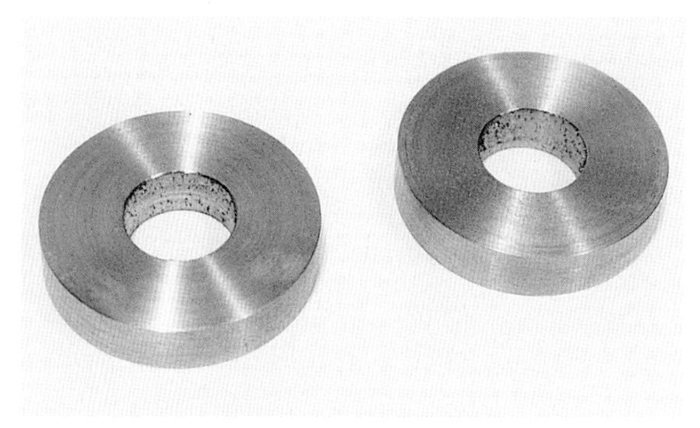

Fig. 323 Cutter blank.

To make a circular cutter, any carbon tool steel is suitable that can be hardened and hold a good cutting edge: cast tool steel, silver steel, gauge plate, and old files make good cutting tools. The blanks illustrated have been turned to size and drilled and reamed; the bore is 12mm (½in) diameter to suit the mandrel on the form-relieving tool.

Mount the cutter blank on a mandrel held in the lathe chuck and form the radii. Next the twelve teeth are gashed; this is carried out on the milling machine with an equal-angle cutter offset to form the radial cutting face. The dimensions for making the form tools can be obtained from the chart in the Appendix.

Another excellent article on cutter making was described in the *Model Engineer* by D. J. Unwin;[19] in this, complete details were given on making the form tools to produce the cutter profile. As can be seen, the cutter to form the radius is obtained by the use of hardened buttons; an alternative method would be to grind a form tool to a template, as previously described. In the illustration of the Eureka form-relieving tool mounted between the lathe centres, note the slotted lever anchored by a pin protruding from the form tool holder.

Fig. 324 Cutter blank radii being formed.

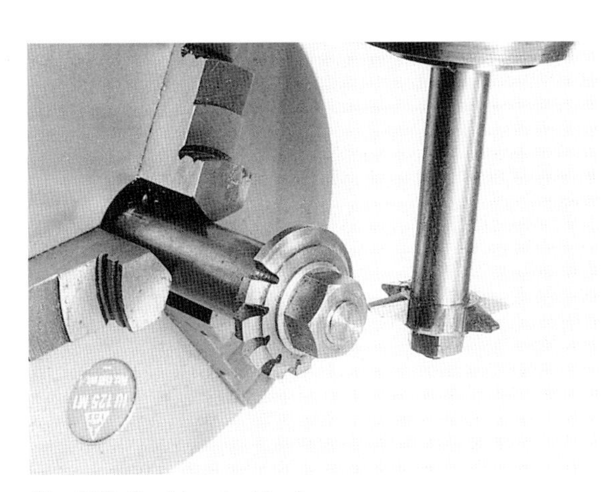

Fig. 325 Gashing the blank.

Fig. 326 Blank and cutter blank gashed.

The lathe is operated in back gear with slow revs. It is fascinating to see the tool in action. Each tooth in turn moves towards the tool and then away in a continuous motion, and then automatically indexes to the next tooth. A new cut is applied after the blank has covered a complete revolution. When the form is near completion, a large cutting area is in contact. A good cutting lubricant is therefore required, and a very fine feed. The tool proved to be extremely rigid in operation. The set-up shows a 1.5 module cutter being produced.

Once the form is complete, the tops of the teeth have to be machined to form relief. The completed cutter now only requires heat treatment, and the cutting face sharpening. Cutters

Fig. 327 Form tool and completed cutter.

Fig. 328 Eureka form-relieving tool in action.

Fig. 329 Relieving the side of the cutter form.

Fig. 331 Machining the outer lobes of a four-tooth form-relieved cutter.

Fig. 330 Relieving the top of the cutter form.

Fig. 332 The jig for holding the cutter blank.

with four teeth form-relieved are extremely satisfactory in use, and can be manufactured with the use of quite simple jigs. A blank is prepared approximately 32mm (1¼in) in diameter, and reamed 8mm (5/16in) in diameter. Four small 2mm diameter location holes are drilled on a 16mm (⅝in) pitch circle diameter; this is to enable the cutter blank to be accurately indexed.

A body to hold the cutter is machined. The diameter is approximately 48mm (1⅞in) with a smaller diameter to fit in the three-jaw chuck. An eccentric pin is fitted 16mm (⅝in) from the centre as shown, with a hole tapped 2BA for a socket-head cap screw to secure the cutter blank. When the blank is held in place, drill through into the body 2mm (0.078in), then a small pin can be used to locate the cutter blank in four positions. Machine the outer lobes, then form the radii either side of the blank. The previous methods of producing the form tool can be used.

Once the form is complete on all four lobes, it is necessary to set up for gashing the teeth. The blank is held on a mandrel either on the milling machine or lathe. A standard end mill is used to cut out the four sections as shown to produce the cutting edge. Once the cutter is completed, all that is then required is for it to be hardened and tempered, and the cutting faces ground and lapped.

Whilst this method has been describing the making of wheel cutters, similar methods can be used to produce pinion cutters. To obtain the

Fig. 334 The jig showing the blank with all sides machined.

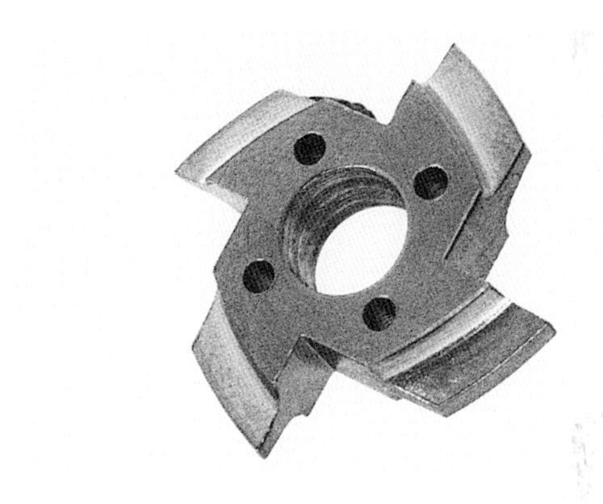

Fig. 335 A completed four-tooth form-relieved cutter.

Fig. 333 Forming the sides of the cutter.

cutter flank angle, the lathe top slide would be required to be set over to half the included angle shown in the table in the Appendix.

When using wheel and pinion cutters, the rate of feed depends on various factors:

a) the rigidity of the machine and the set-up;
b) the surface finish required;
c) the materials being machined.

Feeds of 0.15–0.25mm per rev for steel and 0.3–0.4mm for brass give good results consistent with economic cutter life.

9 CROSSING OUT WHEELS AND MOUNTING

Now that the cutting of wheels and pinions has been covered, the next stage is assembly to their respective arbors: wheels have to be 'crossed out' – that is, wheel spokes must be produced to make the wheel lighter in order to reduce inertia when the train is running. Crossings vary so much on different types of clock: they can be four, five, six or eight, and either parallel or tapered. Whilst crossings for longcase clocks do not have to be absolutely accurate, high grade regulator clocks and good quality skeleton clocks that are more aesthetically pleasing and with nice proportions, should have finely finished crossings. On extremely high grade work, wheels should be balanced.

MARKING OUT PARALLEL CROSSINGS

Parallel crossings are the easiest to mark out, but this requires the lathe to be set up with some form of dividing arrangement for the lathe spindle. The scribing block is then used on the lathe bed, and measurements taken with a rule mounted in a block. Alternatively, the wheel blank could be marked out using a dividing head on the surface plate, once again utilizing the scribing block to carry out the actual marking. With either of these methods, there is still a problem with tapered spokes; with this in mind, the following simple jig was constructed.

Not only will this jig cover all types of wheel crossing, it will also be useful for marking out equal radial spacings for pitch circle diameters on many small components if they have a central hole. Basically, the jig consists of a circular blank with a series of equally spaced holes for shouldered pegs of various diameter. The wheel blank is located on the centre peg and held in position with two strips of double-sided Sellotape. Originally it was thought it would be necessary to clamp the wheel blank to the jig baseplate with some form of screw fixing, but this would have complicated matters and entailed shouldered screws of various diameter, whereas simple pegs are used and the Sellotape is perfect.

The plate shown opposite is 115mm (4.5in) in diameter and has four rows of equally spaced 3mm (⅛in) diameter reamed holes: these are eight, six, five and four, with a central hole also. When using the pitch circle shown, this arrangement will cover most requirements on small components – but of course, there is no reason why the baseplate could not be scaled up, if required, to enable larger circular components to be accommodated. Assuming that the wheel to be marked out is for a good quality skeleton clock with six crossings with a spoke width of

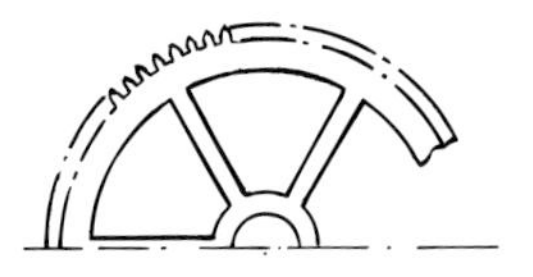

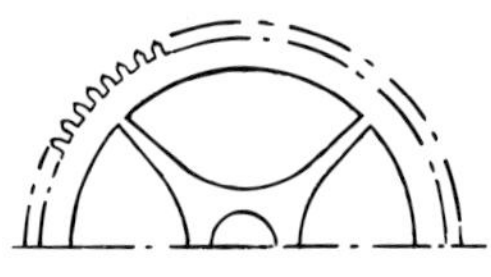

Fig. 336 Various types of wheel crossing.

Fig. 337 Marking out wheel crossings on the lathe.

Fig. 338 Wheel-crossing marking-out jig.

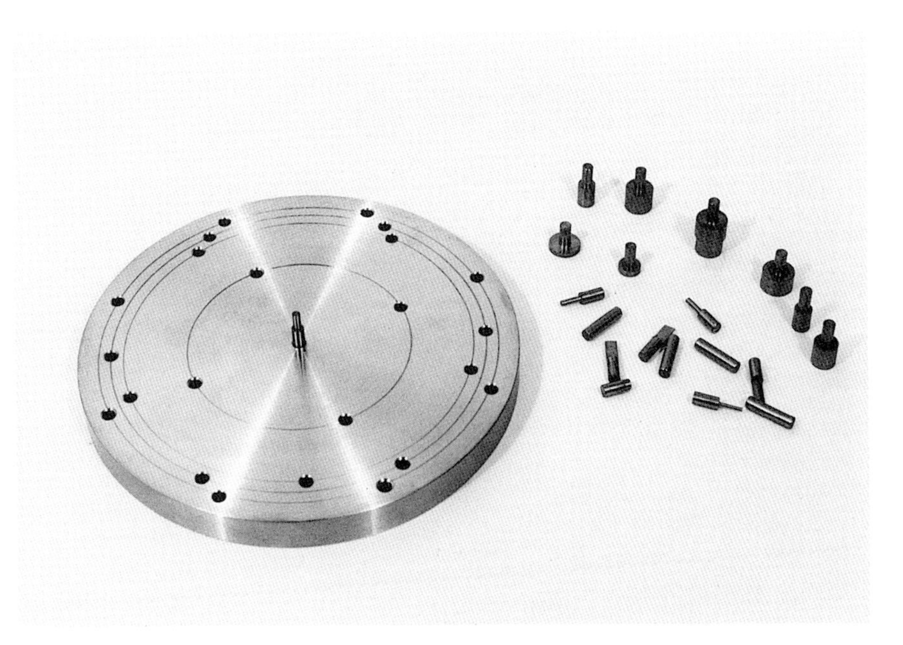

1.5mm ($^1/_{16}$in), cover one side of the blank with marking-out fluid – a wide-tipped felt marking pen which has black ink is ideal; there is no mess and no brush to clean out after use.

Select a pin to suit the bore in the wheel blank, and locate this in the centre hole of the plate. Cut two pieces of double-sided Sellotape and position these on either side of the central locating pin underneath where the wheel blank will be located. Several pins of different diameter are required to suit standard bores, for example, 4.75mm ($^3/_{16}$in), 6mm ($^1/_4$in), 8mm ($^5/_{16}$in) and 9.5mm ($^3/_8$in). All these pins have a very small centre dot: this is for locating the dividers when scribing the boss and outer rim circles.

When the circles have been scribed, select two 1.5mm ($^1/_{16}$in) dia. pins, and position in any two opposing holes on the six-hole circle. Place a rule

Fig. 339 Marking out a six-spoke wheel for a skeleton clock.

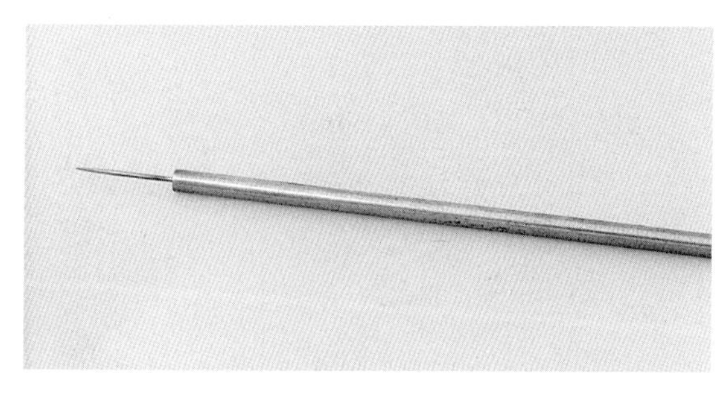

Fig. 340 Fine-pointed scriber.

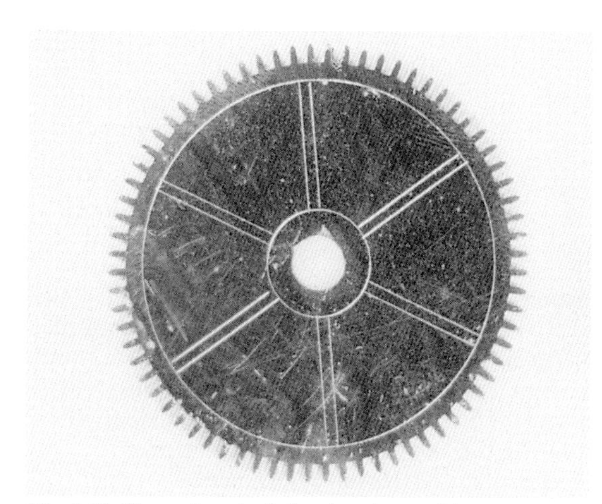

Fig. 341 Six-spoke wheel showing the marking out of crossings.

or straight-edge against the two pins, as shown, and with a very slim and sharp scriber, scribe a line to join the inner and outer circle for both spokes. Then place the rule on the opposite side of the two pins and scribe a parallel line. This will complete the marking out of the first two spokes. Now move the pegs to the next position and repeat the procedure until all the spokes are complete.

NOTE: An excellent scriber can be made using a sewing needle let into a piece of round 4.75mm (3/16in) dia. brass rod. This will give a very accurate and clean outline to work to when filing up and finishing the crossings, making it relatively easy to follow the line. When scribing lines, do not press too hard, because if the lines overrun at any point they take a lot of removing when the final polishing of the wheel takes place.

Tapered Spokes

Tapered spokes can be catered for in a similar manner, but in this instance two different-sized pegs are used to suit the crossing. It is possible to calculate the required pegs, but this does complicate matters, and it is useful to have a number of pins that will cover most types of wheel crossing. When designing a new clock and the layout of the wheel design, extend the spoke lines on the drawing paper to the appropriate PCD of the marking-out jig; this then determines which

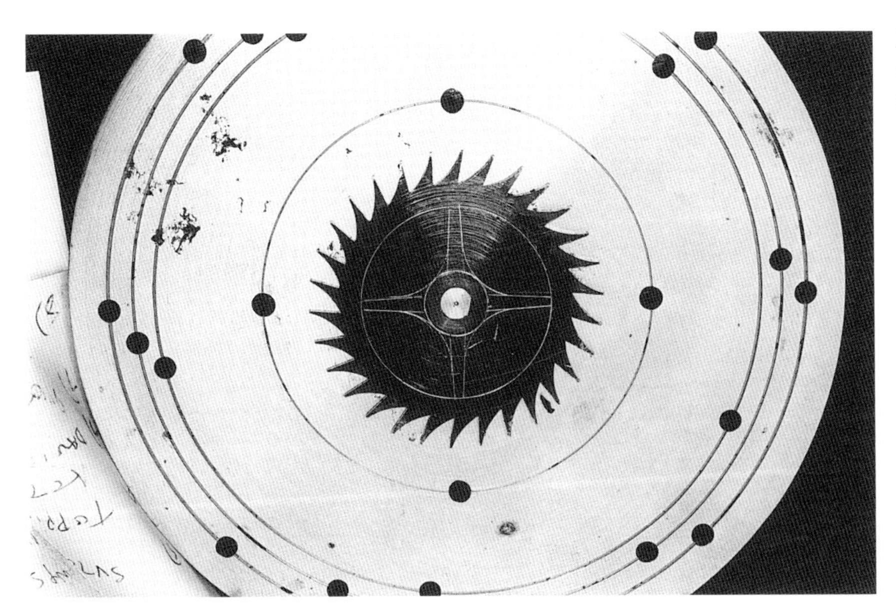

Fig. 342 Escape-wheel crossings for a longcase-clock movement.

diameter pins to use. The illustration shows an escape wheel for a longcase clock; the crossings or spokes have quite a large amount of taper, and to accommodate this, a larger diameter pin is required. These pins take only a few minutes to make, and it is useful if a number of different diameters are produced to cover future wheel crossings.

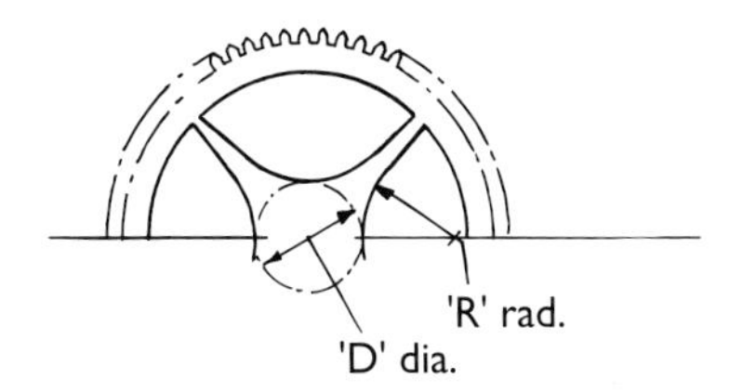

Fig. 343 Wheel-crossing radii.

Marking-Out Jig

The large radius that joins the two adjoining spokes and the central boss is determined approximately by the following formulae; these have been calculated after measuring many wheels on old clocks and give a nicely proportioned crossing. In certain cases they may require amending slightly. When applying these formulae, refer to Figure 343.

$$\text{radius} = \frac{\text{wheel dia.}}{4.5}$$

$$\text{central boss dia.} = \frac{\text{O/dia.}}{4}$$

Figure 344 shows useful items for marking out the wheel crossings; the draughtsman's curves are particularly useful when joining the straight lines with the large radius.

The construction of the jig is quite straightforward. Make the baseplate from a piece of brass, 115mm (4½in) in diameter; steel or aluminium could also be used for the baseplate,

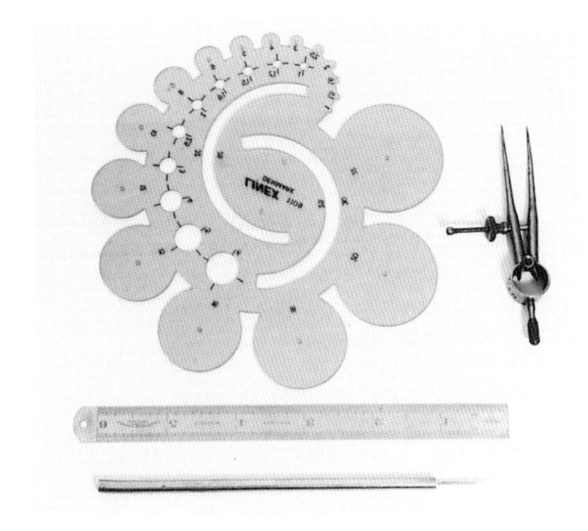

Fig. 344 Various items useful for marking out.

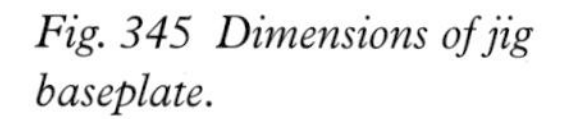

Fig. 345 Dimensions of jig baseplate.

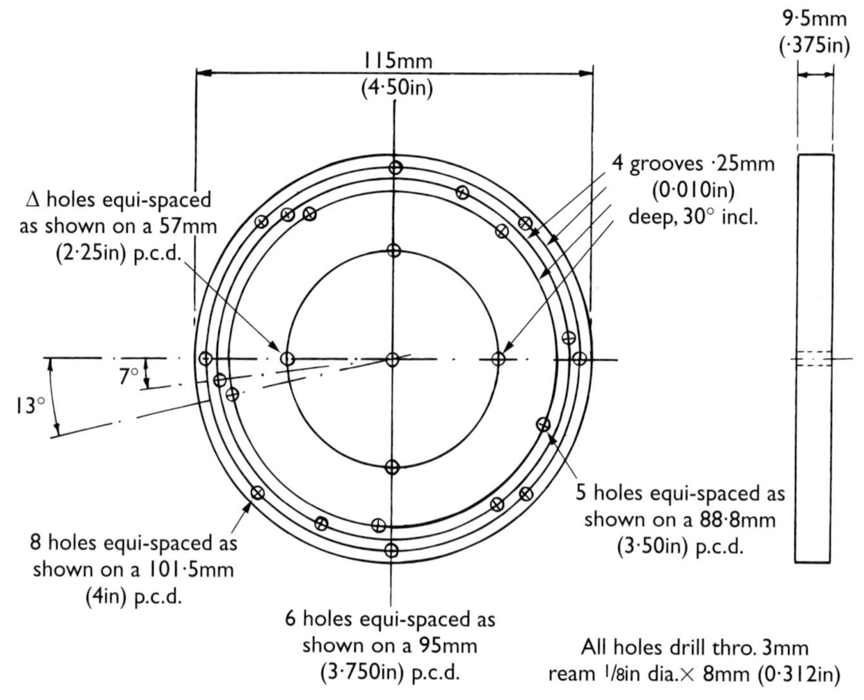

or perhaps cast iron. However, with ferrous materials care would have to be taken to ensure that none of the pegs were left in the jig after use if the jig were stored in a damp cupboard, or it might be difficult to remove the pegs at some later date due to rusting. Brass is to be preferred, if available.

Machine the outside diameter and face. Neither of the dimensions is critical. With a very sharp centre drill, drill a small centre using the tailstock chuck, just deep enough to accept the point of the dividers. Then scribe the four circles to the dimensions shown, using a sharp vee tool, ground to 30 degrees inclusive. Pick up from the circles scribed with the divider, and groove to 0.25mm (0.01in) deep. The circles assist when the jig is in use, by enabling the PCDs to be picked out more clearly.

Once the grooves have been machined, centre with a number 2 centre drill and drill number 31, and ream the 3mm (⅛in) dia. centre hole. Drill right through, but only ream 8mm (0.312in) deep; this is so that the plain 3mm (⅛in) pins or unshouldered pins will not pass right through the plate when in use, but can be removed easily if they become tight or jammed. When machining the four rows of holes on the PCDs, accurate marking out is required. Drill and ream using the drilling and milling spindle on the lathe vertical slide, the spindle being driven from an overhead drive. The dividing of the lathe spindle was carried out by mounting the dividing head to the rear of the lathe. If this facility is not available, then the holes can be marked out with the use of dividers; if care is taken, the results will be quite satisfactory. The drilling and reaming can then be carried out on the drilling machine.

As mentioned previously, there are four rows of holes: eight, six, five and four. The three outer rows are fairly close together. It is important, therefore, that none of the holes coincides with another – although it does not matter in what relationship to each other they are, as long as any hole on any one PCD does not foul. The 3mm (⅛in) reamed holes are machined to the same dimensions as the centre hole. Remove the plate from the chuck, and rechuck and face down the back if you have not already done so. The plate is now complete.

The machining of the various pins is quite straightforward, the material used being silver steel. Mild steel could be used, and would be perfectly adequate. Unless collets are being used, make sure material is selected that is slightly

Fig. 346 Drilling the baseplate.

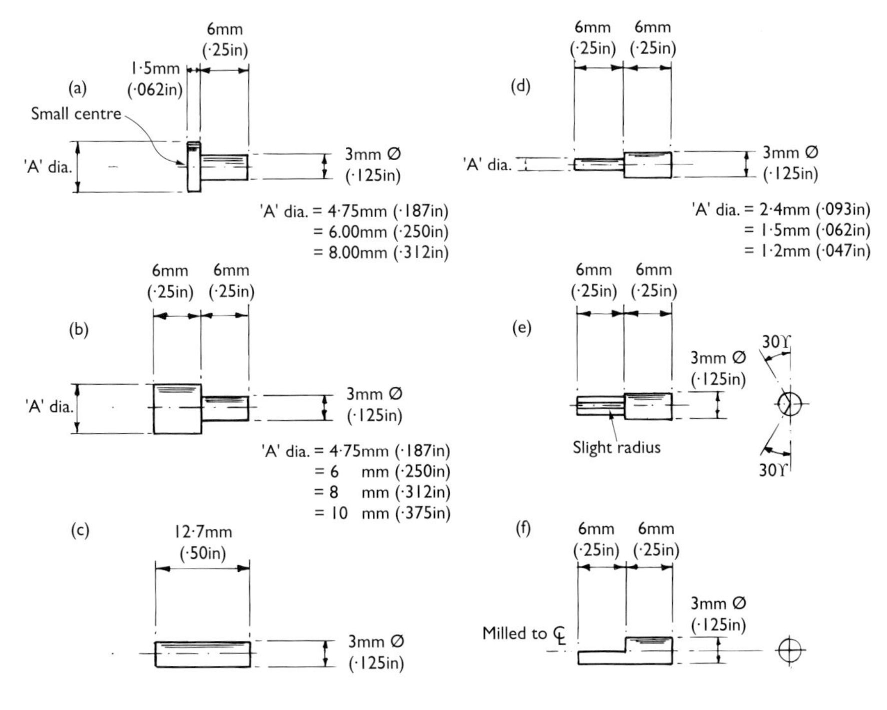

Fig. 347 Dimensions of the pins.

larger than the maximum outer diameter of the pin required, to ensure that both diameters are turned at the same setting, so that the two diameters are concentric. Where pins have flats, these can be filed or milled.

Now that a suitable method of marking out the wheel crossings has been established, perhaps a few notes on completing the wheel would be of interest, and in particular, the use of the piercing saw. When only a few wheels have to be crossed out, then the simplest way is to use a piercing saw. If larger batches are to be considered, then these are often accommodated on the engraving machine or pantograph miller. This does, of course, necessitate making simple holding attachments and profile plates, but the end result

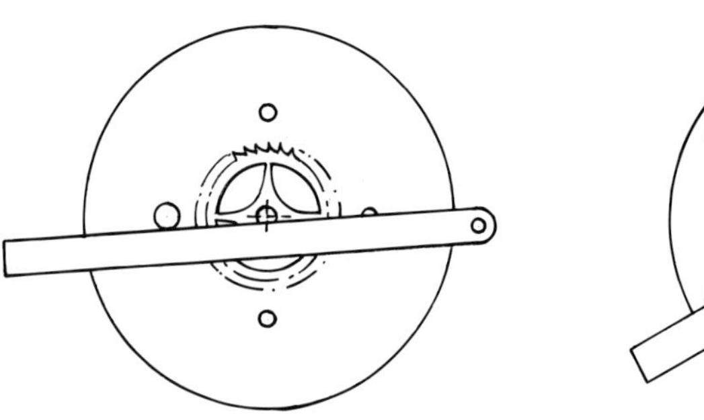

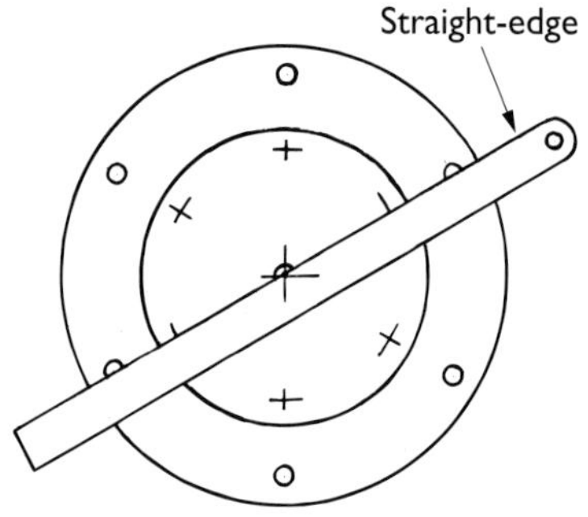

Fig. 348 Marking out wheel crossings.

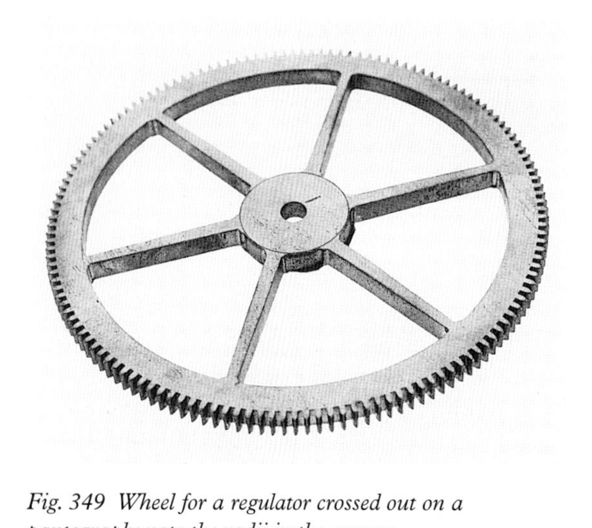

Fig. 349 Wheel for a regulator crossed out on a pantograph; note the radii in the corners.

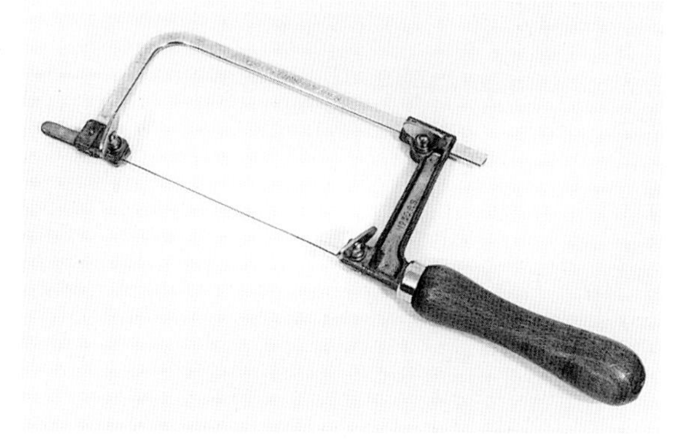

Fig. 350 Piercing saw with an adjustable frame.

Fig. 351 Saw table.

is accurate and speedy. All that is required then is that the corners have to be removed by filing, and the wheel polished. However, normally this type of equipment is not found in the model engineer's or amateur clockmaker's workshop, although the use of this machine will be covered later.

USING THE PIERCING SAW

The piercing saw is always used vertically, with the handle beneath the work. A saw with an adjustable frame is a definite advantage, as tension can be applied to the blade more easily, and also the adjustment allows the use of blades that have been shortened by breakage.

A saw table is used to support the work; this is basically a flat piece of wood with a 'V' cut in it. The one shown has a vertical post that is clamped in the vice, which brings the table up to the correct height for sawing.

Piercing Saw Blades

There are many makes of piercing saw blades, and all have advantages and disadvantages. The blades manufactured in this country are mainly Eclipse; these tend to be softer than the Continental blades, so do not break so easily –

Fig. 352 Piercing-saw blades.

but they do lose their edge more quickly. Get used to the idea that blades are expendable, and be prepared for plenty of breakages – particularly at your first attempt at piercing. Most Swiss or German blades have round backs, so that curves may be accommodated more easily. If fancy scroll work is to be attempted, then the round-backed blades are an advantage; as already stated, these blades hold their edge longer, but tend to snap more easily.

Blades are made in a good range of sizes, some of which are shown in the table below. Only two or three grades are required, however, for the type of work being discussed here. The table lists

Blade size	Length and width	Thickness	Teeth per inch
M4/0	5in × 0.006in	0.018in	80
M3/0	5in × 0.007in	0.019in	80
M2/0	5in × 0.008in	0.021in	60
M1/0	5in × 0.009in	0.023in	60
M0	5in × 0.010in	0.025in	60
M1	5in × 0.011in	0.026in	52
M2	5in × 0.012in	0.027in	44
M3	5in × 0.014in	0.030in	44
M4	5in × 0.015in	0.032in	32
M5	5in × 0.017in	0.036in	32

the size and number of teeth for a given reference number. This reference is a standard, and is the means of ordering the required blade. The complete range is from M8/0 to M6 (8/0 being the finest), but only M4/0 to M5 is listed here, as this will more than cover our requirements.

For the type of piercing work encountered in clocks, blades are required that will accommodate a metal thickness of, say, 0.4mm to 3mm (0.015in to 0.125in), or a maximum of 5mm (0.187in). Breaking this range of metal thicknesses down, the following blades would be all that were required:

Blade M3/0 for 0.4mm to 0.8mm (0.015in to 0.031in)

Blade M1/0 for 0.8mm to 1.5mm (0.031in to 0.062in)

Blade M3 for 1.5mm to 3mm (0.062in to 0.125in)

Blade M5 for 3mm to 5mm (0.125in to 0.187in)

These are recommendations and have been found to be satisfactory, but individual users may find they have slightly different preferences, and they will have to decide which blade best suits their method of working. Excellent piercing-saw blades are manufactured by Valorbe-Glarden.

Preparing to Saw

Using a wheel blank that is already marked out with the required spoke configuration, drill a small hole through each segment or waste portion, just large enough for the piercing-saw blade to pass through. Fix the blade in one end of the frame, making sure the blade is clamped square, and that the teeth are facing down towards the handle. Pass the blade through the drilled hole in the wheel blank and clamp the other end of the blade. Care is sometimes required when tightening the wing nut, as the twisting action will often move the blade out of alignment.

Tension should now be put on the blade with the adjustment on the frame itself. The blade should be nice and taut; if it is left slack it will break in no time at all. The blank is held against the saw table with the left hand, and the piercing saw in the right hand; that is, if you are right-handed. The work is carried out whilst seated, so a stool is required with a suitable height to bring

Fig. 353 Using the piercing saw.

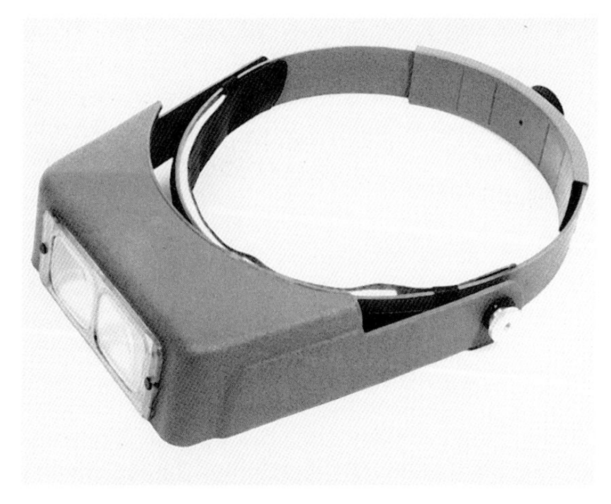

Fig. 354 Head-band magnifier.

the eye level something like 230mm (9in) above the work; a good operating position would be, say, 1.1m (44in) from the floor.

Excellent lighting is also required. A bench light with a halogen bulb gives good results. Another useful accessory is the Optivisor or head-band magnifier. Various magnifications and working distances are available, and these can be used whilst still wearing spectacles.

Sawing Technique

The actual sawing is carried out in a vertical plane – that is, 'up and down' strokes are used, with the pressure exerted on the down stroke. Do not force the blade, or this will result in premature breakage. Try and keep as close to the line as possible, whilst keeping the saw vertical and square to the work. At first this will be difficult, but as you become more proficient with practice, you will find it easier to 'hug' the line – and this, of course, will save time when the wheel or component has to be finish-filed. A spot of beeswax or tallow rubbed on the blade helps to lubricate the cutting. This is particularly beneficial when the blade teeth start to lose their sharpness or set, when any attempt to force the blade should be resisted – just keep an even pressure. As soon as you feel that the blade is no longer capable of any useful work, change it for a new one.

When a number of components have to be dealt with, you may find it preferable to clamp the wheel blank to the saw table with a toolmaker's clamp, to save your hand from tiring – although you may find it easier to saw to the line when actually holding the blank by hand. Whilst cutting is taking place, the sawing chips will tend to obliterate the line, and these should be blown or brushed away as you work.

FILING UP THE WHEEL

When all the crossings have been pierced out, the wheel has to be filed up and polished or burnished. 100mm (4in) and 150mm (6in) crossing files of three different cuts will be required – 0, 2 and 4: 0 cut is the coarsest, and the most useful if it has not been possible to cut close to the line when saw-piercing. The section of the crossing file is in the form of two different radii, one fairly flat and the other more convex; this is to enable the file to accommodate a wide range of curves. A word of warning here: when filing, always use a handle, as one slip with a file without a handle will cause a nasty cut. Apart from that, when a handle is fitted, better control of the file is obtained whilst filing.

For smaller wheels, precision needle files are required; a suitable size is 16cm (6in) long. This is the overall length, including the handle. Regarding files for fine work, it is always advisable to obtain precision files. Whilst these are obviously more expensive than the normal engineers' files, they are far more accurate, have better shape and cut, and will last longer. In addition to the needle crossing file, hand, three-square (triangular) files will be required; these are useful for the straight sides of the wheel spoke on crossing. On some files, grind a safe

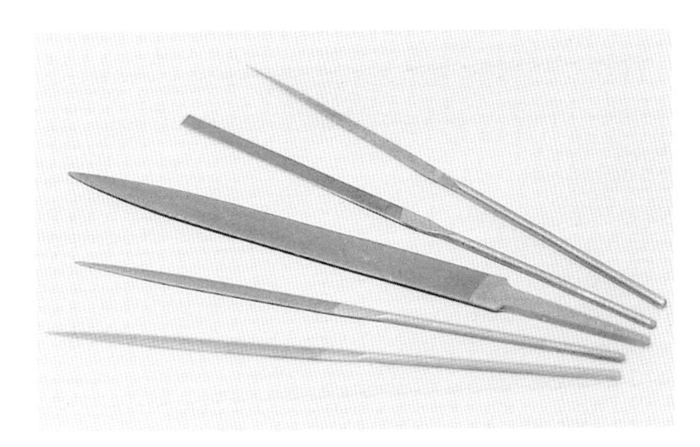

Fig. 355 Selection of precision files.

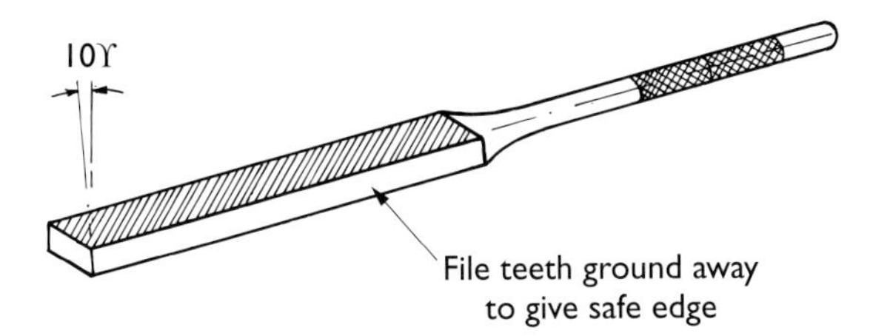

Fig. 356 File with a safe edge.

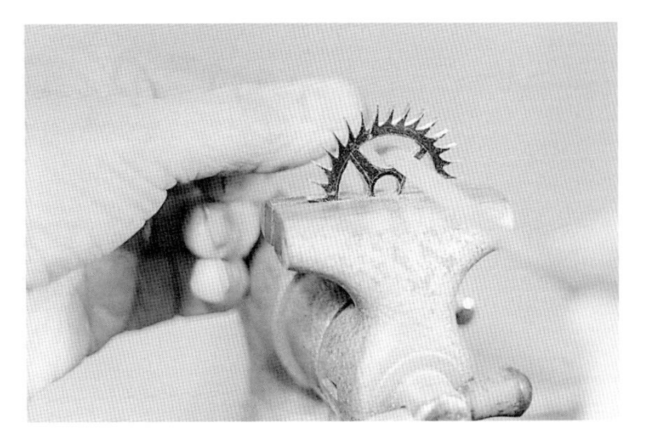

Fig. 357 Filing the wheel crossings or spokes.

edge to enable corners to be filed without affecting surfaces already finished.

The wheel blank is held in a small vice, and the rough filing carried out to each section. It is better to do it this way, because if the guide line has been encroached upon, this can be balanced by filing a little more off each crossing without spoiling the wheel; otherwise, if one section was completely finished, this may not be possible, and the wheel would have to be scrapped. Wheels with nicely cut crossings have a well balanced and pleasing effect. After roughing out work on the corners first, then completely finish with a number 4 cut file of the necessary shape. The straight and curved sections are 'draw-filed', and then finished off with 600- then 1200-grade 'wet-and-dry' paper wrapped around a small file or piece of pegwood flattened on one side. This will impart a very good finish.

FINAL POLISHING

Ensure that the surface is completely free from scratches, and then the final polishing can be carried out with an oval steel burnisher. If a proprietary manufactured burnisher is not available, use either a steel darning needle, or make a suitable burnisher out of an old crossing file. To do this, first remove all the file teeth by grinding, then polish smooth with an emery stick. The polish can, alternatively, be applied by using a hard circular felt pad in the lathe with a suitable polishing compound. Make sure, when the operation is complete, that the lathe slides are cleaned down well to remove all traces of abrasive.

Straight burnishers for polishing pivots are made in a similar manner, but a grain is applied to the active faces by rubbing along an emery

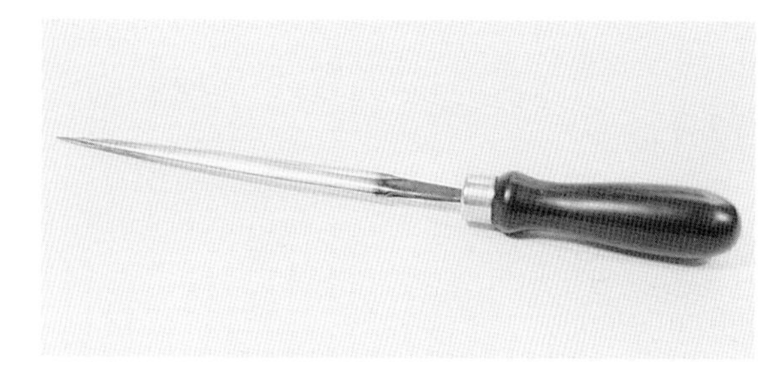

Fig. 358 Oval burnisher.

Fig. 359 Burnishing a regulator great wheel.

Fig. 360 Completed escape wheel.

stick to impart a grain across the width of the blade. This, in effect, acts as a very fine file, which smoothes and compresses the surface of the material being polished. Use the oval burnisher in the same way as when filing – just one or two even strokes on each surface will impart a mirror finish.

Fig. 361 Verge crown wheel.

Ensure the burnisher is held square across the wheel, otherwise the edges will not be nice and sharp. When burnishing brass, no lubricant is necessary.

Verge crown wheels and contrate wheels are more difficult to finish, as the rim has to blend in with the crossing, and these are at right-angles to each other; also the back edge of the rim is chamfered. Care is required when filing, and it is essential to use files with safe edges to prevent damaging the corners where the crossings meet the rim.

CROSSINGS FOR SMALL CLOCKS AND WATCHES

When producing wheel crossings on small, high quality clocks and particularly watches, the spokes or crossings have to be accurately produced. Some form of jig is required so that it will form a filing guide, and with this in mind, the jig shown was designed. It consists of an outer brass body counter-bored to accept a hardened inner ring; this ring is the guide for the inner wheel rim on the crossing. Each different wheel size will require a hardened ring of different diameter, but these are easily produced.

Fixed to the outer brass body by the small screw is a hardened straight-edge. In operation the wheel to be crossed out is shellaced into the hardened disc. This disc has five holes drilled around its periphery, as this is generally the number of crossings in a pocket-watch wheel. The steel disc and brass outer rim are located together with a small pin. After marking out

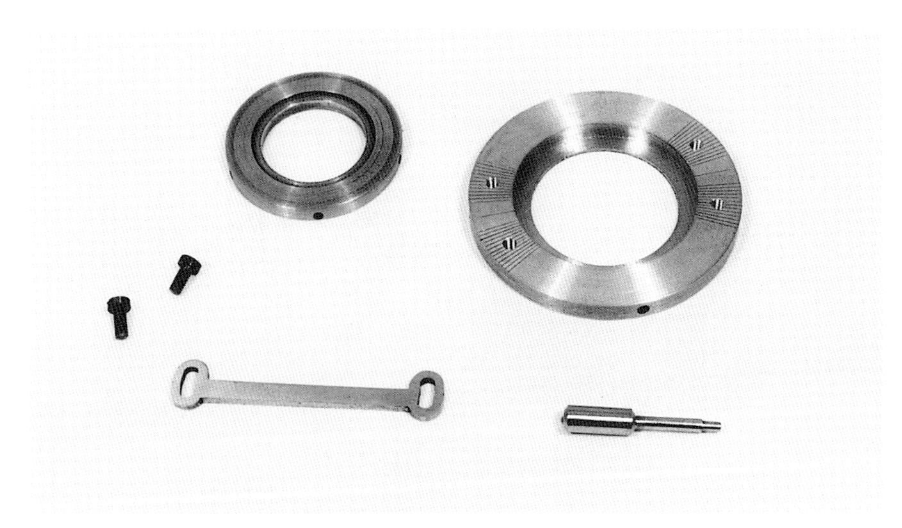

Fig. 362 Jig for crossing watch wheels.

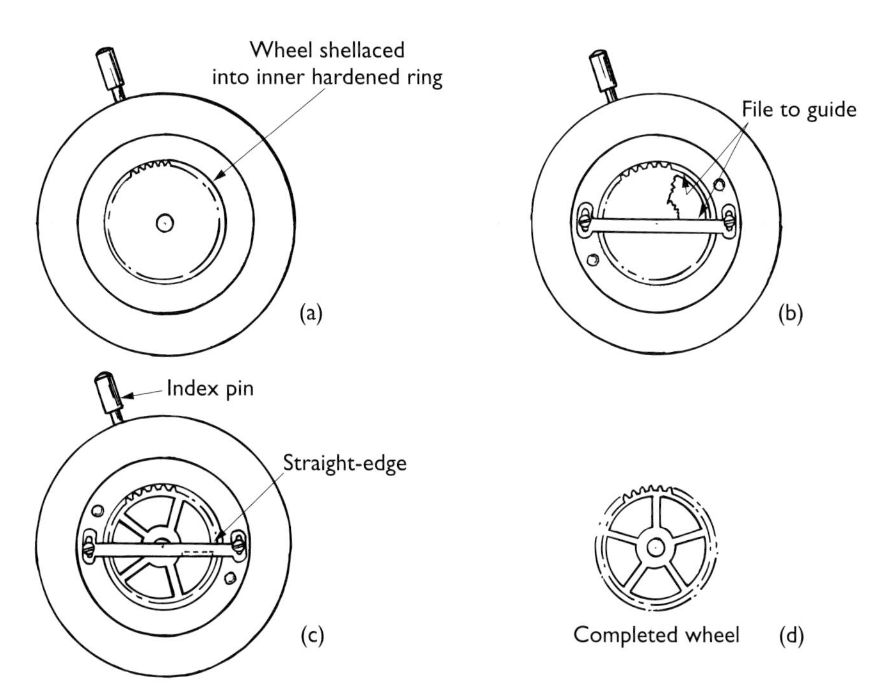

Fig. 363 Various stages in crossing out a watch wheel.

and cutting the surplus material away with the piercing saw, set the straight-edge to the required angle using the engraved angular markings, and tighten the securing screws. File the crossing using the hardened straight-edge and steel ring as a guide. When complete, index the assembly one division.

Carry on until all five crossings are complete. Move the straight-edge to the second set of fixing holes; this will enable the other side of the spokes or crossings to be finished to size. Machine two small silver-steel discs to suit the diameter of the centre boss, drill a central hole for a fixing screw, then harden. Fit to the wheel, and finish-file the central boss. This will produce a very accurate wheel. Now remove from the jig by dissolving the shellac in methylated spirits.

If replacing an old wheel and only the teeth are damaged, it is possible to use the wheel as a pattern for marking out the crossings.

Miniature needle files or escapement files are ideal. When crossing out small wheels, even using

Fig. 364 The jig and finished wheels.

a filing guide, great care is required when finishing; it is essential to keep the file square with the wheel and jig.

CROSSING OUT WHEELS USING A PANTOGRAPH PROFILING ENGRAVING MACHINE

There are a number of machines suitable for this type of work. Sometimes these become available on the secondhand market and can be purchased quite reasonably. To buy new is generally out of

Fig. 365 Model 'K' Mk II engraving and profiling machine.

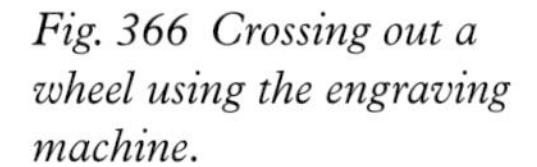

Fig. 366 Crossing out a wheel using the engraving machine.

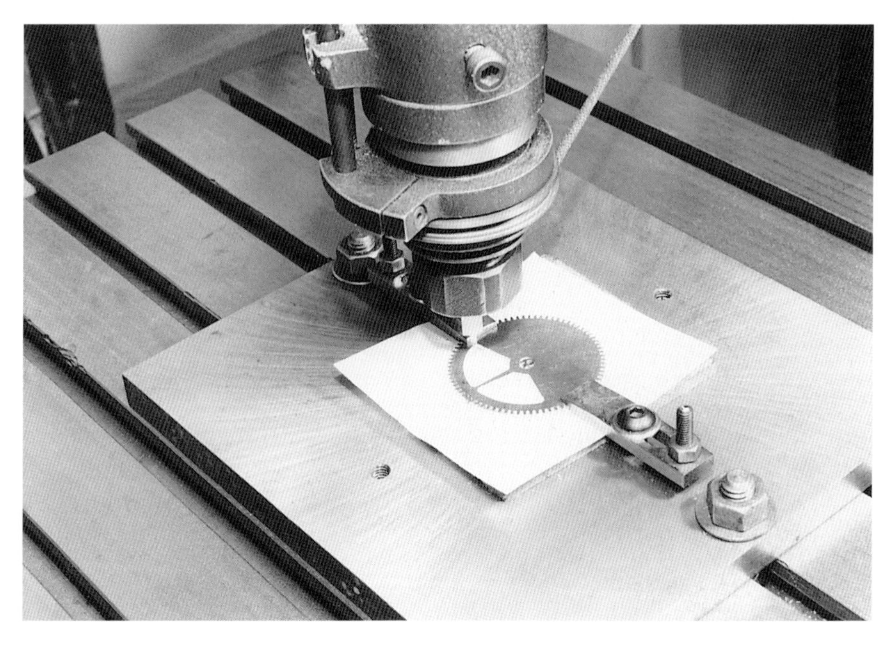

the question for the craftsman in his own workshop, whether he be amateur or professional.

The most likely machines to be found are the Taylor Hobson models D, J, H or, more recently, model 'K' Mk II. The latter is an extremely versatile, two-dimensional machine capable of cutting brass plate 5mm (3/16in) thick with little problem. It will also accommodate steel, if carbide cutters are used with the correct lubricant.

The model K machine has a pantagraph ratio that is adjustable from 1/1 to 50/1, an extremely wide range. The master or template used generally needs to be four or six times the finished work size to obtain an accurate component. The machine is shown set up for wheel crossing, with the wheel blank partially crossed out. Also shown is the master in place on the copy table.

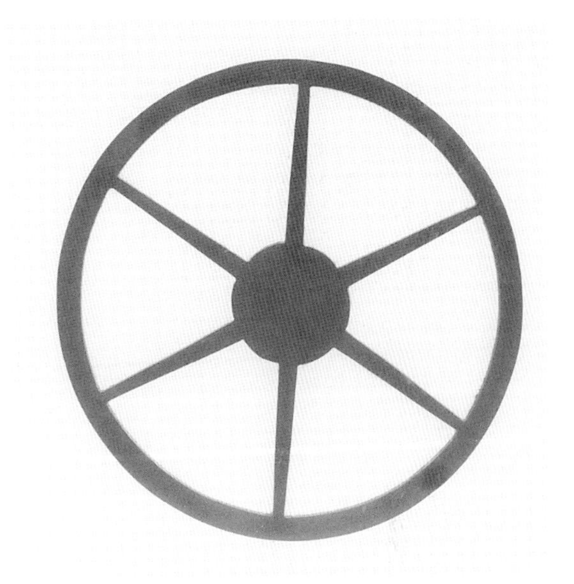

Fig. 367 Solid master template.

Masters

Masters can be made in any suitable material, though plastic sheet can be cut and shaped easily. Two types of master are shown. First, the solid type is made from two thicknesses of perspex, the profile marked out and then cut out with a piercing saw, and the edges filed smooth. This is then bonded to a base.

When using this type of master, the stylus should be the same ratio to the cutter as the master is to the original. For example, if a 4× reduction is being used and the cutter diameter is 2mm (0.080in) dia., then the stylus would be 4 × 0.080in = 8mm (0.320in).

The second type is a line master. This is in the form of a grooved, enlarged profile of the component required, and is produced by first making a drawing at least 4× full size with black ink on white paper to give a good contrast. As in this instance the stylus is a print from the drawing, a negative is produced and then a brass master is produced by using a light-sensitive fluid painted onto the brass master. This is then burnt away

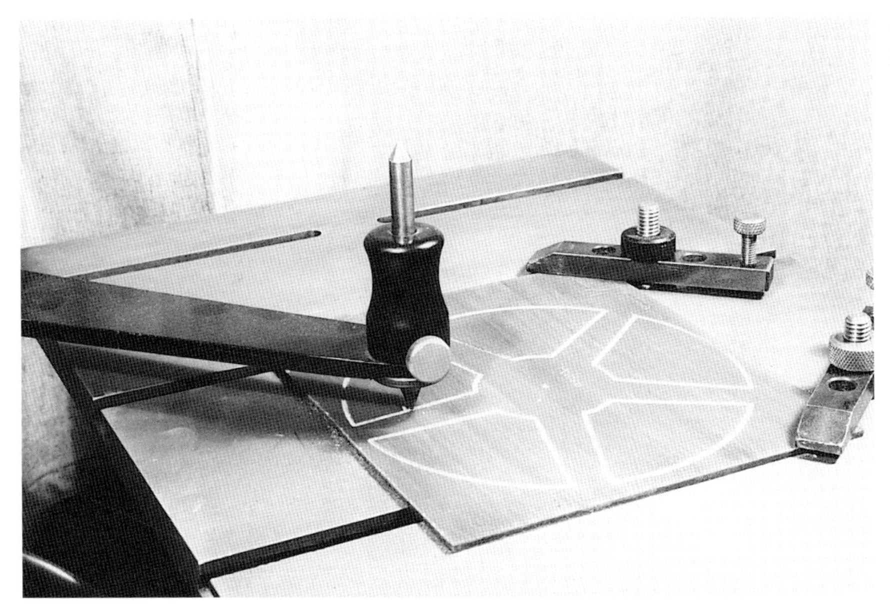

Fig. 368 Line master template and stylus.

with a UV lamp. The line is then etched, which produces a groove for the stylus to follow. If a number of components are required, it is often necessary to make the master more permanent, then a deeper groove can be cut into a perspex sheet using the brass-etched master as the original, the work being carried out on the engraving machine with the ratio at 1/1. Once the ink drawing is produced, a local firm dealing in etching can produce a master quite cheaply.

It is necessary to apply a formula in order to determine the drawing sizes. This is because the ratio is operating from a point, and the stylus is not working from the edge of the master. The

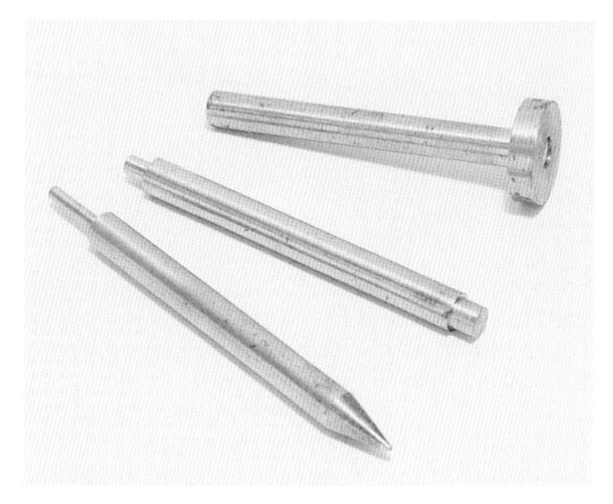

Fig. 369 Various styluses.

Fig. 370 Line master template.

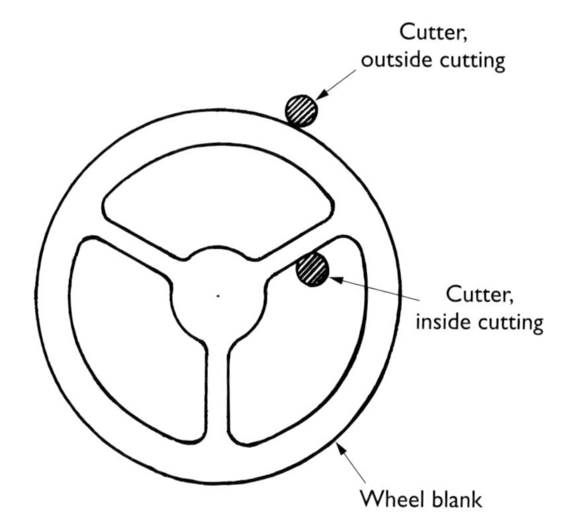

Fig. 371 Inside and outside cutting.

difference between the inside and the outside cutting is shown in the illustration.

Outside cutting
= (size required × reduction)
+ (reduction × cutter dia.)

Inside cutting
= (1/dia. required − 2 × cutter dia.)
× reduction + (reduction × cutter dia.)

Use the above formulae for drawing out a master for a centre wheel, with a boss dia. of 11mm (0.437in), inner rim 43mm (1.687in) dia., thin or narrow end of crossing 1.5mm (0.062in), and wider portion 3mm (0.109in):

Boss drawing dimension
= (11 × 4) + (4 × 1.5) = 50mm
(note outside cutting)

Inner rim drawing dimension
= 43 – (2 × 1.5) × 4 + (4 × 1.5)
= 166mm
(note inside cutting)

Thin end of crossing
= (1.5 × 4) + (4 × 1.5)
= 12mm
(outside cutting)

Thick end of crossing
= (3 × 4) + (4 × 1.5)
= 18mm
(outside cutting)

Alternative calculation in imperial measurement:

Boss drawing dimension
= (0.437 × 4) + (4 × 0.062)
= 2in
(note outside cutting)

Inner rim drawing dimension
= 1.687 – (2 × 0.062) × 4 + (4 × 0.062)
(note inside cutting)
= 1.562 × 4 + 0.250 = 6½in

Thin end of crossing
= (0.062 × 4) + (4 × 0.062)
= 0.500 in
(outside cutting)

Thick end of crossing
= (0.109 × 4) + (4 × 0.062)
= 0.687in
(outside cutting)

Wheels crossed out by pantograph require very little finishing, mainly the radius left by the cutter in each corner has to be removed by filing, then normal finishing applied, as described previously.

With a 6 × master it is possible to cut accurately to within 0.13mm (0.005in) quite easily, and even less in some instances.

Cutters ground to 1.5mm (0.062in), whilst giving a smaller radius in the crossing corners, are liable to breakage. A diameter of 2mm (0.080in) is a good average. Remember that when cutting, it is better to take small cuts often, and that a speed of 18,000rpm is recommended for engraving brass. With a pantograph engraving machine it is essential that a tool and cutter grinder is available, as cutters require to be sharpened quite frequently.

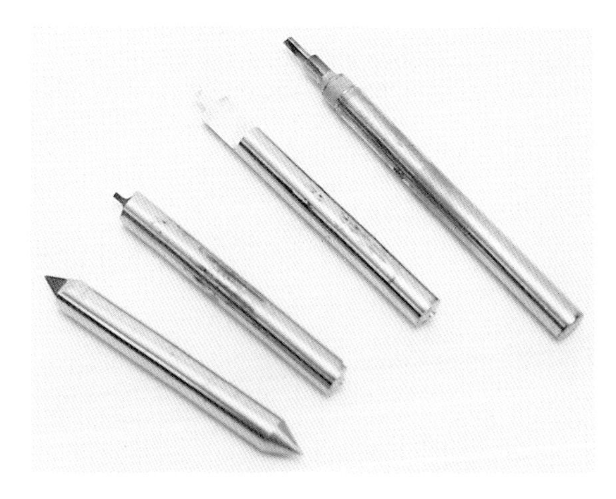

Fig. 372 Various types of cutter, high-speed steel and carbide.

The Alexander machine is ideal. It is a commercial model where all types of cutter can be sharpened, including spherical, angled, and so on. The sharpening and making of cutters can also be carried out on the Quorn tool and cutter grinder, as previously described.

Fig. 373 Alexander tool and cutter grinder.

Fig. 374 Trueing a wheel collet.

MOUNTING WHEELS ONTO COLLETS

Collets should be turned and drilled to be a good location fit onto their mating arbor. If possible, the centre hole should be reamed to give a smoother and more accurate location. The turned register for the wheel bore should be concentric with the bore. In certain cases when fitting a replacement wheel to its arbor, it may be necessary to turn the collet true on its mating arbor, rather than remove it, as the arbors on old clocks are often tapered and quite rough and irregular, and it would seem that their method of manufacture must have taken this form.

Certainly French clock collets would have been machined in position, as the bore of the collet is quite large in comparison to the arbor, and the space is taken up by the use of soft solder. Complete blanks were available, the wheel blank being already mounted on its matching collet, completely assembled onto the arbor with pinions already cut. Great care would be required to finish the wheel to size, then cut the teeth on such a slender arbor. An accurate method of clamping and supporting the wheel blank and locating it centrally would have been required.

When re-mounting a French clock wheel, discard the old collet as it is virtually impossible to solder it back true. Machine a new collet to match the arbor diameter.

Normally collets would be soft soldered onto the arbor, but a more modern method would be the use of Loctite adhesive, either 601 or 638 (the latter having a higher shear strength), particularly if working on modern clocks or building a clock from scratch. Obviously more traditional methods would be used on restoration work, either soldering or, quite often, the collets are driven onto a taper on the arbor. If the tapers are matched correctly by broaching the core of the collet, then an extremely satisfactory result can be achieved. On early English longcase and bracket clocks, the collets were often shallow,

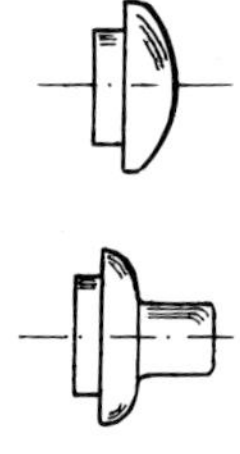
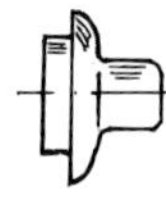
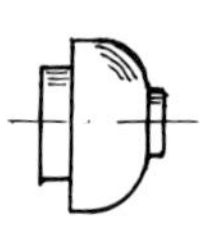
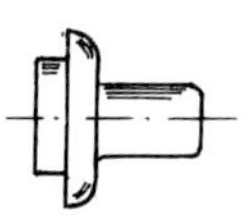
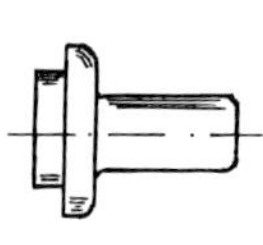

Fig. 375 Various styles of wheel collet.

and had very little area of contact with the arbor. They were therefore brazed to the arbor before shaping and trueing the wheel seating.

Wheels can be either rivetted to their collet or screwed onto them, as in high quality regulators, with small precision screws. It is essential that the bore of the wheel is true with the outside; the chapter on wheel cutting covers methods of trueing the bore. Ensure that all burrs are removed so that the wheel sits down snugly against the collet shoulder; a slight chamfer is permissible.

Some form of staking tool is required to ensure the punch is held concentric and square with the work. In the illustration, the staking tool manufactured by JMW (Clocks) has a most useful feature: in most staking tools, the head which locates the punch is solid and integral with the main body; this makes it difficult for the tool to accept long arbors. With this design, it is possible to lock the punch in the correct position with the corresponding hole in the stake, then swing the arm away clear to enable the long arbor to be accommodated. The arm, together with the punch, can then be accurately repositioned against the stop provided. It is quite often necessary to make up a special punch to suit the work in hand, although a range of commercial punches will cover most requirements.

Use a hollow punch to seat the wheel down hard against its shoulder, then use the rivetting punch to close the collet down onto the wheel, and spread the metal evenly. Turn the complete assembly of the wheel through 90 degrees, each time delivering a light blow to the punch, making sure everything is kept square. Check the wheel is held tight when rivetting has been completed.

In restoration work, when fitting a new collet with a curved shape, as the brass can be quite soft and easily damaged when rivetted, it is easier to

Fig. 376 Wheel screwed to its collet.

Fig. 377 Wheel ready for rivetting; note the slight amount of collet protruding.

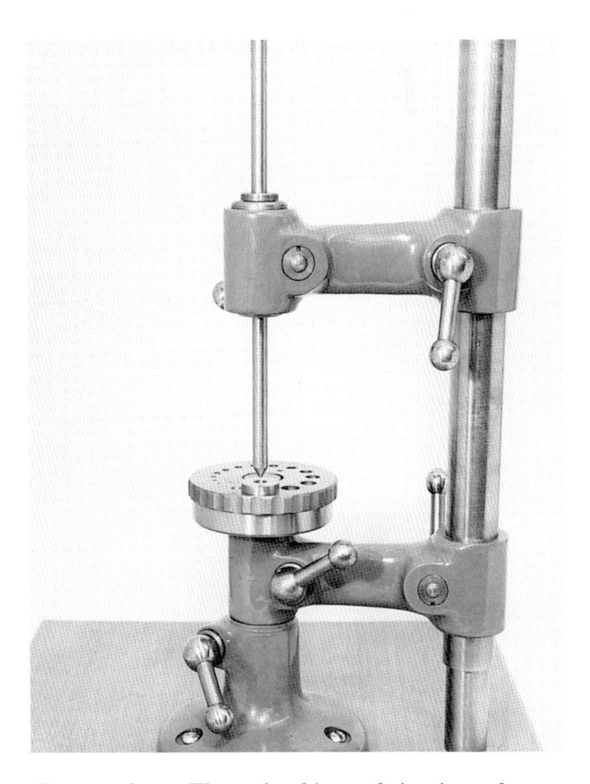

Fig. 378 George Thomas' staking and rivetting tool.

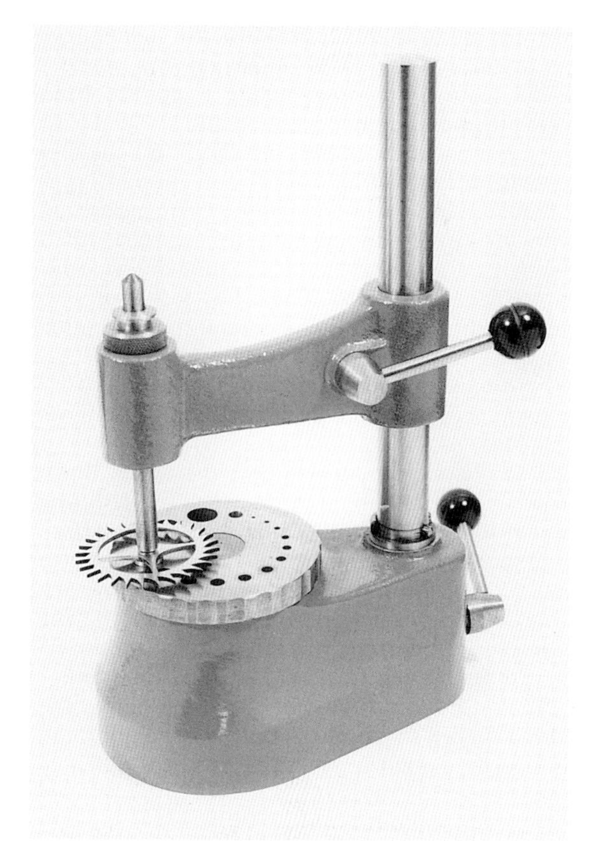

Fig. 379 JMW staking tool.

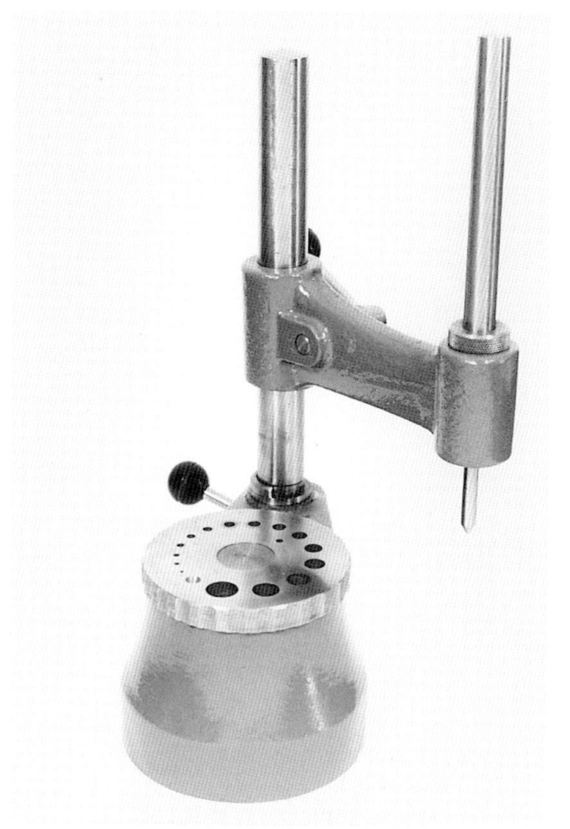

Fig. 380 JMW staking tool with the arm swung away.

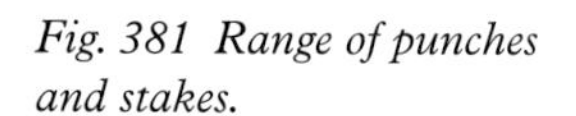

Fig. 381 Range of punches and stakes.

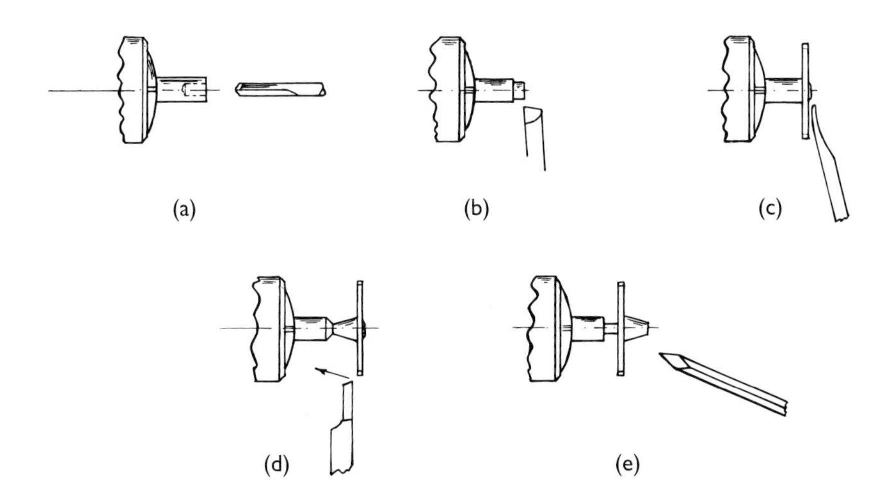

Fig. 382 Mounting a chronometer escape wheel.

machine the collet with a flat seating for rivetting. After rivetting, turn the shape with a graver and hand rest.

MOUNTING A CHRONOMETER ESCAPE WHEEL

When awkward-shaped collets are used for wheel mounting, it is often quite difficult to hold the wheel or collet when finishing and sizing. A method of carrying out this type of work successfully is illustrated above: the material is first held in a collet and carefully centred, then drilled, ensuring that the hole is true. The register diameter is then turned to accept the bore of the wheel. The wheel is then rivetted or burnished in position, as shown at 'c'. With the topslide set over at the appropriate angle, the back of the collet is turned and then parted off. Finally, the wheel and collet assembly is mounted on an arbor for final finishing. To assemble the collet onto its arbor, broach out the hole until it matches the arbor, giving a good interference fit.

Fig. 383 Turning a pinion to accept the wheel.

MOUNTING WHEELS ONTO PINIONS

Where the wheel is rivetted to the pinion, as in an English longcase clock, the pinion is held in a suitable collet in the lathe, and a diameter turned for the wheel seating. This is to have a slight taper from the front, the diameter being such that it will enter the hole in the wheel approximately a third to half its length, then a tapered groove is turned into the face of the pinion to enable the edge to spread when rivetting takes place. The arbor and pinions are then placed in the staking tool for final assembly. When the wheel is pressed home, the remaining portion of the pinion leaves will form internal splines, and key the wheel and pinion to give a positive location. As shown previously, use a hollow punch to completely seat the wheel against the shoulder of the pinion. Complete by rivetting with a tapered punch, and finally with a flat punch.

A wheel from an English high-grade Library clock is shown. The rivetting is well finished in the best English manner, and both ends of the pinion are highly polished. Many pocket-watch pinions are finished in a similar manner.

Fig. 385 Seating the wheel.

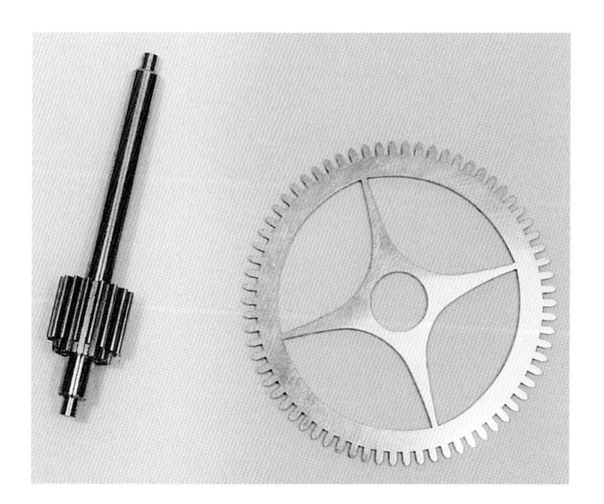

Fig. 384 Wheel and pinion ready for assembly.

Fig. 386 High-grade wheel and pinion from an English Library clock.

10 FINISHING AND REPLACING WORN PIVOTS

Now that the cutting of pinions and wheels, and the crossing and mounting of wheels to their respective arbors has been covered, the machining and finishing of pivots will be described. It is essential that an extremely good finish is imparted to the pivot to reduce friction in the wheel train.

MACHINING PIVOTS

Grip the arbor in a suitable collet. When extreme accuracy is required, it is necessary to mount the work between centres; in normal circumstances an accurate lathe collet is quite suitable. Use a sharp lathe tool with a honed cutting edge and try to obtain a good finish, as the better this is, the less burnishing will have to be carried out. Once the pivot is reduced to the approximate size, just leave a slight amount for finishing. The precise diameter is not important, as the pivot hole in the clock plate can be opened out to match the finished pivot.

When turning small pivots for French clocks, ensure the tool is exactly on the centreline of the work, otherwise there is a good chance that when the final diameter is arrived at, the pivot will snap off! With pivots as small as 0.25mm (.010in), particularly fly pinions, it is better to finish these off by using the hand graver. Whilst a fair amount of skill and practice is required to carry out this type of work, it is possible to turn much smaller diameters. The tool for sharpening previously referred to in Chapter 8 will produce an excellent cutting edge on the graver, and consequently this is transferred to the work; with a nicely sharpened graver, blued pivot steel will cut like butter!

Various instruction books are referred to in the Bibliography at the end of this book, which show the use of the graver.

Fig. 387 Turning a pivot using the lathe slide-rest.

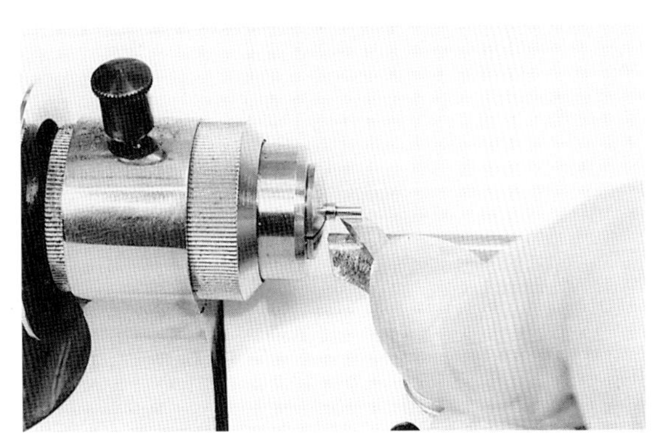

Fig. 388 Using the graver to turn a fine pivot.

FILING AND BURNISHING

To finally finish the pivot, a pivot file and burnisher are required. These are available in both left-hand and right-hand design: some operators prefer to work overhand when they would use the left-hand burnisher; if working underhand, a right-hand file burnisher would be used. The advantage of working underhand is that the work can be viewed whilst burnishing is in progress. Note that the edge of the burnisher is quite sharp to get right into the corners of the work. It is important that no radius is left between the pivot and shoulder, as this could eventually stop the wheel train. Only one cut of file is available, and that is extremely fine, around cut eight.

There is a large step between the finish imparted by the file, and that by the burnisher. Once the file has been used, it is often preferable to use an Arkansas stone to improve the finish before imparting the final finish with the burnisher. Ensure the Arkansas stone is flat; it can be cleaned and shaped by rubbing on wet/dry paper. When polishing with the stone, use a light smear of oil.

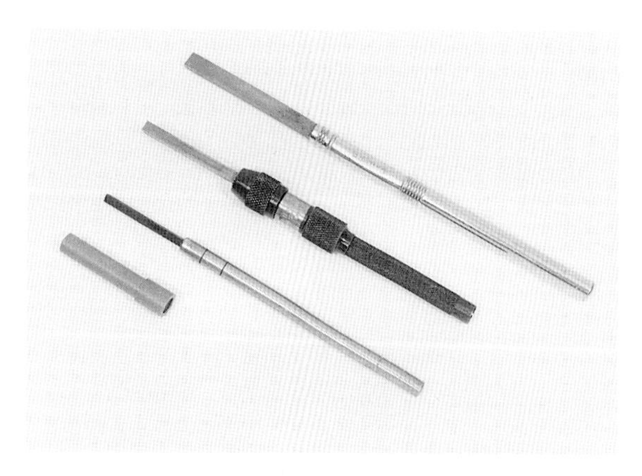

Fig. 389 Various types of burnisher.

Before carrying out any burnishing, it is necessary to prepare the burnisher. A cross-grain has to be imparted to the blade to act as a type of file, but in actual fact this is the process of burnishing where the hardened burnisher will close the surface grain of the metal by pressure and impart a mirror finish. Prepare a curved wood block as shown, and fit coarse emery over it; the burnisher is now drawn sideways to impart the cross-grain.

The sharpening of the blade can also be carried out on a diamond lap such as an Eze-Lap. Always ensure the burnisher is kept flat when sharpening.

When driving the work, it is not necessary to have a large amount of torque available. In fact if the work is held in the chuck and driven by the lathe motor, it may come out of the chuck, with serious consequences, perhaps damaging or breaking a pivot.

Rotating by Hand

The simplest method of rotation is to hold the work in a pin chuck and revolve it by hand on a wooden block with vees held in the vice. Practice is required in using this method, otherwise flats

Fig. 391 Sharpening the burnisher.

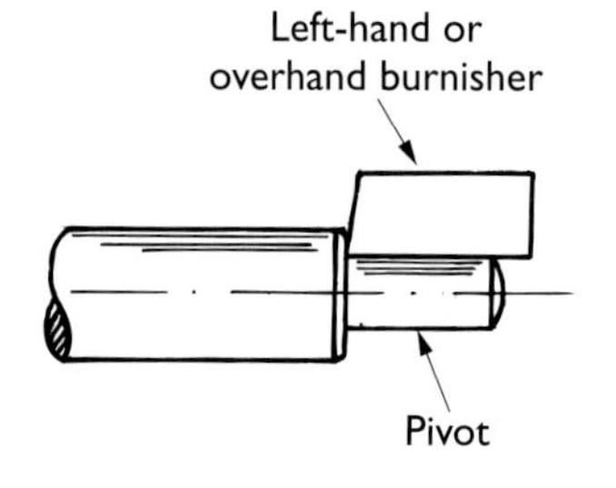

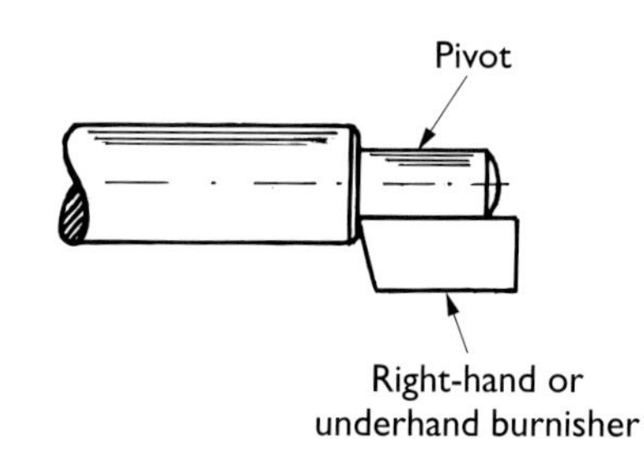

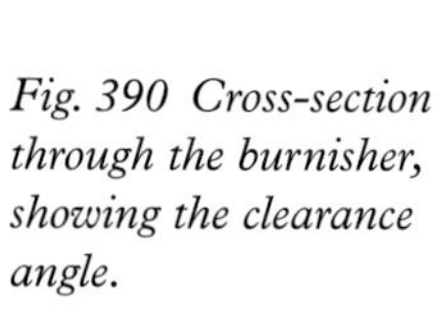

Fig. 390 Cross-section through the burnisher, showing the clearance angle.

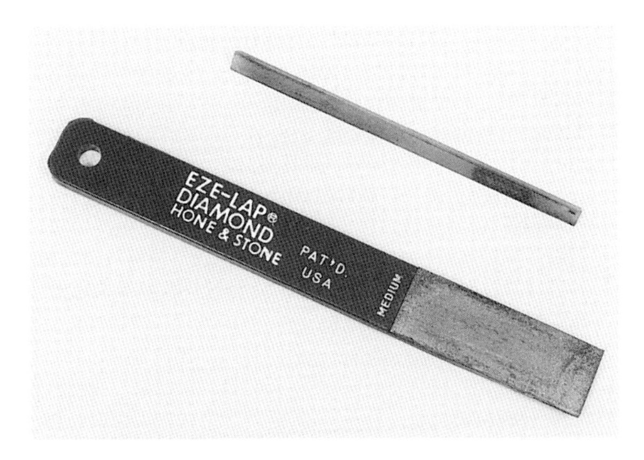

Fig. 392 Diamond lap and Arkansas stone.

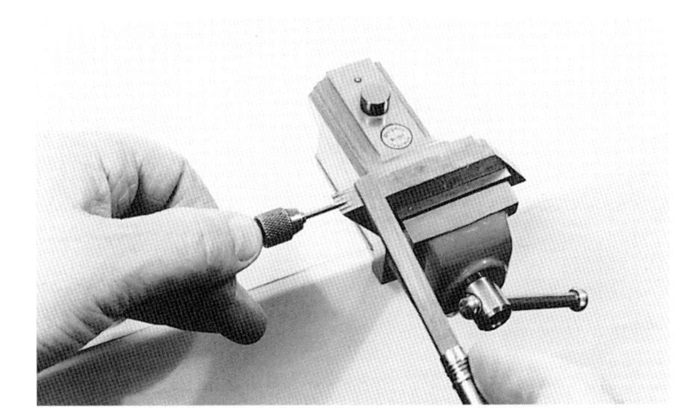

Fig. 393 Finishing the pivot by hand.

will be formed on the work. Although considerable control can be achieved as everything is under the guidance of both hands, care is required to keep the pivot file flat to give a parallel pivot. Likewise when using the burnisher. Place a slight amount of oil on the burnisher, which will allow it to slide over the work and help to prevent scoring. With the left hand, rotate the work towards you whilst the right hand is pushing the burnisher forwards, with a fair amount of downward pressure being applied onto the work. After a time the action becomes second nature.

Rotating on the Lathe

When using the lathe or turns to carry out pivot burnishing, more control is achieved on the actual burnishing if the work is supported on a jacot drum; bi-directional rotation is then possible by the use of the bow. The bow shown is made from the tip of a carbon-fibre fishing rod, and the handle is an adjustable plastic file handle; equally, a wooden handle could be used. If burnishing watch or very small pivots, the bow

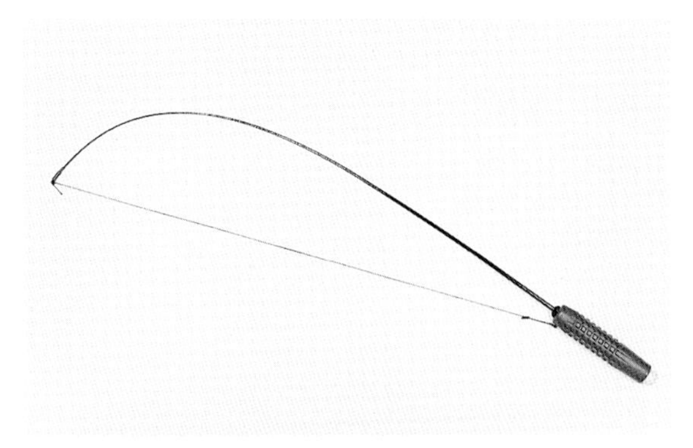

Fig. 394 Bow made from carbon fibre.

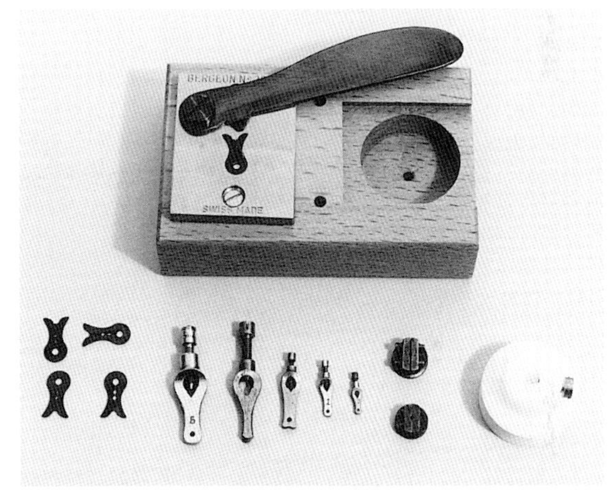

Fig. 395 Bergeon carriers and opening tools, also various carriers.

would be used with ferrules or lathe carriers, often supplied with the lathe. An excellent range of spring carriers is supplied by Bergeon; these are available complete with opening tool to assist in assembling the carrier to the work. For larger work these can be made from any suitable material. The one shown mounted on an arbor is made from white nylon or plastic rod.

To fit the bow to the ferrule, with the bow handle facing away from the body and the driving line under the ferrule, grip the line and nip its two sides together around the small pulley or ferrule. Now twist the complete bow around until the handle is in the correct working position. This takes only seconds, and becomes second nature after a while. To assist in keeping the file and burnisher parallel, a filing roller can be used, or an adjustable support that is held in

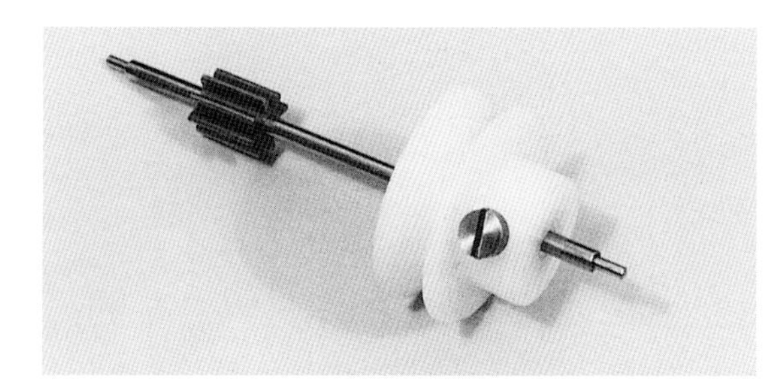

Fig. 396 Home-made carrier or ferrule.

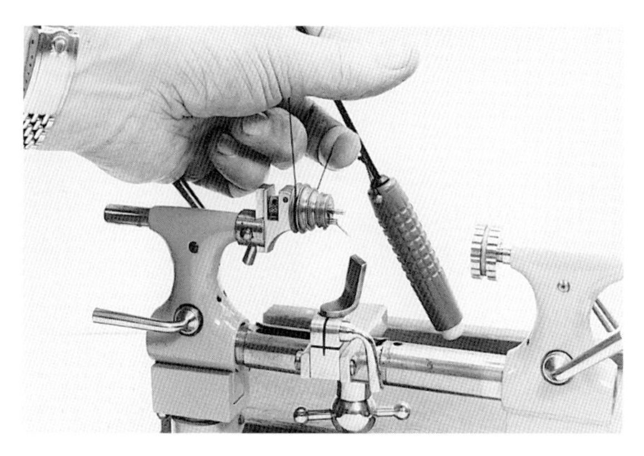

Fig. 397 Method of fitting the bow to a ferrule or pulley.

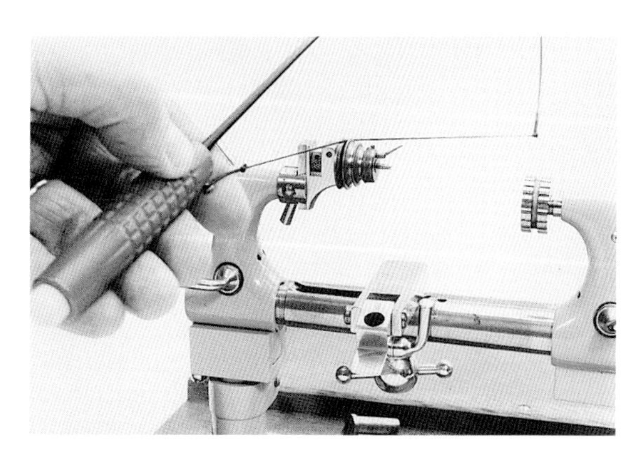

Fig. 398 The bow ready for use.

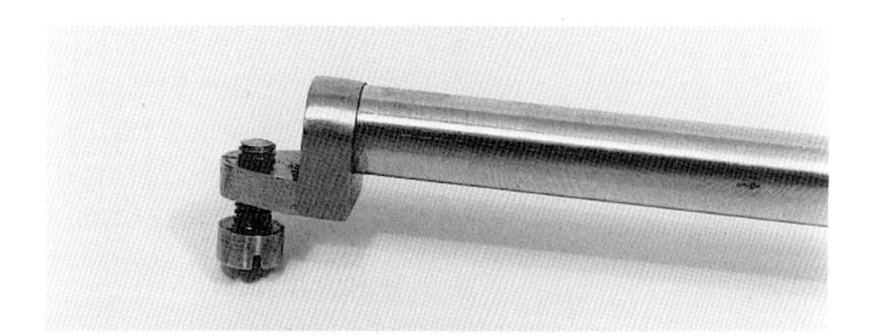

Fig. 399 Adjustable support for the burnisher.

the tailstock of the lathe. Furthermore, a simplified version of a jacot drum can easily be made; this is quite often one of the standard accessories of a good watch or instrument lathe outfit. It is hardened and polished, and in some cases there are two available, covering both watch- and clock-size pivots.

The Steiner turns are excellent; the model shown here is held in the vice and has a hand-wheel to rotate the work. The turns are supplied

Fig. 400 Lathe jacot drum to support the work whilst it is being burnished.

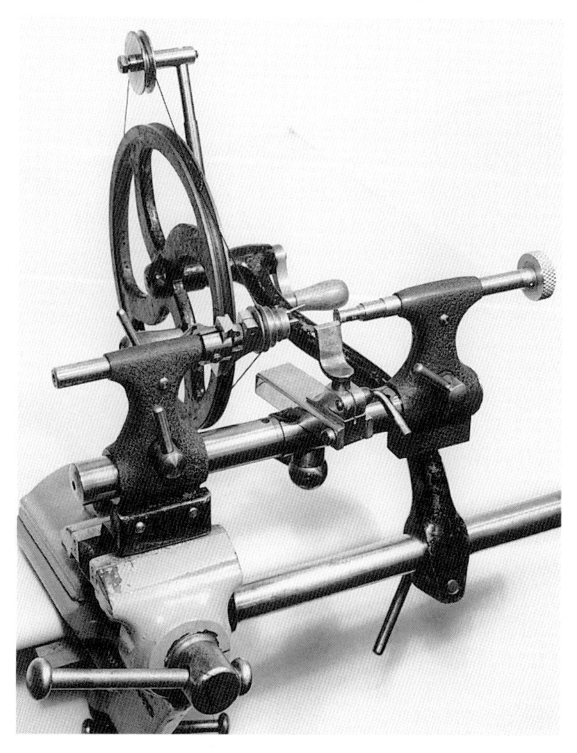

Fig. 401 Steiner turns.

Fig. 402 Accessories for the Steiner turns.

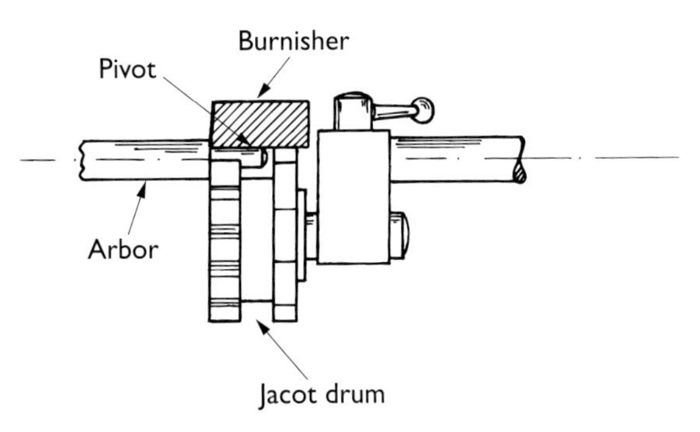

Fig. 404 Jacot drum and burnisher.

Fig. 403 Burnishing a pivot using the bow.

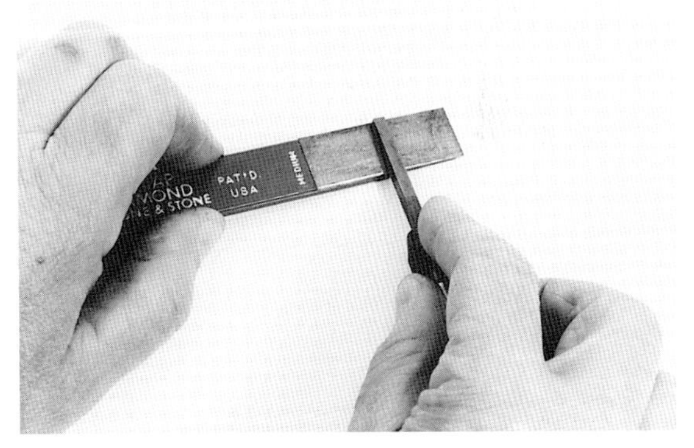

Fig. 405 Sharpening a carbide burnisher on a diamond hand lap.

with many types of runner, and centres for both clock and watch work. Methods of driving the work can be either as shown with the handwheel, or with the bow, or motorized. The box of accessories shows centres, drill plates and various jacot drums.

When using the bow, control of the rotation of the work is excellent. Pull the bow towards the body in a slightly downward motion, at the same time pushing the burnisher forwards. Keep pressure on the work only on the forward stroke. After a few strokes, a mirror finish will be imparted to the work. Ensure the burnisher is kept well sharpened.

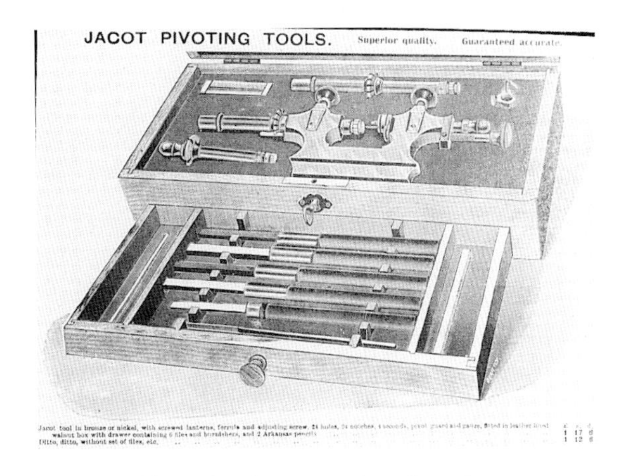

Fig. 406 Illustration from an early tool catalogue showing jacot tool and burnishers.

OTHER TYPES OF BURNISHER

There are other types of burnisher available. One of the most useful is carbide, which can be held in a pin chuck. A diamond lap is essential for sharpening this. If a fine Eze-Lap is used, it is possible to achieve a 'black' polish to the work, particularly if the pivot material is hardened and tempered, for example blue pivot steel or similar. This type of polish is essential for watch work.

When using the jacot tool or lathe for burnishing, it is essential the correct vee-bed is

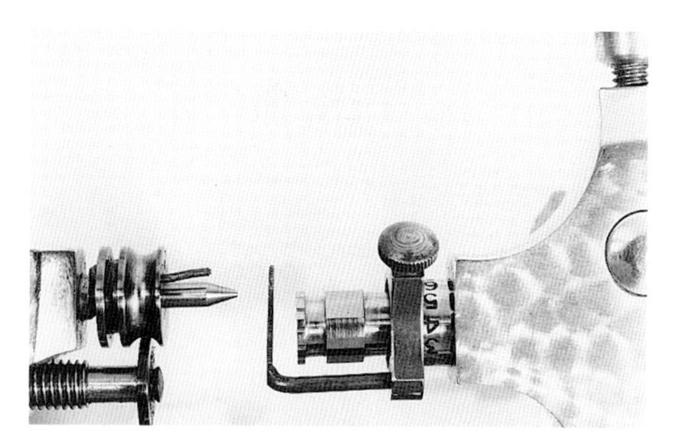

Fig. 407 Jacot tool with a guide for the burnisher.

selected for the pivot being worked on. The jacot drum is made with varying sizes of vees to support the work. These are often numbered, which is most useful when a pivot has been reduced in size but has not been completely cleaned up; then it is an easy task to select the next lower numbered vee-bed. Burnishing is now carried out as previously described, until the pivot has been reduced and the burnisher is being guided solely by the jacot drum. Always ensure the work is kept clean, and remove metal and oil particles frequently.

When working on small pivots, a guide for the burnisher is often necessary to prevent the burnisher slipping sideways and damaging either the wheel or the arbor. This is particularly important if burnishing balance-staff pivots.

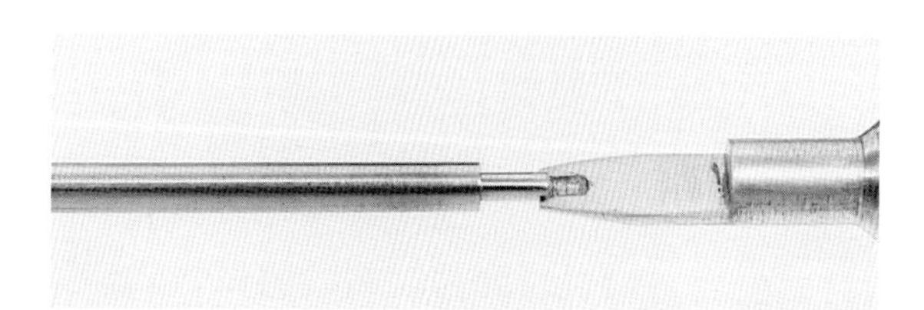

Fig. 408 Tailstock runner supporting a pivot.

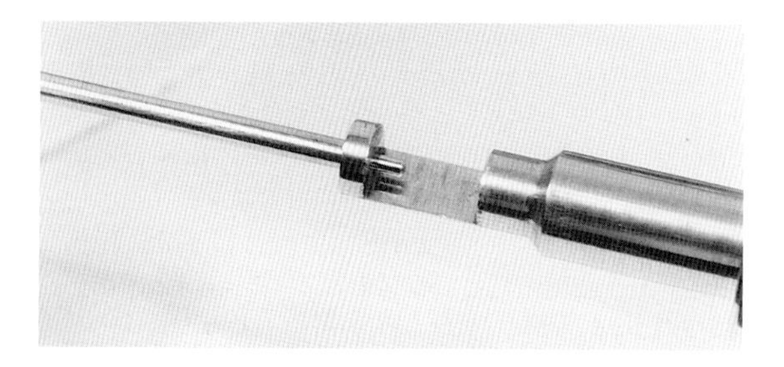

Fig. 409 Runner supporting a pivot whilst the pivot end is burnished.

Special attachments can be made to support the pivot whilst it is being polished. These can be made from silver steel and hardened; they can be held in the tailstock drill chuck, or made to fit directly into the lathe tailstock, as the one shown.

DAMAGED PIVOTS

When clocks have been in service for a long time, the oil will dry out and dust will deposit, forming a grinding paste that will reduce the pivots unevenly. In the worst case this could cause the pivot to shear. In the illustration below, the pivot is nearly cut through, leaving only its end attached by a slender piece of metal; if this had sheared whilst the clock was in service, serious

Fig. 410 Severely grooved pivot.

Fig. 411 Slightly grooved pivot.

Fig. 412 Re-finished pivot.

damage would have resulted. In this instance, only replacement of the pivot would be acceptable, as to reduce the diameter of the pivot would not be possible.

When a pivot is grooved only, it is possible to reduce the diameter with the pivot file and burnisher, as previously described. Then, of course, the clock plates would have to be bushed to reduce the hole size to match the reduced pivot diameter.

REPLACING WORN PIVOTS

When a pivot is worn nearly through, there is only one solution and that is to cut off the old pivot, face the end of the arbor square, and drill and fit a new pivot using blue pivot steel. This is not quite as easy as it sounds, however, particularly when a French clock pivot has to be replaced. Here again, a good watch- or clockmaker's instrument lathe would have pivoting attachments that mount in the tailstock. The arbor is supported by the coned drill plate. Select the appropriate hole and locate this centrally by using the pointed runner. Place a small amount of oil or thin grease between the work and the coned, shaped hole in the drill plate.

Fig. 413 Locating the correct cone support with the centre.

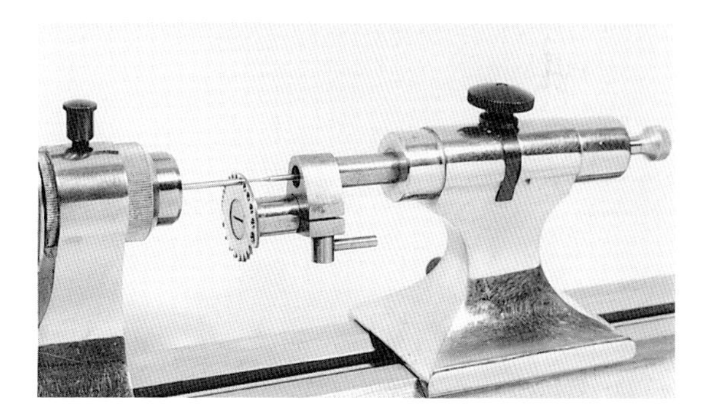

Fig. 414 Drilling an arbor to accept a new pivot.

Fig. 415 Striking a centre with the graver.

Problems with Drilling

The only problem with this attachment for the watchmaker's lathe is that the holes in the drill plate do not often match a standard drill diameter, as the centring is meant to be carried out with the graver and tee-rest. This is not easy, and requires considerable practice. One way of overcoming this is to lap out each hole to match the standard diameter drills available. This can be carried out quite easily with diamond paste and either brass or copper wire. Once a centre is struck or centre-drilled, then it is not too difficult to drill the arbor, as long as suitable drills are available.

Of course, the arbor has to be softened. A Bergeon tool is suitable for small pivots, and brass plugs of various sizes can be drilled to accept the end of the arbor. Heat is applied to the brass until the arbor to be drilled turns blue locally; the material should then be soft enough to drill. Maintain the flame on the brass, then remove it gradually so the metal cools slowly. Sometimes it is necessary to repeat the procedure. If the arbor is still not machinable, try and let the heat down more gradually.

Eureka drills are the most suitable for drilling, as they have two straight flutes which make the drill extremely stiff, and therefore ideal for this type of application. Carbide drills will work on very tough steel – though if one should break whilst drilling, it is extremely difficult to remove the broken drill, and sometimes impossible. In this case the end of the arbor will have to be removed with the portion of broken carbide drill inside it, and the arbor extended.

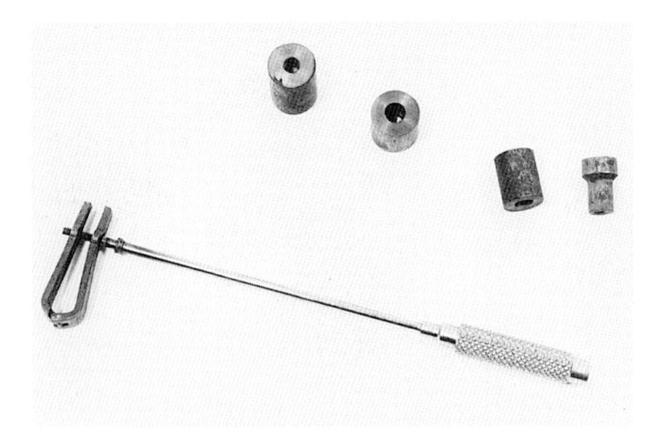

Fig. 416 Various items to assist in annealing arbors.

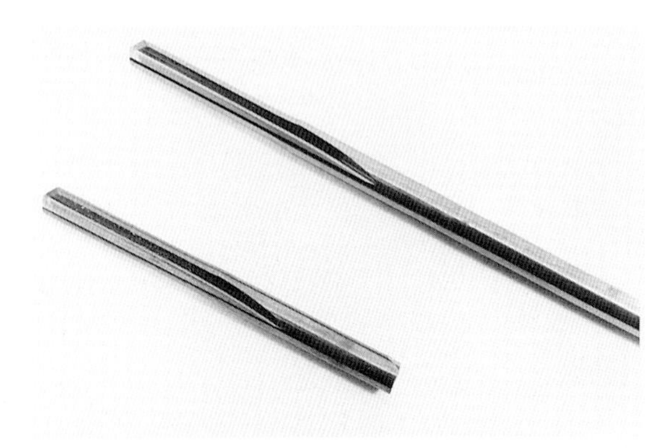

Fig. 417 Eureka pivot drills.

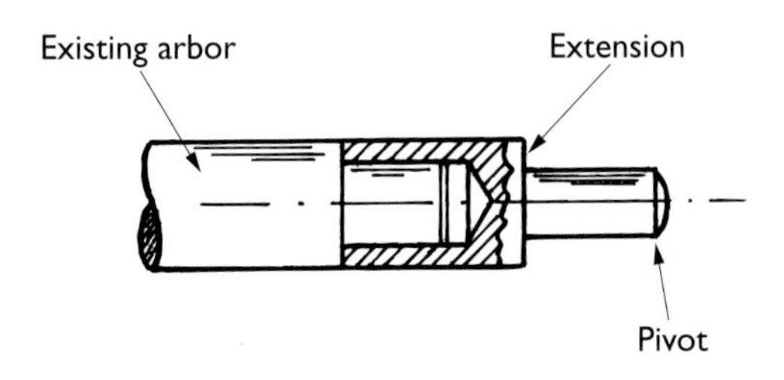

Fig. 418 Extending an arbor.

Range of Drills

There are surplus carbide drills on the market. These are usually of the spiral flute design used mainly for the drilling of PCB boards, a type of fibre insulation. They are used at extremely high speeds and are intended for drilling soft material and clearing the swarf quickly; they are not suitable for drilling pivots in tough steel, as they snap so easily. There are Eureka-type drills available manufactured in carbide which will last a lifetime. Whilst they are expensive, it is money well spent, as their use can often speed up the job tremendously if a particularly tough arbor is encountered. The range of this type of drill is not extensive, although smaller-diameter, spade-type carbide drills are available, and these are equally successful in drilling tough material.

Drilling Methods

Depending on the relationship of the pinion arbor and wheel, it is sometimes necessary to remove the wheel, as it is not possible to hold or support the work. A method is shown here whereby the wheel can be left in place. A brass collar is placed over the pinion, and this is supported in the lathe fixed steady. Centring is carried out from the tailstock using a small centre drill, then use the appropriate pivot drill.

The illustration below shows a pivot and jacot tool manufactured by JMW (Clocks). In this case the hardened drill plate is made to match a standard-size drill diameter, and the cone on the other side of the plate locates the arbor true

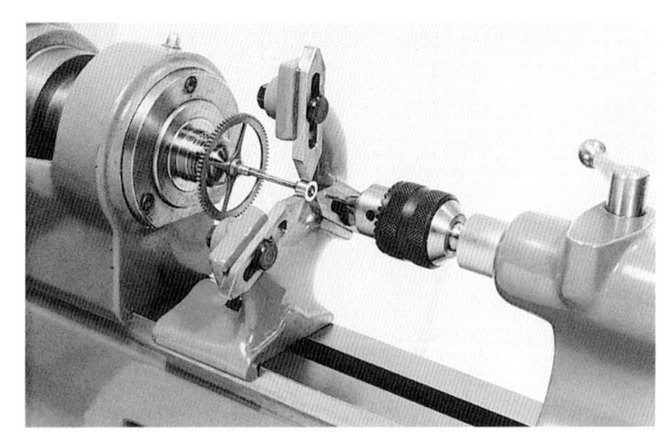

Fig. 419 Supporting the arbor for drilling using the fixed steady.

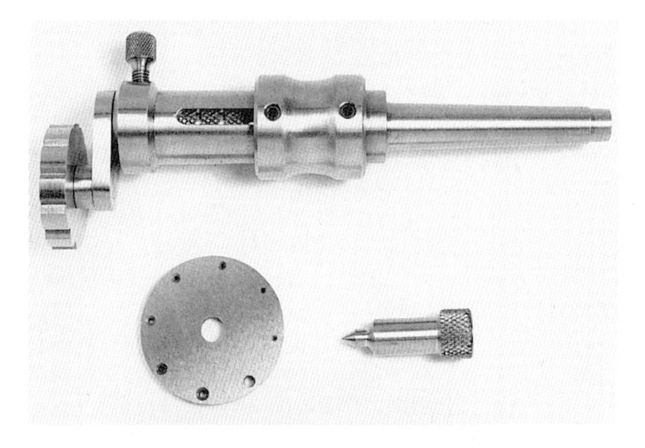

Fig. 420 Commercial pivot and jacot tool.

Fig. 421 Burnishing a pivot, supported by the jacot drum.

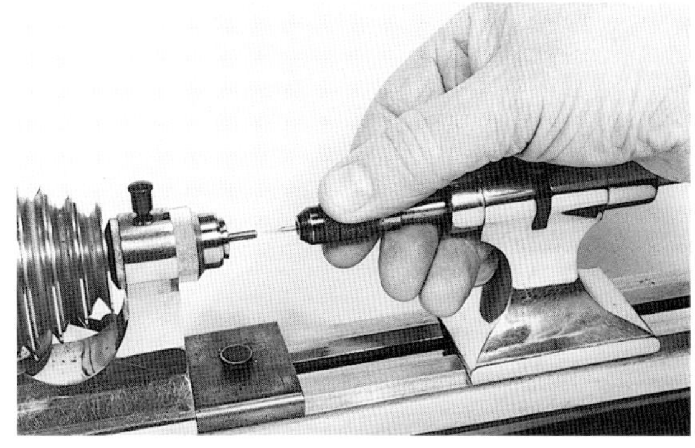

Fig. 423 Drilling small holes with the pin chuck and runner.

whilst drilling takes place. With this tool a larger drill may be used with the matching hole in the drill plate to centre the work, then a smaller drill fitted to suit the pivot steel available. The plate is provided with a standard reamed hole to enable guide bushes to be made to suit any particular size drill.

Drill the hole at least three diameters deep. Always ensure that the drill is sharp, and do not force it into the work or there will be a greater risk of breakage. As soon as the metal chips appear in the drill flutes, clear the drill from the hole and clear both the hole and drill from swarf. Use a suitable cutting lubricant to assist when drilling; one formulated with a 50/50 mix of camphor and turpentine is ideal. After drilling, clean the hole with pegwood.

Another method is to mount a pin chuck on a piece of round steel in the form of a runner, in this case an 8mm (5⁄16in) dia. rod; this is a loose fit in the Boley & Leinen tailstock. After centring the arbor, the drill is applied by hand pressure. With this method, extremely small holes can be drilled, if care is taken.

MAKING THE PIVOTS

Select a suitable piece of blued pivot steel slightly larger than the hole, put it into the lathe collet, and with a fine file or pivot file, put a slight taper on the pivot steel so that it will enter the hole by approximately one third.

Cut off the blued steel at the appropriate length, and square the end. It now only remains for this to be driven into the hole. Make sure that when pressing it home, the pivot is square and true with the arbor. One method of carrying this out is the set-up as shown: a bush that fits in the tailstock takes a standard staking-set punch; arbor and pivots are then kept completely in line, and light hammer blows are applied to the punch. Finally, radius the pivot end and burnish the pivot – and the job is complete.

Fig. 422 Drilling a pivot with the JMW pivot and jacot tool.

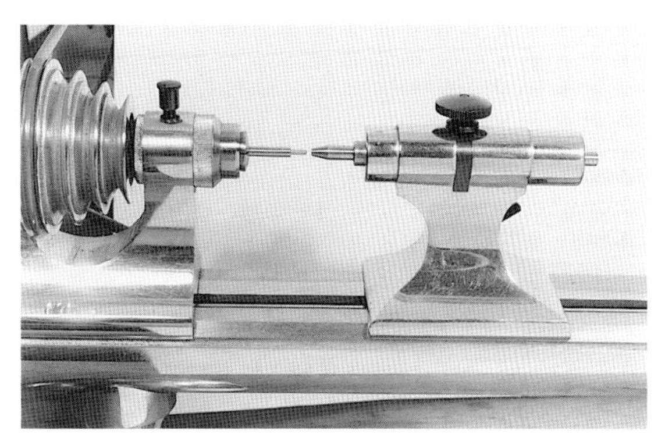

Fig. 424 Using a staking punch to drive home a pivot.

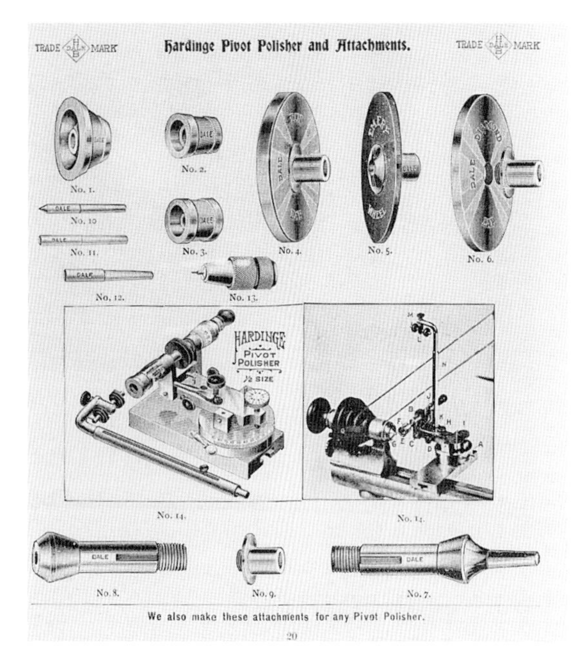

Fig. 425 Hardinge pivot-polishing attachment.

POLISHING/BURNISHING

Specialist machines have been developed for carrying out pivot polishing or burnishing, also separate attachments for mounting on the lathe. An early model made by Hardinge to fit the American-style W. W. bed lathe is shown; this is from an early Hardinge catalogue *c.* 1901.

The machine shown (*see* Figure 426) is a Hauser bench model mainly for production work, where batches of pivots of the same size are required to be finished. Also shown is a Pivofix machine. This is extremely compact and well designed. It is for smaller pivots; carriage clock pivots and watch pivots can be burnished quite easily. Both machines incorporate a carbide wheel that is driven either by a motor or by hand to

Fig. 427 Bergeon pivot-polishing attachment.

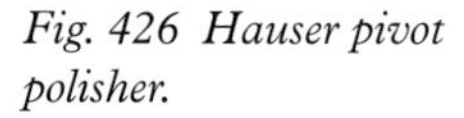

Fig. 426 Hauser pivot polisher.

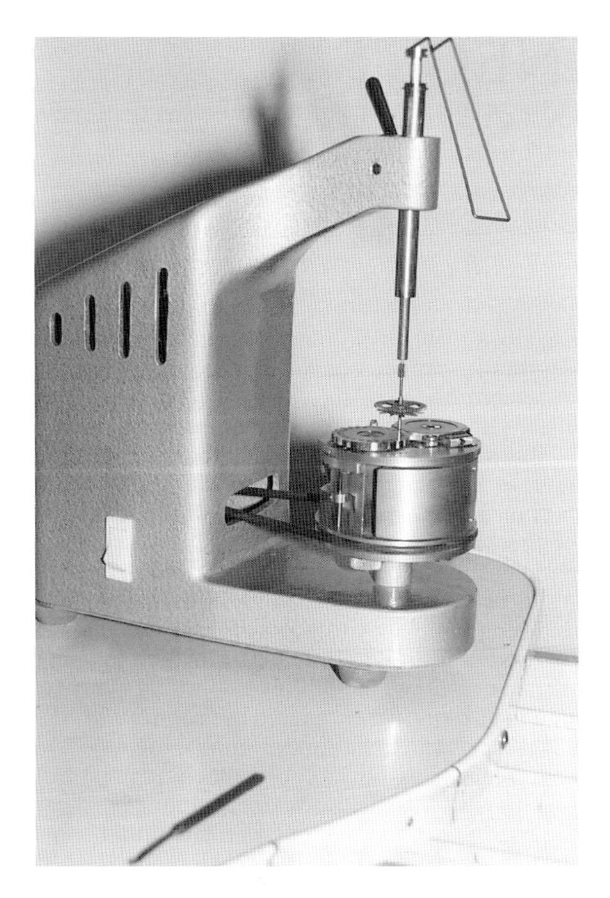

Fig. 429 Drum-type polisher, bench-mounted.

Fig. 428 Pivofix pivot polisher.

Fig. 430 View of carbide wheel and jacot drum.

impart the finish to the work. This type of machine ensures a completely round and true pivot.

Another attachment is shown (*see* Figure 427); this is by Bergeon and mounts in the lathe tailstock. Again it has a carbide wheel, in this case driven by the overhead drive. It can be fitted to most small lathes, as the arbor can be easily interchanged.

Finally, various views are shown of a specialist pivot polisher, which can be either bench-mounted or fitted in the lathe. The design is quite ingenious: the whole head with jacot drum and carbide wheel revolves, but the work remains stationary.

Fig. 431 The unit mounted in the lathe.

11 MISCELLANEOUS OPERATIONS

DEPTHING THE TRAIN

Whilst the theoretical centres of wheels and pinions can be calculated, as we have seen in Chapter 5, this does not always give the optimum rolling action between meshing wheel teeth and their matching pinion leaves; a depthing tool is therefore essential. Similar tools have been shown in early French books on horology since the early eighteenth century.

In the late nineteenth and early twentieth century, horological tool catalogues offered many sizes of tools, from those suitable for depthing watch trains to those that would accommodate large clock trains. It is essential that the tool being used is accurate, and that the runners are parallel and true. The body is usually two brass castings machined and hinged together at the base. The steel runners have male points for scribing, and at the opposite end, female runners for locating the pivots of the wheel arbors. Locking screws are provided for the runners, and a larger screw situated near the base of the tool provides infinitely variable adjustment for the depthing of wheel and pinion centre distance. Normally a return bow spring is fitted along the length of the tool. Various runners are provided, as shown: plain runners as previously described, also a cone runner to locate from larger holes in the clock plates, and a lantern runner, useful for depthing escapements. When large arbors have to be accommodated, for example longcase barrel arbors, special trumpet runners are used.

It is always advisable to work to one end of the tool, as it is virtually impossible to retain accuracy of the centres over 250mm (10in). Methods of checking a depthing tool are illustrated; basic parallelism can be checked with a micrometer.

To use the tool, proceed as follows. Open the depthing tool with the adjusting screw, and place the first wheel and arbor between the matching female centres of the first two opposing runners. Use the locking screw to hold the runner with the slightest of end shake, just to enable the wheel to spin freely. Now position the matching pinion to line up with the wheel already positioned. Adjust the meshing of the two components until it is as near as can be achieved by eye, so that the pitch circles of the pinion and wheel coincide: this is the theoretical engaging position. With the adjusting screw, slightly deepen the engagement, and test the rotation of wheel to pinion; if they run lumpy and tend to jam, slightly reduce the engagement.

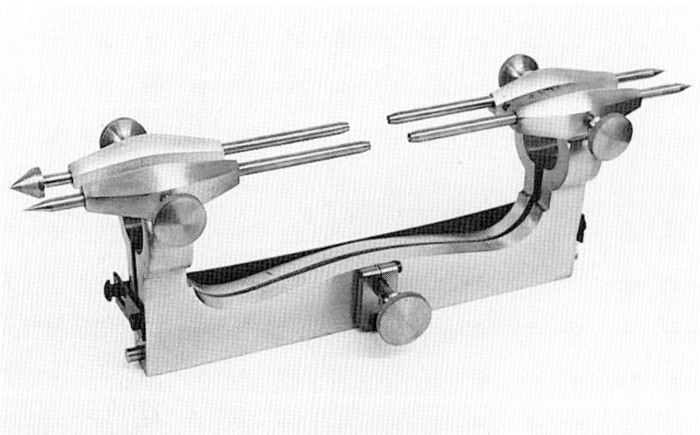

Fig. 432 Clockmaker's depthing tool.

Fig. 433 Depthing tool shown in Ferdinand Berthoud, 1763.

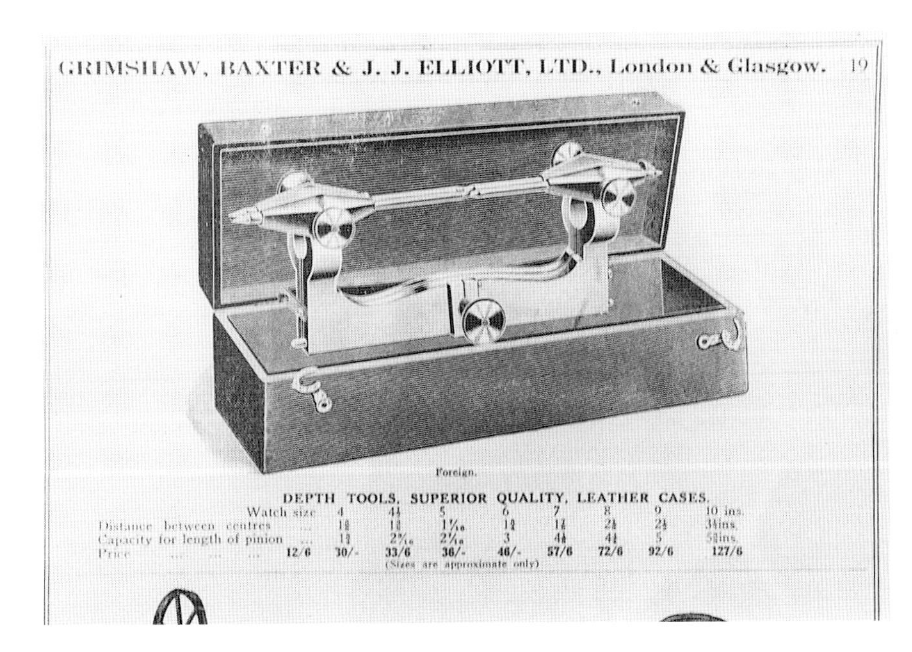

GRIMSHAW, BAXTER & J. J. ELLIOTT, LTD., London & Glasgow. 19

Foreign.

DEPTH TOOLS, SUPERIOR QUALITY, LEATHER CASES.

Watch size		4	4½	5	6	7	8	9	10 ins.
Distance between centres ...		1⅜	1⅝	1⁷⁄₁₆	1¾	1⅞	2⅛	2½	3¼ins.
Capacity for length of pinion ...		1¾	2⁵⁄₁₆	2⁷⁄₁₆	3	4⅛	4½	5	5⅜ins.
Price	12/6	30/-	33/6	36/-	46/-	57/6	72/6	92/6	127/6

(Sizes are approximate only)

Fig. 434 Various sizes of depthing tool from an early tool catalogue.

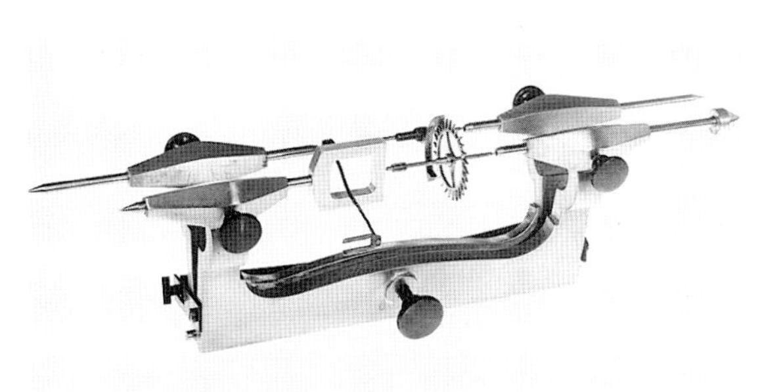

ABOVE: *Fig. 435 Lantern runner in use.*

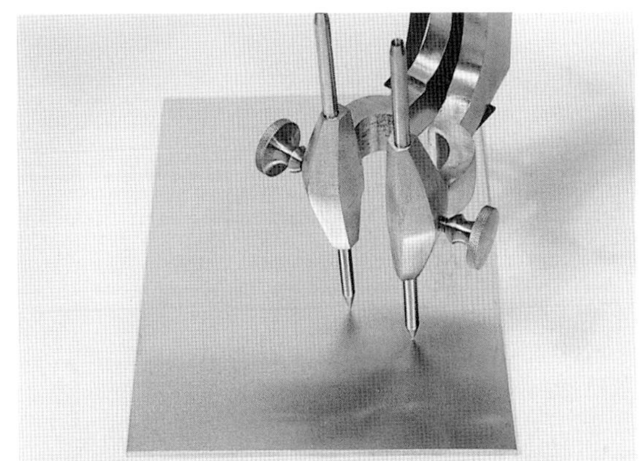

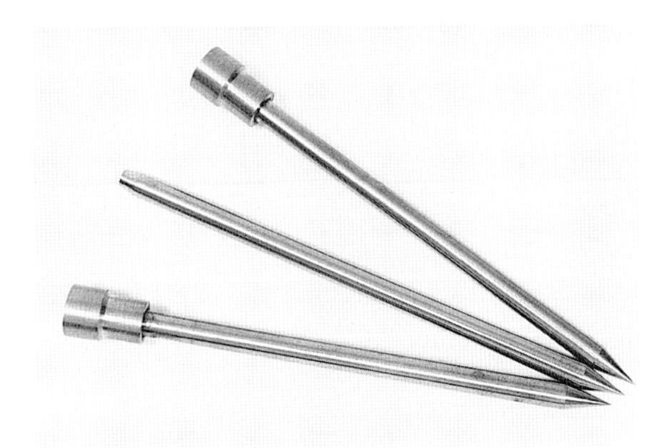

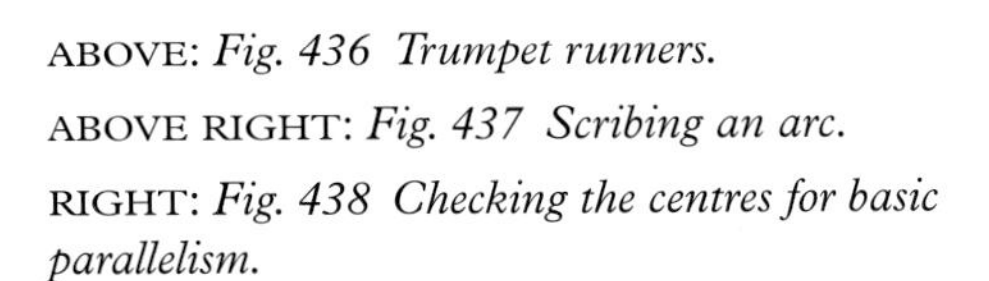

ABOVE: *Fig. 436 Trumpet runners.*

ABOVE RIGHT: *Fig. 437 Scribing an arc.*

RIGHT: *Fig. 438 Checking the centres for basic parallelism.*

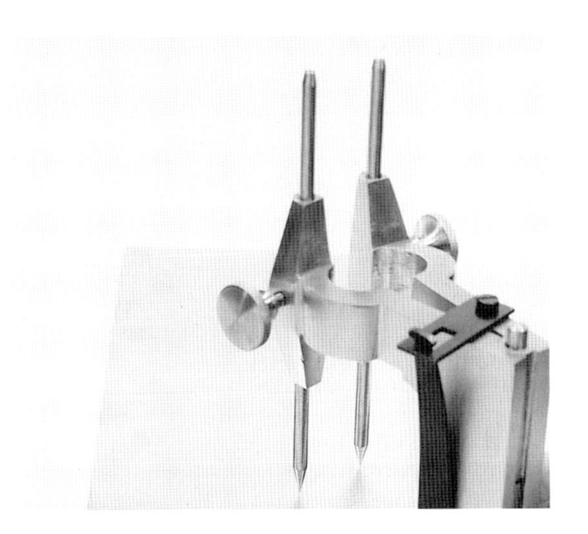

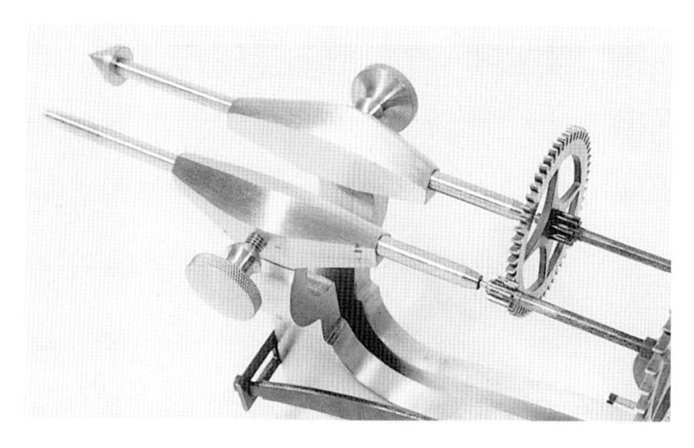

Fig. 439 Adjusting the meshing.

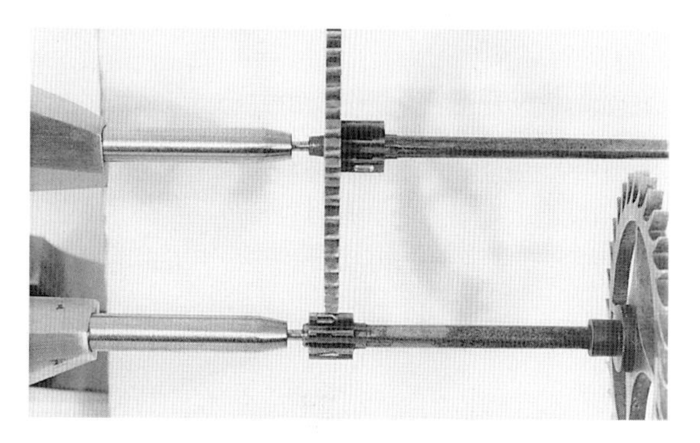

Fig. 440 Close-up of the wheel and pinion in engagement.

Fig. 441 The cone runner in use.

Experience will tell you when the depthing is correct, but you can check with the eyeglass. Viewing can take place from the end of the depthing tool – there are circular apertures provided for just this purpose. Spin the wheel and listen to the sound: it should spin freely and noiselessly. Whilst the optimum position is the theoretical centre distance of wheel and pinion (and will give engagement just before the line of centres), this does not always work out in practice, which is why a depthing tool is used, to locate the most efficient working engagement. A cone runner is shown being used to locate the tool from a winding hole in the clock plate.

As we have seen in the chapter on wheel and pinion theory, if the centre distance is too great, the teeth will engage before the line of centres, and this will increase friction. Conversely, if the meshing is too deep, the teeth and leaves will jam, and will not run smoothly.

Now that the wheel and pinion centres have been ascertained, these can be transferred to the clock plates. When rebushing an antique clock and the centre distance is incorrect, it is necessary to plug the hole completely and then depth the wheel and pinion correctly. The tool should be held vertically when scribing, otherwise results will be inaccurate. Once the marking out is complete, use a fine centre punch and eyeglass to position the centre pop exactly. Alternatively, an optical centre punch can be used: this is an excellent method of obtaining the most accurate position for drilling. Use a small centre drill to start the hole – sometimes a number one is too large for very small pivot holes, in which case use a miniature centre drill. If a Quorn tool and cutter grinder is available, 'D' section, 60-degree cutters can be produced; these are excellent for producing a fine centre, and are ground from small, round, high-speed steel tool bits. They can also be produced in the lathe using silver steel, a filing rest or milling attachment being necessary to produce a flat halfway through the diameter of the cutter. When the cutter has been finished and polished, then harden and temper.

When drilling holes for the pivots, always select a drill slightly smaller than the pivot. A

Fig. 442 Trumpet runners locating a longcase barrel arbor and centre pinion.

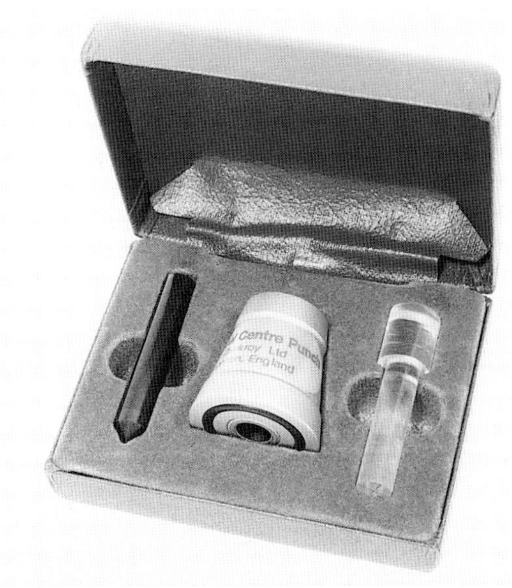
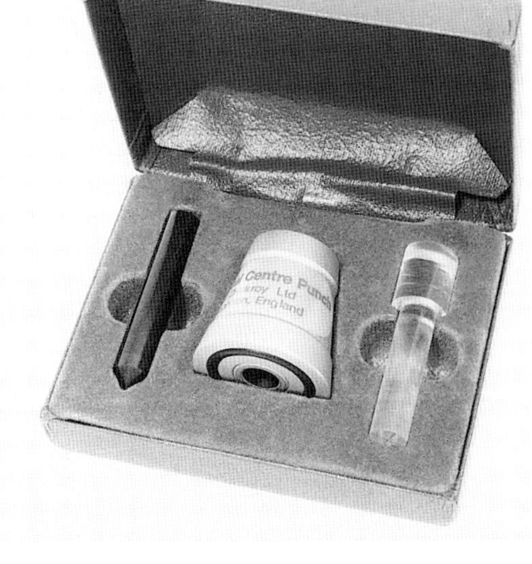

Fig. 443 *Optical centre punch.*

five-sided cutting broach is then used to open out the hole; this is used from the inside of the plate. Finally, a smoothing broach is used with oil, again from the inside, to burnish the bore. Some restorers advocate broaching from each side of the plate; however, it is entirely a matter of preference which method is employed. Just ensure that all burrs are removed from the edge of the hole. A number of clocks have oil sinks, such as French clocks: these can be produced with roller countersinks, or home-made from silver steel, as shown.

When one-piece arbors and pinions are being used, then only a conventional depthing tool will suffice. If pinion heads are used, it is possible to use a pinion-head depthing tool (illustrated); this is a far simpler tool, but obviously has limitations. The wheel and pinion centres can be varied with the knurled adjusting screw located at the

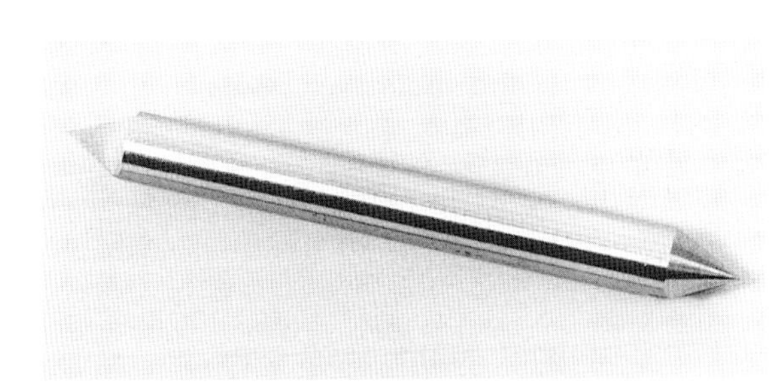

Fig. 444 'D'-section centre drill.

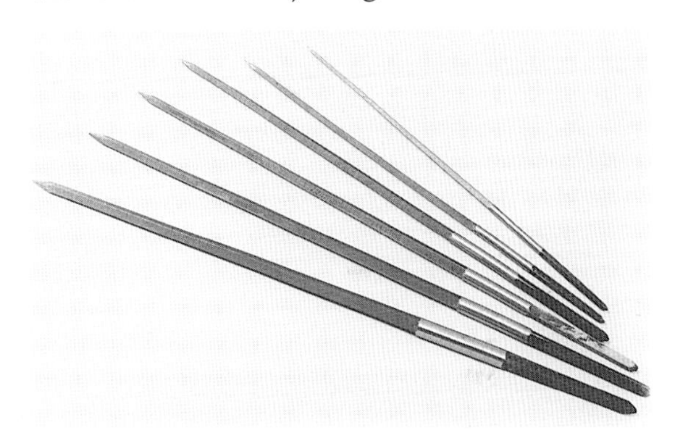

Fig. 445 Range of taper-cutting broaches.

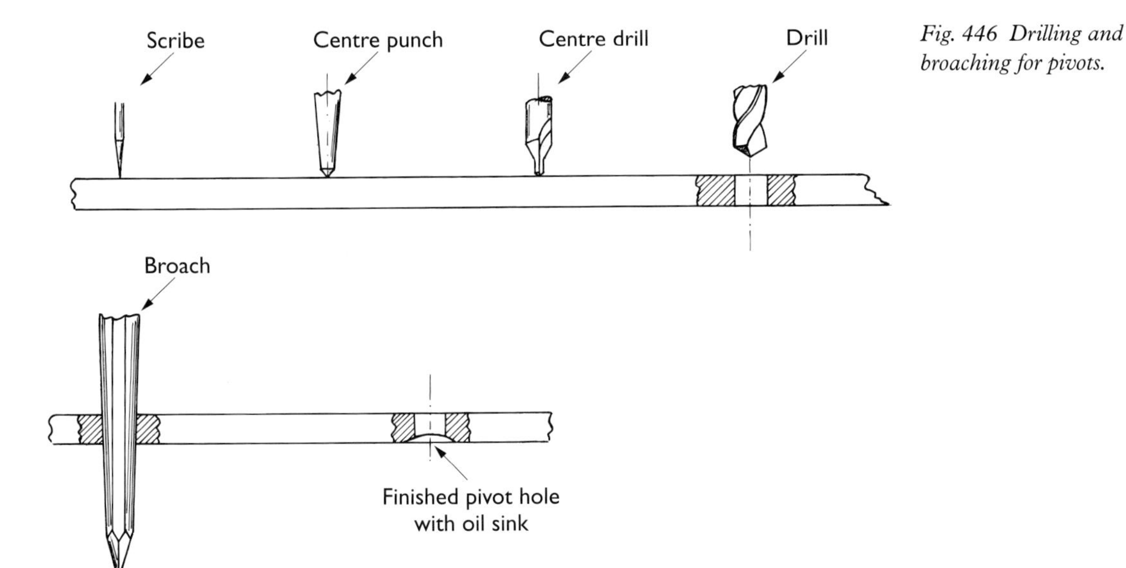

Fig. 446 Drilling and broaching for pivots.

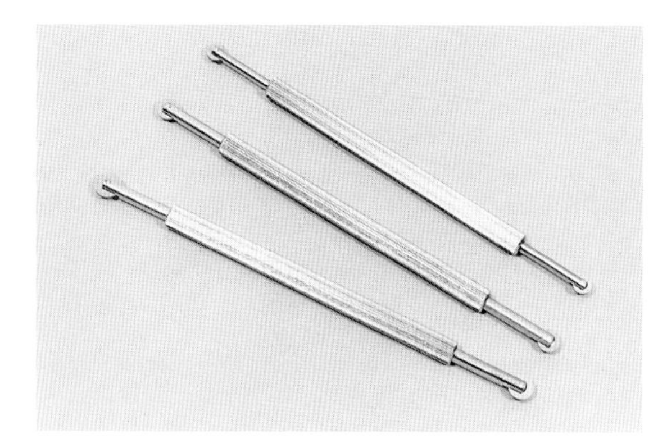

Fig. 447 Wheel countersinks.

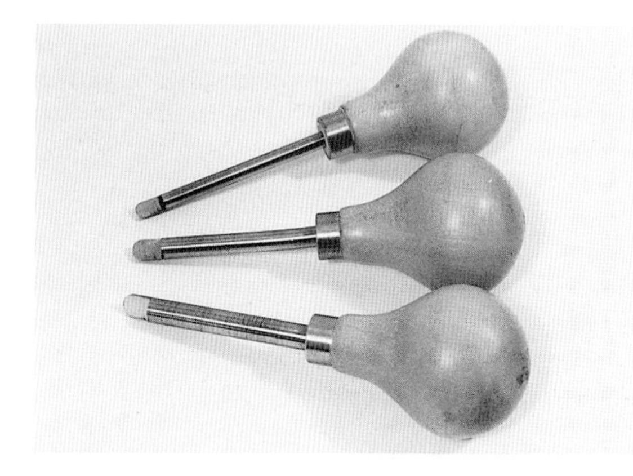

Fig. 448 Home-made countersinks for producing oil sinks.

Fig. 449 Pinion-head depthing tool.

end of the tool, and the matching wheel and pinion head are located directly onto the depthing-tool runners.

For checking contrate-wheel engagement, a special right-angle bracket is used. A special depthing tool for this operation was offered in the Wyke tool catalogue, but it would appear that none has survived.

The bracket is attached to the standard depthing tool with two securing screws. It can be put in use quite quickly if required, then removed, when the tool reverts to its conventional form. Note the special runner used to locate the crown-wheel arbor.

When the contrate-wheel and crown-wheel pinion are in correct engagement, measurements can be taken to check depths.

Also provided is a special runner for locating the verge staff in conjunction with the crown wheel. This method is most useful. A guide to the centres of engagement and the proportion of the verge flags for English bracket clocks and verge watches is as follows:

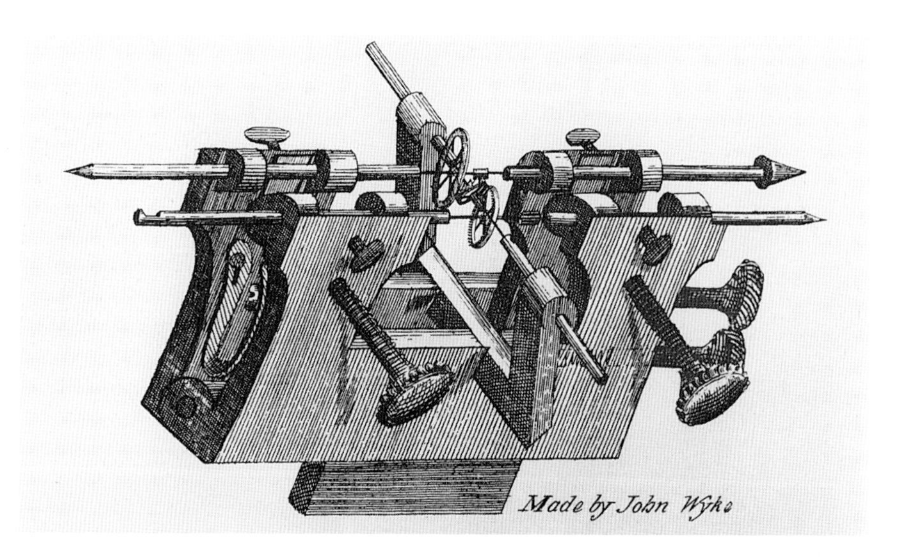

Fig. 450 Contrate-wheel depthing tool from John Wyke tool catalogue.

Fig. 451 Commercial depthing tool with contrate wheel attachment.

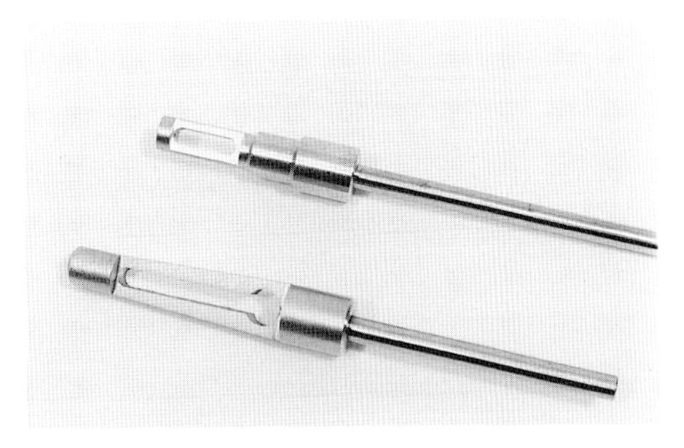

Fig. 453 Special runners.

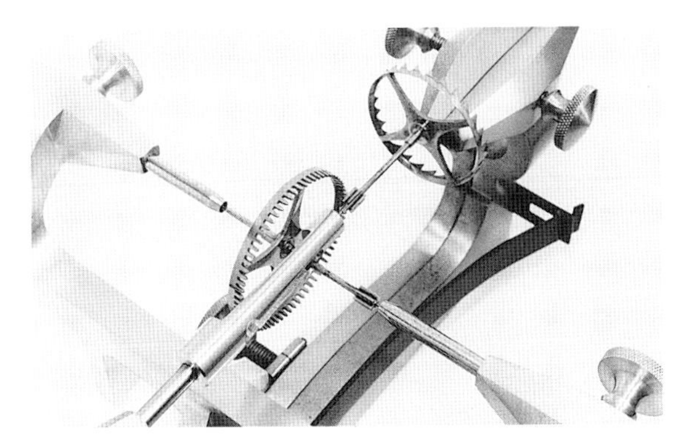

Fig. 452 Close-up showing contrate wheel and pinion engagement.

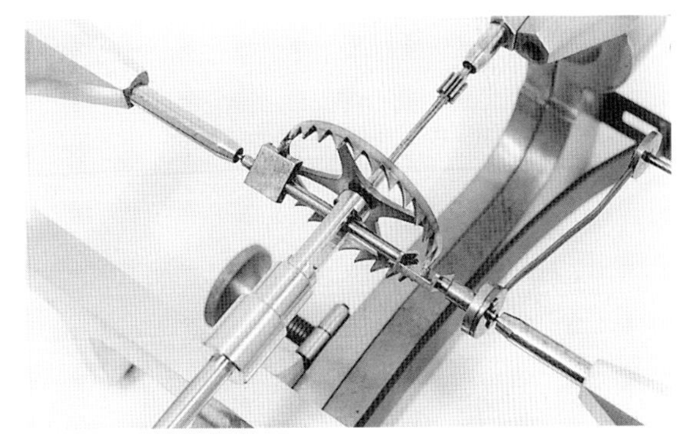

Fig. 454 Adjusting the verge escapement.

Angle between flags = 90 degrees

Height of staff above wheel teeth

$$= \frac{2 \times \text{pitch}}{3}$$

Length of pallet $= \frac{6}{5}$ of pitch

For clocks with long pendulums, e.g. French Comtoise clocks, the pallet angle is approximately 60 degrees.

Very small depthing tools were produced for watch work, and it is essential that these are absolutely accurate. They can often be found quite cheaply at clock and watch fairs, representing excellent value considering the cost of producing such a tool today.

For special pocket-watch escapement and platform escapement work, an escapement matching

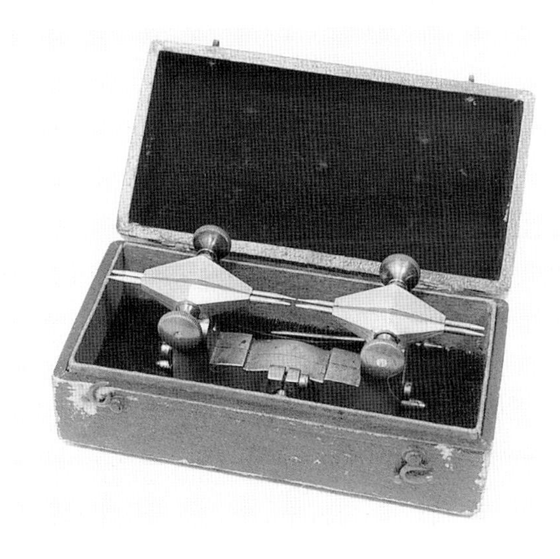

Fig. 455 Watchmaker's depthing tool.

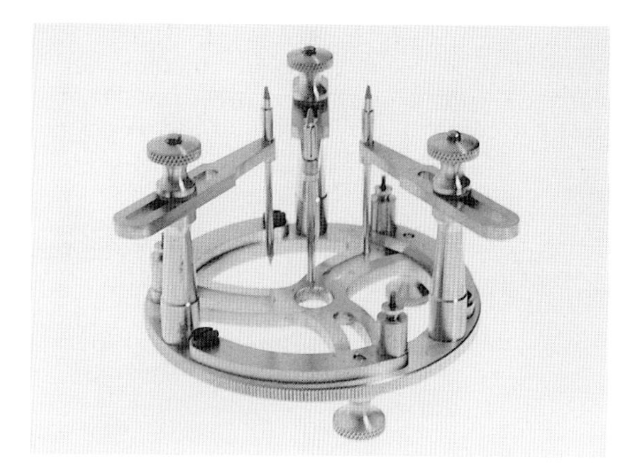

Fig. 456 Escapement matching and depthing tool.

Fig. 457 Escapement tool in use.

Fig. 458 Longcase movement with calendar wheel missing.

and depthing tool was produced, with three separate arms, each with its own runner.

MISSING CALENDAR WHEEL, LONGCASE MOVEMENT

Typical of many longcase movements that require restoration is the situation where the stud is still intact, but the twenty-four-hour wheel is missing. Fortunately it is not too difficult to calculate this module and the dimensions of the missing wheel, and replace it.

When working on antique clocks, it is preferable to restore the movement using the appropriate material; in this instance, use a cast-brass wheel blank. This is 70/30 brass, and will generally be a good match in colour; in its 'as cast' condition, the brass is quite soft and requires work-hardening by hammering. This can be carried out on a small anvil or steel plate. Hammer both sides of the wheel-blank casting, then grip in the three-jaw chuck and machine all over. Drill and ream a hole for the wheel mandrel.

Fig. 459 Yellow cast-brass rods and discs.

Fig. 460 Hardening a cast-brass wheel disc.

Fig. 461 Completed sixty-tooth wheel meshing with existing thirty-tooth calendar wheel.

Because the old clockmakers did not work to an acceptable standard with their tooth form and size, it is not always possible to cut the matching wheel the correct size first time, when using standard formulae for the wheel cutters available. It is sometimes necessary, therefore, to cut the wheel and test for depth, then mount it back on its mandrel to recut the teeth smaller. The mandrel shown previously in the chapter on wheel cutting is ideal for this purpose. A small locating pin is used to position the blank to the arbor accurately. When the wheel is complete, the hole can be plugged if required.

The calculation for the missing wheel was as follows:

The existing wheel on the hour pipe was thirty teeth

Centre distance of centre arbor and stud
$= 38.3\text{mm}$

We require to find the wheel O/dia. and module of the sixty-tooth wheel

From previous formula,

$$\text{c/distance} = \frac{(N + n)}{2} \times M$$

$$\text{Therefore } 38.3 = \frac{(60 + 30)}{2} M$$

$$\text{Therefore } M = \frac{38.3}{45}$$

$$= 0.85 \text{ module}$$

O/dia. of 60 tooth wheel

$$\text{O/d} = (N + 2.76)\text{module}$$

$$\text{O/d} = (60 + 2.76)0.85$$

$$= 53.35\text{mm } (2.1062\text{in})$$

Therefore a wheel blank is machined to 53.58mm (2.109in), and sixty teeth are cut using a wheel cutter with a module of 0.85.

Fig. 462 Calendar wheels for an early longcase clock.

These calculations actually did work, without having to recut the wheel. There is, of course, more latitude with calendar wheels than train wheels; furthermore early clocks had calendar wheels with extremely coarse teeth and this would give more margin when depthing.

To finish the job, all that is required is to turn a suitable collet from cast-brass rod – again, hammer the material before machining. The collet is rivetted in place, and then a steel pin has to be fitted to drive the calendar ring.

MAKING THE RACK AND SNAIL FOR STRIKING WORK

The front-plate layout for a rack-striking longcase clock illustrated is typical, but quite often these parts are either damaged or missing and have to be replaced.

Laying out Snails

Racks and snails can cause problems due to wear or poor previous repairs. Moreover in some instances, the striking had not been set out correctly when the clock was first manufactured. To lay out a snail for a longcase clock, proceed as follows:

Determine the travel of the rack tail from its rest position to twelve o'clock. Draw two circles, one to represent the rack tail at rest, and the other at twelve o'clock. Draw a downward vertical line to represent the starting position, and from this, draw twelve radial lines, each one being 30 degrees. Then divide the vertical line into twelve equal spaces as shown. The rack-tail pin is shown on '0', the rest position just above the one o'clock on the snail; each of the steps can then be marked off from the divisions on the vertical line.

Fig. 463 Eight-day longcase rack-striking movement.

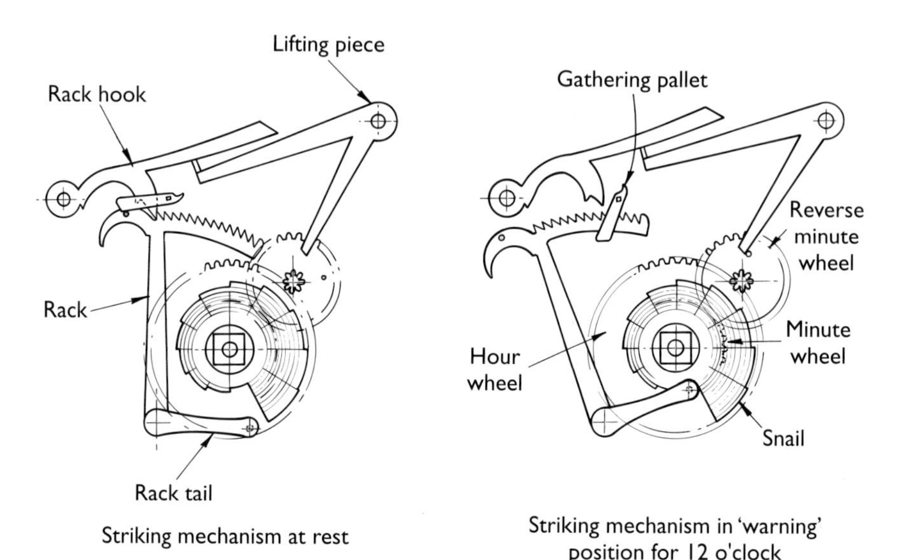

Fig. 464 Rack-striking mechanism.

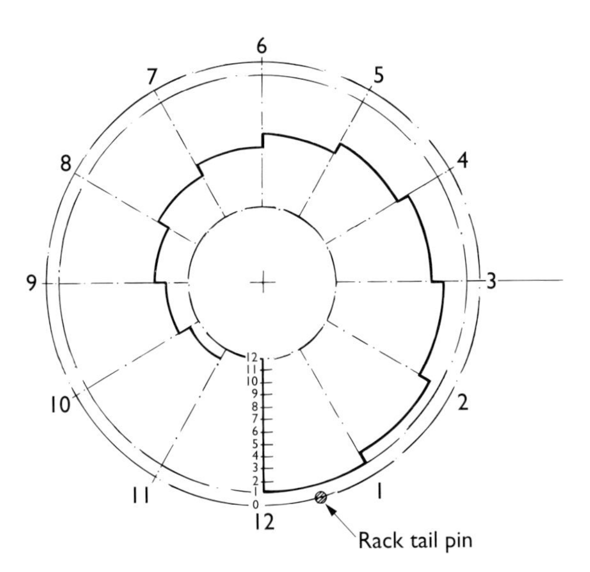

Fig. 465 Layout of stepped snail.

Fig. 466 Scribing the radial lines for a snail.

Fig. 467 Milling the steps on a snail.

Note that the first circumferential line is non-existent: it simply represents the position of the rack-tail pin ready to drop to the one o'clock position.

The snail is made from brass sheet approximately 1.5mm (0.062in) thick. There are eleven steps and twelve arcs, onto which the rack-tail pin drops to give the correct number of blows when the clock is striking. The steps can be marked out on the lathe or surface plate. In the example, it was determined that the steps equalled 1.29mm (0.051in), and it is essential that these are marked out accurately. Fit a division plate or dividing head to the lathe mandrel, and set for twelve divisions; the radial lines can be marked using the scribing block as shown. For the twelve arcs, set the scriber to a rule at the maximum step, rotate the lathe headstock by hand, and scribe the first step. Move to the next radial space, reset the scribing block, and continue until all steps are complete. The snail can then be cut out with the piercing saw, and filed to size.

A more precise method is to mount the blank in the lathe chuck or collet on a suitable arbor. Remove some of the surplus metal, and mount the dividing head onto the lathe mandrel as previously shown for wheel cutting. The dividing head is required to be set for twelve divisions. Set up the milling spindle as shown, with a 2mm (.078in) end mill. Commencing with the largest step, mill the circumference for 30 degrees (one twelth of a revolution). Take care when approaching the end of the cut. When the first cut is complete, lower the cutter the calculated amount for one step, and repeat the procedure until all steps are completely milled. With the worm and wheel dividing head, complete control can be achieved when milling of the snail steps takes place.

Take care at the end of each cut not to mill into the edge of the adjacent step. When milling is complete, all that is required is to remove the small radius left by the cutter with a needle file. In most longcase clocks, the large step is cut back as shown to give the rack tail more clearance at twelve o'clock; it can be either milled or filed. Also, this deep step is often chamfered on its radial face to enable the pin on the rack tail to pass over the step if the striking mechanism jams or has not been wound.

The tool for milling stepped snails illustrated is useful: there are twelve holes drilled in the mandrel body, and a pin locates one of these from a hole drilled in the outer body. An adjustable stop is provided at the rear of the tool

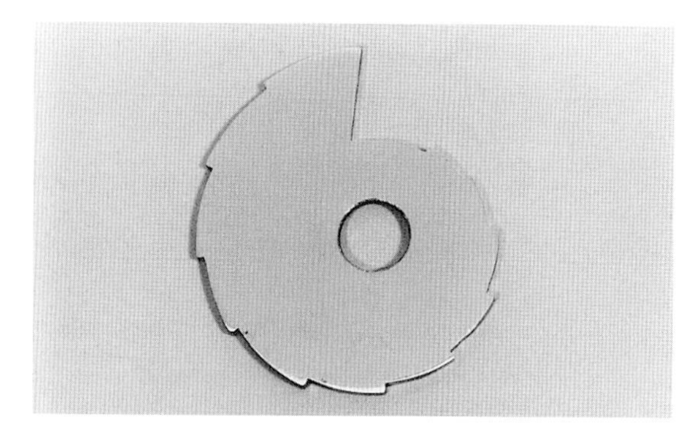

Fig. 468 Completed snail.

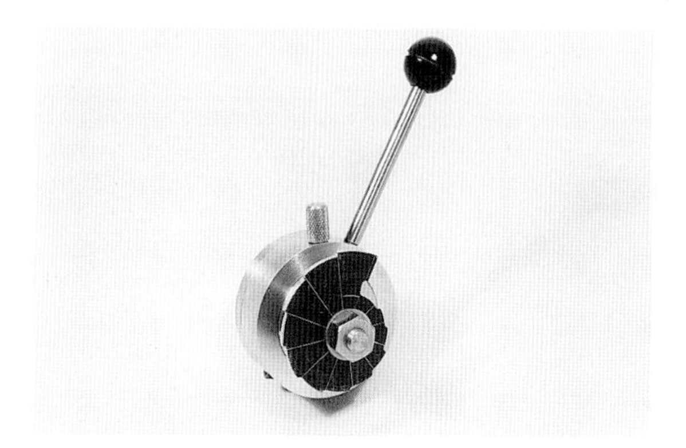

Fig. 469 Special tool for milling snails.

to enable the length of cut to be controlled, which is necessary when cutters of different diameter are used. A ball handle gives rapid and easy control of the traversing, and the stops previously mentioned give a positive cutting length.

Marking out on the lathe with the scribing block is quite useful, and fairly accurate work can be achieved measuring from a rule mounted vertically, and the readings taken off with the scriber pointer. However, a far more accurate method was developed and is shown here: this is a spring-loaded scriber made from 3/16in silver steel, which fits into a 5mm (3/16in) reamed hole in a brass body 12mm (1/2in) square × 50mm (2in) long. The scriber can be accurately positioned by use of the centre micrometer.

Clamp the scriber block in the lathe tool holder, ensuring it is parallel with the lathe axis. Set the micrometer to half the diameter of the scriber and, with the micrometer head vertical, adjust the tool-holder height so that the scriber just touches the face of the micrometer anvil. Now rotate the micrometer head through 90

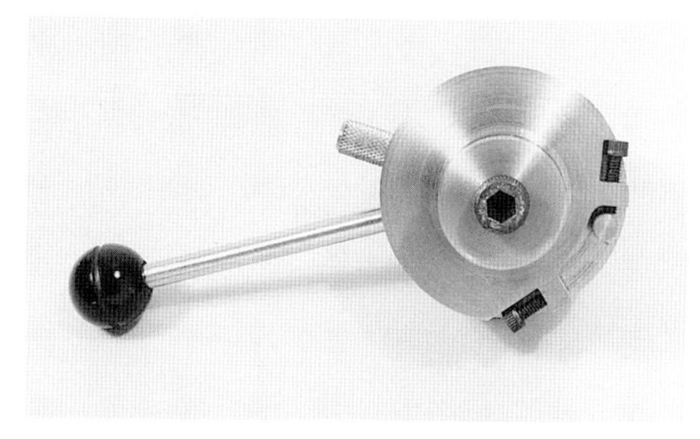

Fig. 470 View from the rear showing adjustable stops.

Fig. 471 The component parts for a snail milling tool.

Fig. 472 The snail milling tool set up in the lathe.

degrees, and with the lathe cross-slide handle, bring the scriber to bear against the micrometer anvil, lock the cross-slide – and the scriber point is now exactly on the lathe centreline. Using the lathe cross-slide graduated thimble, the twelve radiused steps can be accurately marked out.

French clock snails differ from English clocks, there being no individual steps, just a gradually

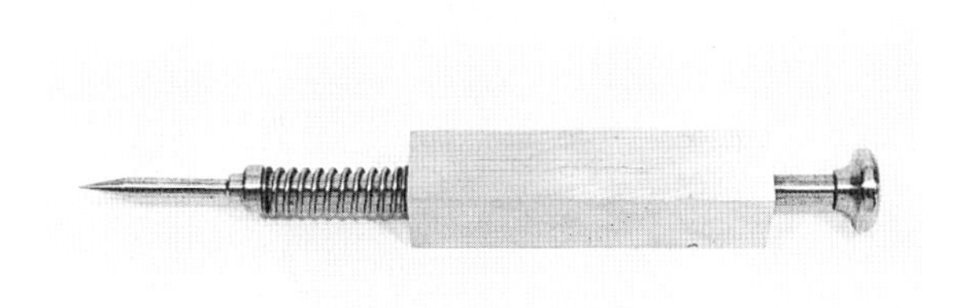

Fig. 473 Spring-loaded scriber.

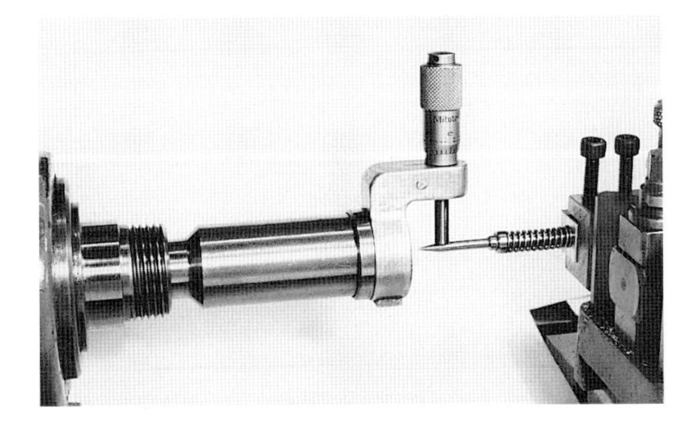

Fig. 474 Setting the scriber vertically.

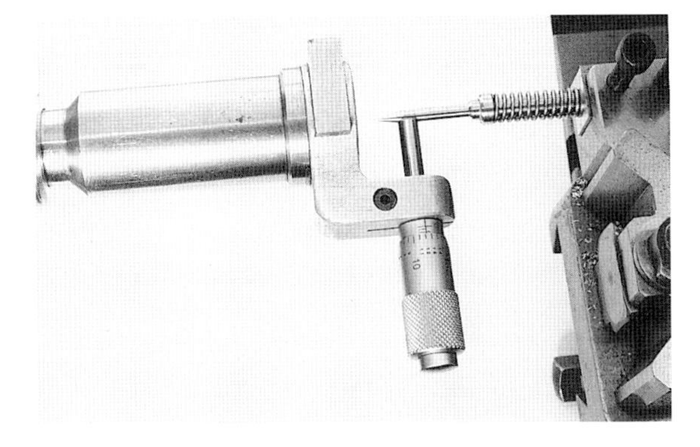

Fig. 475 Setting the scriber in the horizontal plane.

increasing radius from the smallest to the largest step (see next page). These are more difficult to mark out and machine than previous methods.

To mark out a snail for a French clock, first determine the travel or drop of the rack tail from its rest position to twelve o'clock. Draw two circles, one to represent the rack tail at rest, and the other at the twelve o'clock position. Draw a downward vertical line to represent the starting position, and from this draw twenty-four radial lines, each one being 15 degrees apart. Divide the bottom vertical line between the inner and outer circle into twenty-two equal divisions, then

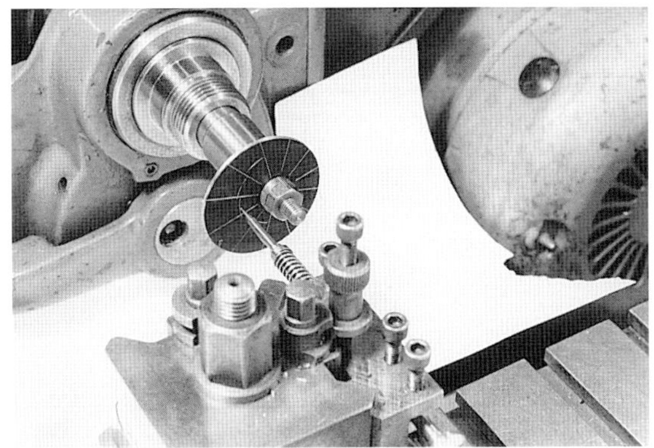

Fig. 476 Marking out a snail with the special scriber.

add one division inside to the inner circle, and one division to the outer circle, giving twenty-four divisions in total. Number the divisions 0–24 as shown. Also starting at the base of the vertical line, number each radial line 1–24.

Note that the rack-tail pin drops on all the even numbers: this is the mid-point for each 'step'. When marking out, start with radial line number two, as the first division is already positioned. When all points have been plotted, use French curves to join up to form a progressively reducing cam.

It has already been noted that the rack-tail pin drops onto the even-numbered radials, therefore we see here the reason for the two extra divisions at each end of the vertical line: they allow for the start and finish of the progressive curve in practice. If desired, the curve can follow the circumference line at position twenty-four for the last 15 degrees; this will look more aesthetically pleasing. To finish the snail, cut out with a piercing saw, then carefully file to the line.

If the rack is present and the snail is missing or damaged, it is possible to mark out the snail from the existing rack. This is an advantage, as any error in the pitching of the rack teeth can be accommodated, and the snail steps will then match each individual pitch of the rack teeth. Obtain a suitable brass blank, and bore it out to suit the hour-wheel pipe. Draw twelve radial lines to represent each step or segment. Replace the blank onto the movement. Replace the rack-tail pin with a small pin with a sharp point to act as a scriber. With the rack hook holding the rack at one o'clock, scribe an arc between two radial lines on the snail blank. Repeat the procedure for each pitch of the rack teeth until all the twelve

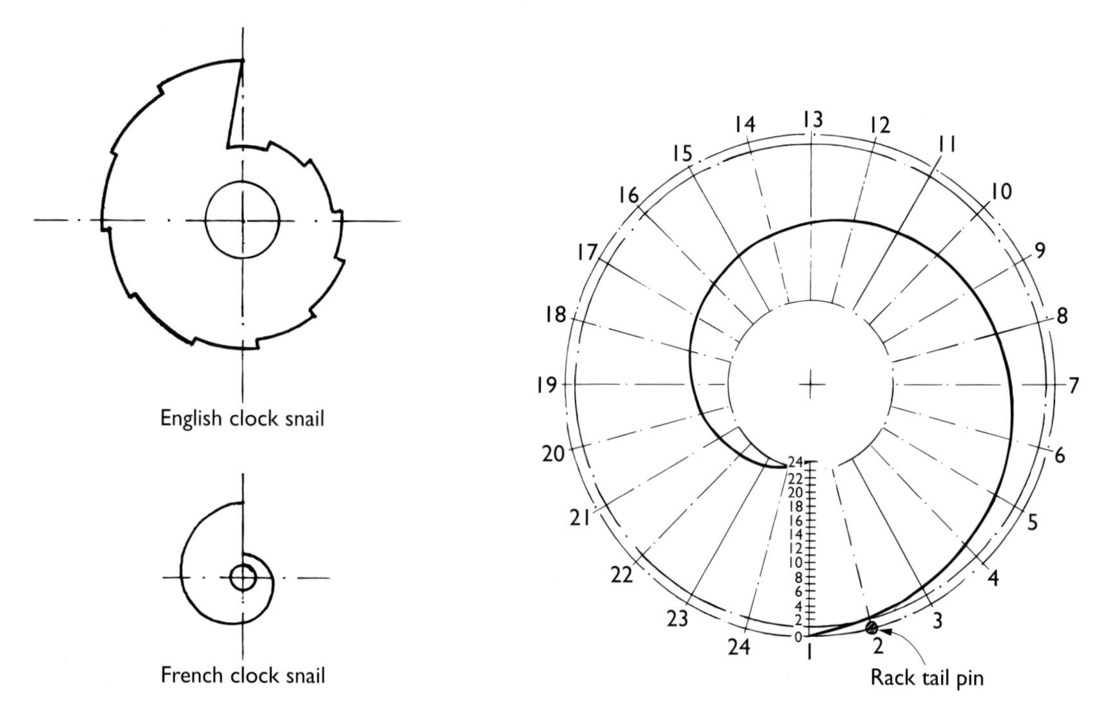

Fig. 477 French and English snails.

Fig. 478 Layout of French snail.

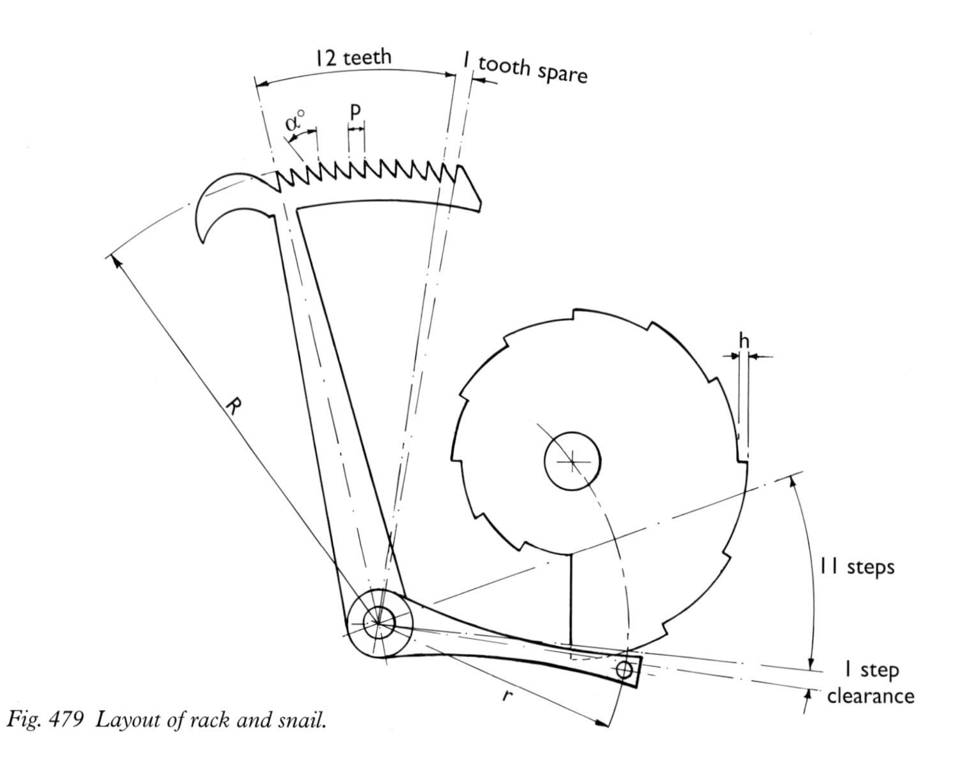

Fig. 479 Layout of rack and snail.

steps have been scribed. The snail can then be machined as previously described.

Replacing Racks

Racks have to be replaced quite often, and the following information should be useful if a new clock is being constructed, or a missing or damaged rack on an old clock has to be made. There is a definite relationship between the steps on the snail and the pitch of the rack teeth and the length of the rack tail; however, if the following notation is used, calculations for both rack and snail can be carried out.

R = radius of rack

P = pitch of rack teeth

r = radius of rack tail

h = height of snail step

Therefore $\frac{R}{r} : \frac{P}{h}$

Therefore $P = \frac{R \times h}{r}$

When setting out a rack and snail, there are often constraints in the design due to the position of various components of the under-dial work, such as winding arbors, bridges, pivots, and so on. Some trial and error is often required, therefore, before actual calculations can be determined.

When sizing a longcase rack, ensure that the pitch of the teeth is not too small; larger teeth give less chance of error when the gathering pallet picks up a tooth: the smaller the tooth, the smaller the gathering pallet. After checking many longcase movements, there appears to be a wide variation between the length of the rack tail and the length of the rack, the ratios varying from 1.5:1 to 2:1. The pivot position of the rack should be such that the arc formed by the rack-tail pin passes through the centre of the snail.

Using the above information, an example of a rack was calculated to match the snail already cut:

Rack radius, R = 60.13mm (2.375in)

Rack-tail radius, r = 37mm (1.460in)

Snail step, h = 1.29mm (0.051in)

Therefore pitch of rack tooth,

$$P = \frac{R \times h}{r}$$

Therefore

$$\frac{60.13 \times 1.29}{37} = 2.10\text{mm } (0.083\text{in})$$

It is necessary to find the number of teeth in a full circle to enable the rack teeth to be cut:

$$\text{Circumference} = (60.13 + 60.13) \times \pi$$
$$= 378\text{mm } (14.925\text{in})$$

Therefore circumference divided by the pitch:

$$\frac{378}{2.10} = 180 \text{ divisions}$$

It may be useful to obtain the angular relationship of each tooth. This can be achieved as follows: using the proportions as before, let $x°$ be the angle in degrees for one tooth:

Circumference = 378mm

Pitch = 2.10

$$\frac{360°}{378\text{mm}} : \frac{x°}{2.10}$$

Therefore

$$x° = \frac{360 \times 2.10}{378} = 2°$$

With this information it is then possible to mark out the rack blank for the position of the first and last tooth. A line is drawn from the centre of the rack pivot point vertically upwards, which should pass through the first acting tooth of the rack. A suitable piece of mild steel sheet 1.5mm (0.062in) is mounted on a wooden faceplate, as shown; this is located in the centre with a small steel pin, and the material fixed to the backplate with small wood screws. The outer rim is turned to size, and a line is marked for the position of the first tooth, using a square mounted on the lathe bed. The lathe is then set up to make 180 divisions.

Rack teeth can now be cut. Recoil escapement cutters are ideal for this purpose; a 3E cutter was used in this instance. Alternatively, a slitting saw can be used, and the surplus material cut away with either a piercing saw or a file.

Fig. 481 Cutting the rack teeth.

Fig. 480 Rack blank mounted on a wooden faceplate.

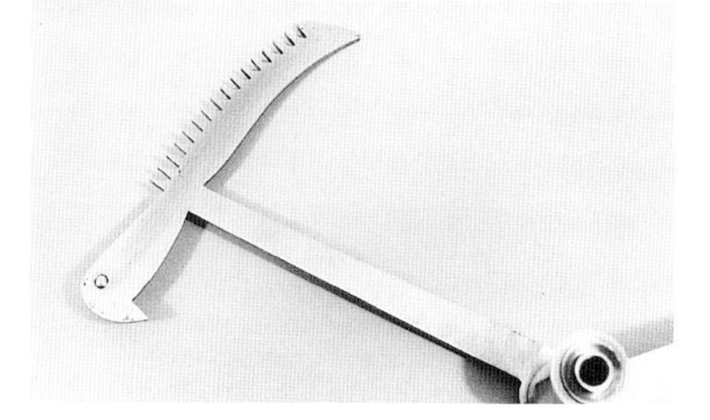

Fig. 483 The completed rack.

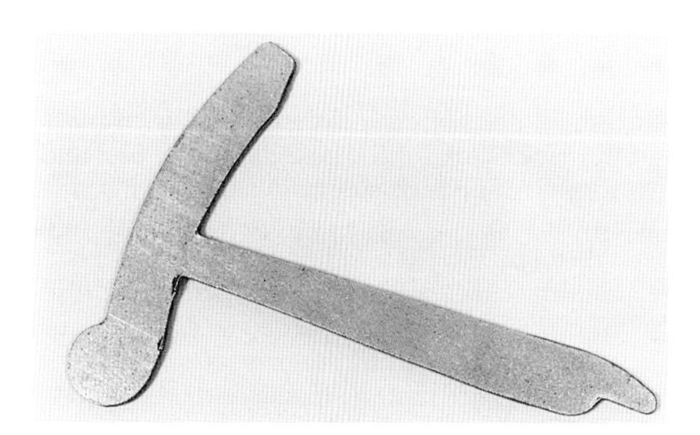

Fig. 482 Commercial rack blank.

If a recoil cutter is used, feed down until only the slightest of lands is produced at the tip of each tooth. The speed of the cutter spindle should be around 250rpm, with a good supply of cutting lubricant. Commercial rack blanks are available, which saves cutting away a lot of waste material when preparing a blank.

The finished rack illustrated has been designed around a completely new snail. When restoring an old clock with an existing snail, it is always necessary to mark out the rack teeth from the snail itself, using each individual step to mark each rack tooth. As variations often occur in the snail, it is not unusual to see examples that have been altered by filing.

Further information on rack-striking geometry can be found in *Clockmaking Past and Present* by G. F. C. Gordon, dated 1928.

ENGLISH COUNT WHEEL OR LOCKING PLATE

Count-wheel striking is generally found on lantern clocks and country-made longcase clocks. The locking plate is a disc with slots cut into its outer periphery, and it provides control for the number of blows struck in each hour.

To make a locking plate, the process of milling is used. This can be carried out either on

Fig. 484 Seventeenth-century lantern clock.

Fig. 485 Rear view of lantern clock, showing the locking plate.

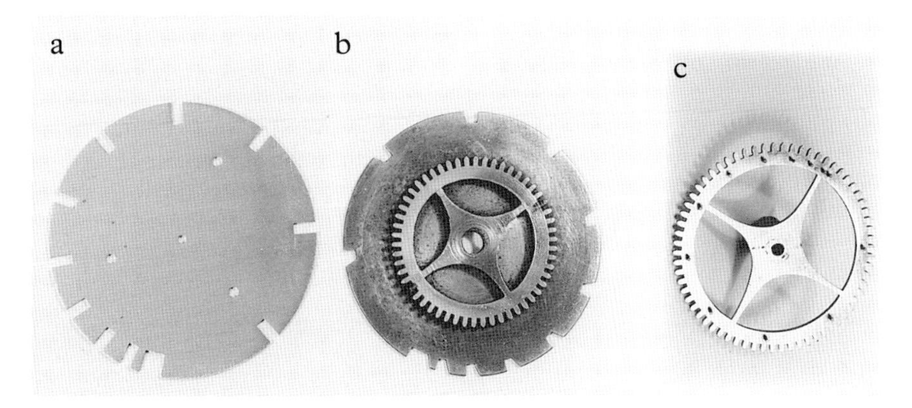

Fig. 486 Three types of locking plate.

Fig. 487 Locking plate or count wheel with ratchet teeth.

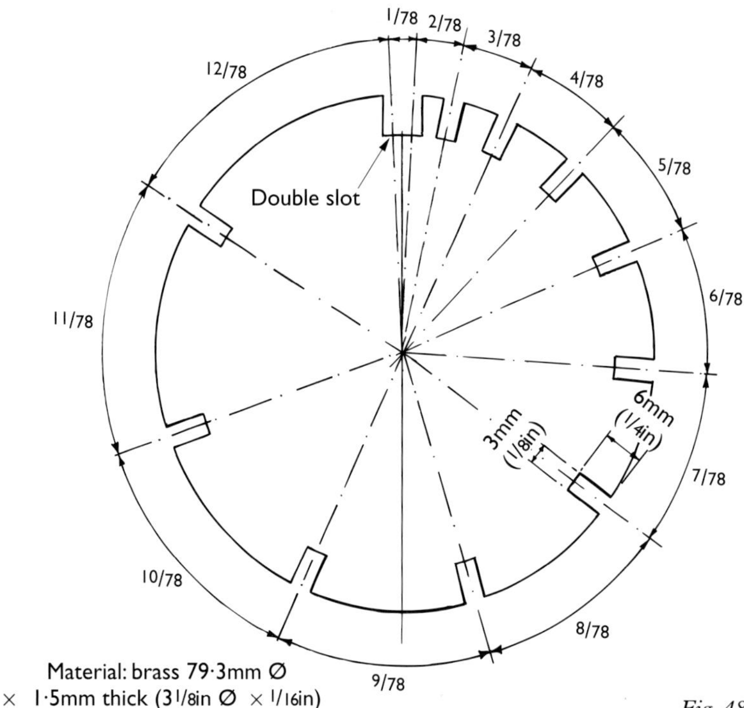

Fig. 488 Layout of divisions for a locking plate.

the lathe, as shown, or on a milling machine. The layout is not difficult, but a few notes on the methods used may be of interest when carrying out this type of operation.

There are three basic types of locking plate, as illustrated in Figure 486: 'a' has straight slots and is usually found on lantern clocks and early provincial longcase clocks; 'b' has slots with one straight side and one tapered, to enable the locking lever to ride up out of the slot; 'c' has steel pins in place of the slots. Another variation of this has seventy-eight ratchet teeth on the outside of the locking plate and is advanced by a gathering pallet, rather than the normal method of a pinion driving a wheel attached to the inside of the locking plate.

There are eleven slots in type 'a' locking plate; one is a double slot. As there are seventy-eight blows or strikes in twelve hours, the locking plate is divided out in this manner.

A brass blank is required; in this instance it was 79.3mm (3.125in) × 1.5mm (0.062in) thick. Mount the blank onto a wood chuck or backplate with wood screws; the fixing holes can be positioned so that later they can be used for rivetting the locking plate to its driving wheel. Alternatively, the locking-plate blank can be mounted on the centre hole, but ensure there is adequate means of support before any cutting takes place.

Set up the division plate or dividing head for seventy-eight divisions, and mark out as shown. Remove the chuck from the lathe and hatch in the slots to be cut; this will make it easier to see when cutting takes place. Re-mount the chuck and position the work. A fly cutter is required, 3mm (0.125in) wide. Set the depth of cut. Note the rotation of the cutter, which tends to force the work against the supporting wooden backplate. With a slow feed and the spindle revolving at 4,000rpm, one pass only is required with the

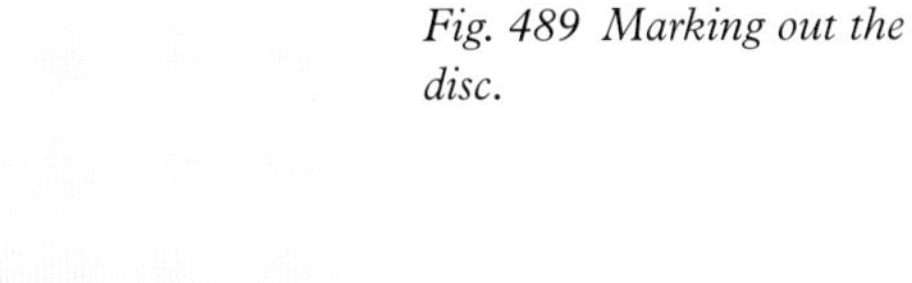

Fig. 489 Marking out the disc.

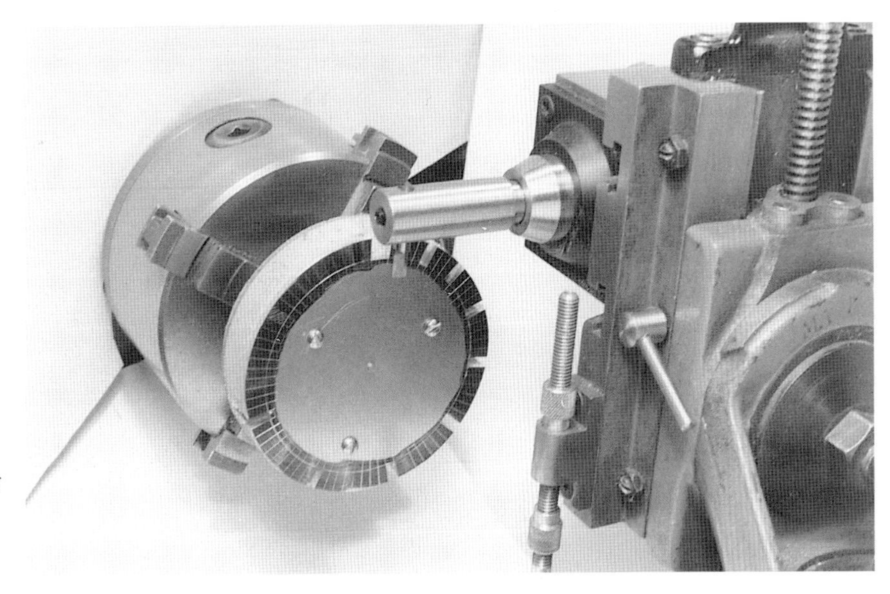

Fig. 490 Cutting a locking plate or count wheel in the lathe with the milling spindle.

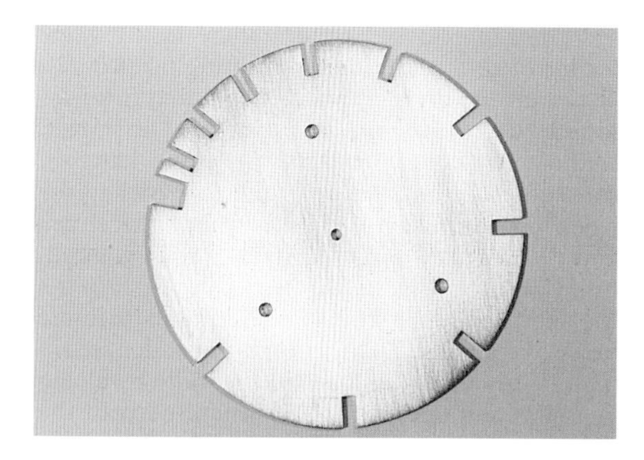

Fig. 491 The completed locking plate.

fly cutter. Index until all slots have been cut. Note the double-width slot previously mentioned; two cuts are taken side by side to produce this.

Type 'b' locking plate is dealt with in a similar manner, except the fly cutter is ground to match the tapered slot. Type 'c' locking plate is cut using methods as previously described when cutting escape or ratchet wheels. Set up for cutting seventy-eight teeth; the pins are positioned by dividing out in exactly the same way as for type 'a' locking plate.

FRENCH COUNT WHEEL OR LOCKING PLATE

As with the previous English system, the French locking plate will revolve once in twelve hours, though in this instance, half hours also are struck. Thus an additional twelve blows are struck, making 78 + 12 = 90.

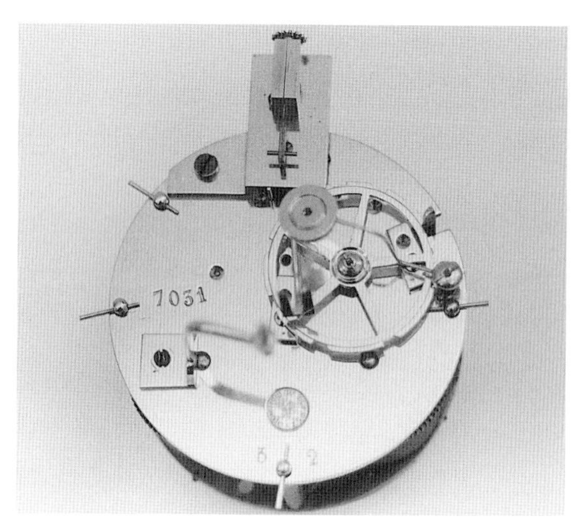

Fig. 492 French clock movement with locking plate striking.

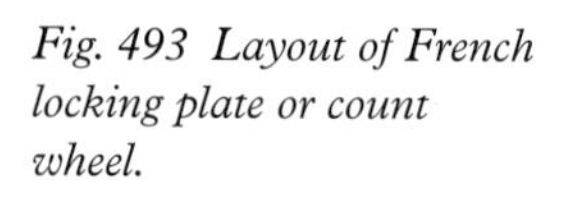

Fig. 493 Layout of French locking plate or count wheel.

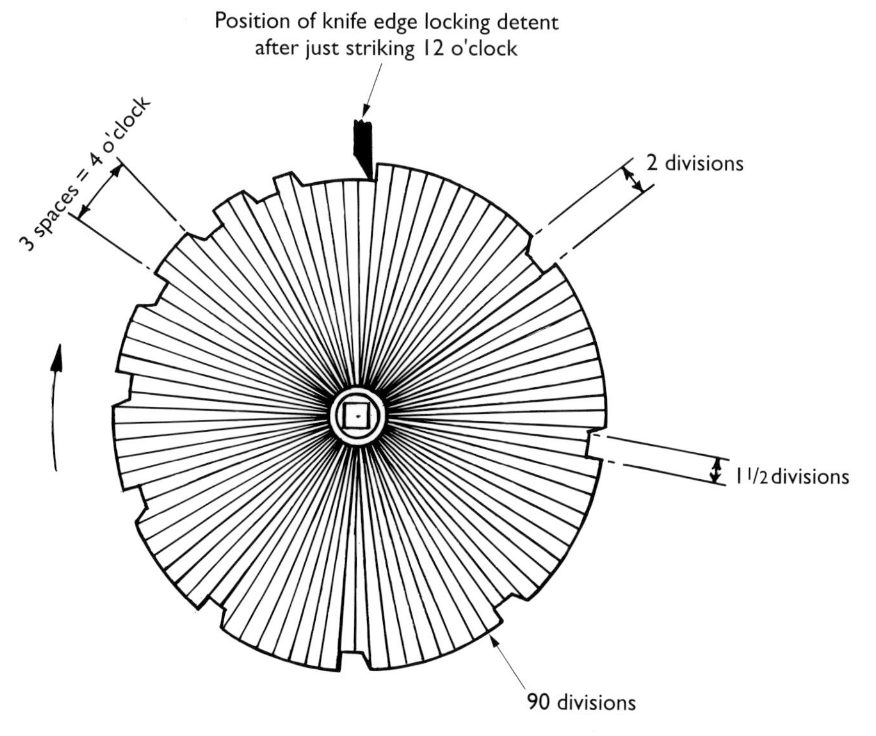

Fig. 494 Completed locking-plate wheel.

Fig. 495 Geneva stop work fitted to Swiss pocket-watch bar movement.

So when dividing out the plate, ninety divisions are required. Note the slots are cut with a slope, to assist the knife-edge locking detent to rise when the clock commences to strike each hour. The illustration shows the general layout and the position of the knife-edge locking detent just after striking twelve o'clock. There has to be a slight amount of clearance between the side of the locking detent, as shown. This also applies to the tapered side of the detent, when it is in its rest position after striking the half hour.

GENEVA STOP WORK

Geneva stop work, or the Maltese Cross, is used on high quality carriage clocks, higher grade pocket watches, and music boxes. It is sometimes found on good quality speciality clocks made in the Victorian period.

Its function is to prevent over-winding, and it allows the centre part of the mainspring with the most uniform power to be transferred to the wheel train. Generally, the most common types of stop work are the ones that limit the number of turns to four, but five and six turns are also employed.

Quite often when a carriage clock is stripped down it is found that one or both of these components are either broken or missing. Figure 496 shows the action of the Maltese Cross and the finger piece when the clock or watch is being wound. The finger fits onto a square on the winding arbor, and the female piece or Maltese Cross is mounted adjacent to the finger piece on the barrel, as shown. Now referring to Fig.496 a, b, c, d, when the barrel is wound, the finger

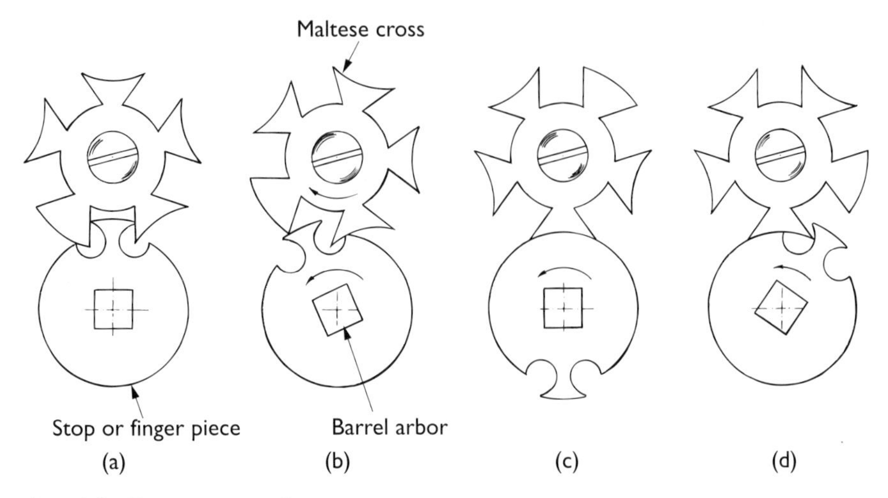

Fig. 496 The action of the Geneva stop work.

Fig. 497 Two sets of stop work as fitted to carriage clocks.

revolves anti-clockwise, as shown, and turns the Maltese Cross clockwise. Each complete revolution of the finger moves the Maltese Cross one tooth until it hits the stop as shown at 'd'; this prevents the mainspring being wound any further.

To make Geneva stop work, various calculations are required before the actual manufacturing can proceed. Assume that a stop work is required that will control four complete turns of the mainspring barrel, and the centre distance between finger piece and Maltese Cross equals 10mm (0.394in). The Maltese Cross and finger piece have a definite relationship to each other and to their centre distances.

Finger Piece

1) O/diameter of blank
   = CD × 1.20 (where CD = centre distance in mm)
2) Finished diameter of blank
   = CD
3) Diameter of holes for finger piece
   = CD × 0.20
4) Two holes are at 38 degrees to each other
5) Radial distance of holes from blank centre
   = CD × 0.448
6) Radial distance of holes from outside of blank = CD × 0.152
7) Width of finger
   = CD × 0.29

The following calculations are based on a centre distance of 10mm (0.394in):

O/diameter of blank
= 10 × 1.20
= 12mm (0.472in)

Finished diameter of blank
= CD
= 10mm (0.394in)

Diameter of holes
= 10 × 0.20
= 2mm (0.079in)

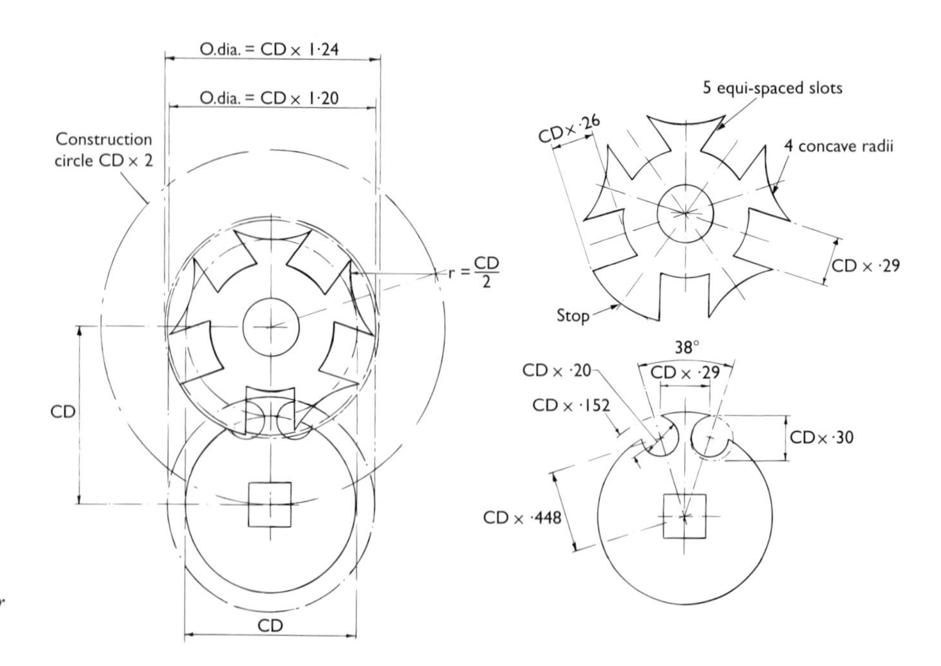

Fig. 498 Layout of Maltese Cross and finger piece – four turns.

Holes are at 38 degrees to each other from the centre line of blank:

Radial distance of holes from centre of blank
= 10×0.448
= 4.48mm (0.176in)

Width of finger at end
= 10×0.29
= 2.9mm (0.114in)

Maltese Cross

A) O/diameter of blank = $CD \times 1.24$

B) Construction circle diameter = $CD \times 2$

C) Concave circle diameter = CD

D) Width of slots = $CD \times 0.29$

E) Depth of slots = $CD \times 0.26$

F) Radius of concave = $\frac{CD}{2}$

O/diameter of blank
= 10×1.24
= 12.4mm (0.488in)

Construction circle diameter
= 10×2
= 20mm (0.787in)

Concave circle diameter
= 10mm (0.394in)

Width of slots
= 10×0.29
= 2.9mm (0.114in)

Depth of slot
= 10×0.26
= 2.6mm (0.102in)

Radius of concave
= $\frac{10}{2}$
= 5mm (0.197in)

Now that all the dimensions are available, the making of the two components can proceed.

Making the Finger Piece

First the finger piece is made: use silver steel, as the component will be hardened when complete. Machine a blank in the lathe, as shown, to a diameter of 12mm (0.472in).

Fig. 499 Drilling the finger piece.

If the radial centre distance is used, then the two holes can be positioned by angular dividing using the dividing head to set the 38°. The position of the two holes can also be set by the use of the cross-slide and vertical slide. In this instance, trigonometry will be required to calculate the horizontal and vertical centres.

By calculation, the horizontal centre distance or pitch of the two holes is 2.918mm (0.115in), and the vertical centre distance equals 4.236mm (0.167in).

Position the centreline of the milling and drilling spindle in line with the centre of the finger-piece blank. With the vertical slide, set the vertical distance of 4.236mm (0.167in), and lock the slide to prevent any movement taking place. Then with the cross-slide, position for the first hole to be drilled by moving the cross-slide half the calculated amount, i.e. 1.459mm (0.057 in). Carefully centre-drill the blank, then drill through 2mm (0.078in) diameter.

The cross-slide should then be moved 2.918mm (0.115in) to the second hole position, and drill as before.

The width of the finger equals 2.9mm (0.114in). With a small slitting saw, cut down either side of the centreline into the space provided by the drilled holes. The waste material can now be milled away using the worm and wheel attachment to produce the finished diameter of the blank, which is 10mm (0.394in). In the absence of a dividing attachment, hardened steel discs can be used as filing guides. These can

Fig. 500 Milling away the surplus material.

Fig. 501 Finger piece ready for finishing the two lobes.

be made from silver steel. Turn to 10mm (0.394in), drill a hole through the centre for a small securing screw, then harden. Drill the centre hole for the winding square, and part off. All that remains is to finish the two lobes with small radii, and produce the square hole for the barrel arbor; this can be either filed or broached.

Making the Maltese Cross

To make the Maltese Cross, select a piece of silver steel to finish at 12.4mm (0.488in). Mount the dividing head or division plate on the lathe mandrel, and set for five divisions. Note that for four turns, five slots are required. On a sixty-hole circle of a division plate, use every twelfth hole, which equals an angular movement of 72 degrees. If a 60/1 worm and wheel dividing is used, then twelve complete turns of the crank handle are required. A slitting saw is required, 2.9mm (0.114in). Mount the cutter central with the lathe spindle, and set the depth of cut to 2.6mm (0.102in); it is better to take two cuts for each slot. A slitting saw will produce a flat-bottomed slot, which is quite acceptable but not as aesthetically pleasing as a radiused slot. Alternatively, a fly cutter can be used, ground to the shape required, or hardened steel discs can be used to correct the shape of the base of the slots after the component has been parted off.

Now that the slots are complete, the concave radii can be machined. Index the lathe mandrel through 36 degrees, or six holes on a sixty-hole circle, or six turns of the crank on a 60/1 worm and wheel dividing head. This now positions the Maltese Cross blank directly between each slot ready for the concave radii to be produced. The radius required is 5mm (0.197in), therefore a cutter diameter of 10mm (0.394in) is used. This can be either a suitable end mill, or a boring bar set correctly for the radius being cut. Check to ensure the cutter is central with the work.

The bottom of the concave radius is equal to the centre distance, which is 10mm (0.394in). The depth of cut is 1.2mm (0.047in). Set the depth on the vertical slide, and machine the first radius. Complete by indexing the remaining four positions. Note the fifth position is not

Fig. 502 Slotting the Maltese Cross.

Fig. 503 The five slots complete.

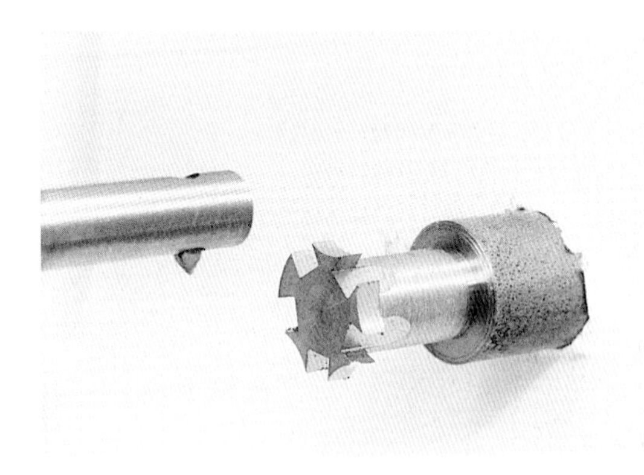

Fig. 505 Milling the four convex radii.

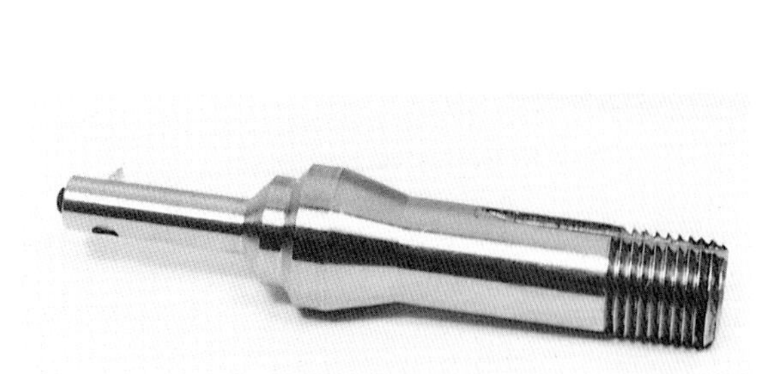

Fig. 504 Miniature fly cutter.

Fig. 506 The Maltese Cross complete, ready for parting off.

Fig. 507 Stop piece and Maltese Cross complete.

machined, as this forms the stop. To complete, drill the centre hole and counter-bore for the fixing screw. Before hardening, check that both the finger and the Maltese Cross work well together, and that the items are well polished and all burrs and sharp edges removed.

CONSTRUCTION OF A FIVE-TURN GENEVA STOP WORK

The geometry is shown below for a Geneva stop-work design for five turns of winding. In this case, six slots are required. As can be seen, the proportions are different to the previous example and the following formulae are used:

Finger Piece

1) O/diameter of blank
   = CD × 1.16
2) Finished diameter of blank
   = CD
3) Diameter of holes for finger piece
   = CD × 0.175
4) Two holes are at 30 degrees to each other
5) Radial distance of holes from centre of blank = CD × 0.484
6) Radial distance of holes from outside of blank = CD × 0.096
7) Width of finger at end
   = CD × 0.25

Maltese Cross

A) O/diameter of blank
   = CD × 1.19
B) Construction circle diameter
   = CD × 2

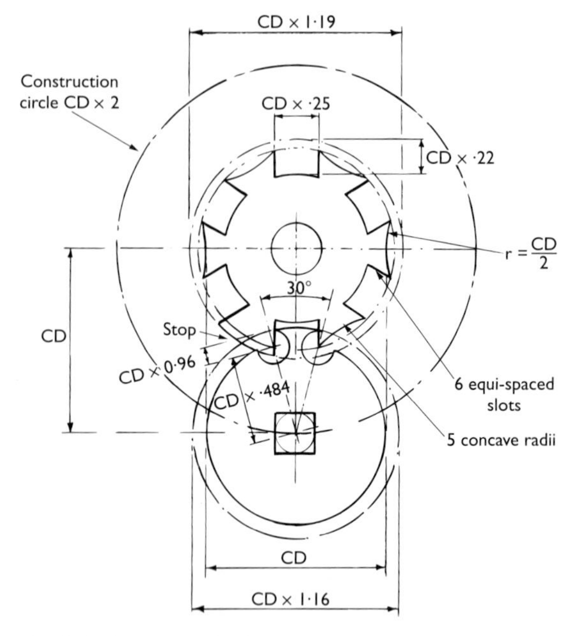

Fig. 508 Layout of the Geneva stop work, giving five turns.

C) Concave circle diameter
   = CD
D) Circle enclosing points of concave arms
   = CD × 1.10
E) Width of slots = CD × 0.25
F) Depth of slots = CD × 0.22
G) Radius of concave = $\frac{CD}{2}$

When a clock or watch is stripped down for repair, it is more likely that the finger piece is either missing or broken when the stop work is examined. If careful measurements of the Maltese Cross are taken and the methods described previously are used, then a suitable matching component can be produced.

Setting up Stop Work

Quite often the reason that parts of the stop work are missing when a clock is stripped down is that the previous repairer did not know how to set it up correctly. The following few notes should assist if this problem is encountered.

With the stop work disengaged, wind the mainspring fully, carefully let it down, then wind it fully once more. Each time count the number of complete turns. If this is six and the stop work on the movement is for four turns, that means there are two turns spare. Now let the mainspring down completely, then wind up one turn, and set the stop work in position with the convex stop piece on the Maltese Cross butting against the shoulder of the finger piece (*see* Figure 496).

If the spring is now fully wound, only four complete turns will be allowed due to the stop work being in operation. There is now one spare turn of mainspring at the beginning of the wind and one spare turn at the end. This allows the centre part of the mainspring with the most uniform power to be utilized.

This section on the design and making of Geneva stop work was based on an excellent article by Archie B. Perkins that appeared in the *Horological Times* (USA) January/February 1988.

CUTTING THE MORE UNUSUAL ESCAPE WHEELS

Brocot Dead-Beat Escape Wheels

A Brocot escapement is normally found on the better quality French clocks and is often mounted on the front of the dial and classed as a visible escapement. It is also found on cheaper German and American mass-produced clocks. It is a dead-beat escapement, similar to the Graham type shown previously, though whereas the Graham escapement normally has solid pallets, the Brocot has jewelled pallets shellaced in place; these are normally semi-precious stones such as red Cornelian.

Fig. 509 Close-up of the Brocot escapement.

When the wheel is cut, a similar process to cutting a dead-beat or recoil escape wheel is used, except that two cutters of different shape are employed. Referring to the drawings overleaf, it will be seen that the first cut is a shallow one which cuts the radius at the back of the tooth, and the second cut is down to the root; use backing washers to support the work whilst cutting takes place. Ensure the fly cutters are sharp and have a good finish. Some wheels have been noted that have escape-wheel teeth with the acting face radial with the centreline. Most, however, are undercut, as is the Graham escape wheel.

Chronometer Escape Wheel

A ship's chronometer by Parkinson & Frodsham is shown, complete in its gimballed brass-bound box. Also shown is the movement. This is a two-day chronometer; others ran for eight days,

Fig. 510 French clock movement with visible Brocot escapement.

Fig. 511 Completed escape wheel.

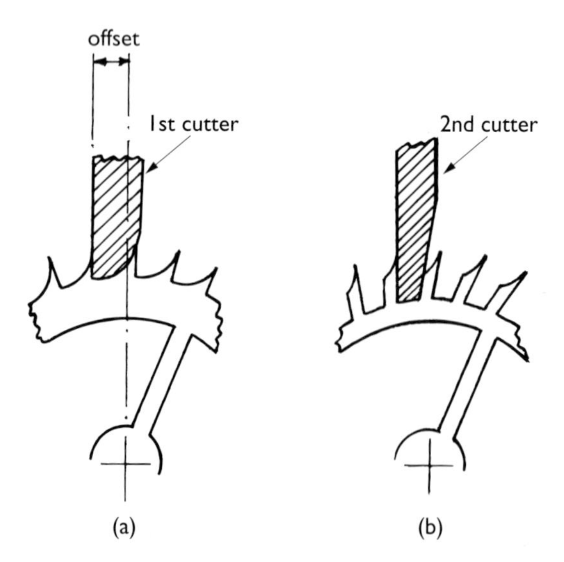

Fig. 512 Method used in cutting Brocot escape wheel.

though these are more rare. Chronometers were manufactured until and during World War II; Mercers of St Albans, England, and Hamilton of the USA produced many thousands for navigation purposes for the war effort.

A chronometer is an extremely accurate large watch, and components get worn and damaged and require replacement. One of the main problems is the escape wheel and detent. The escape

Fig. 513 Ship's two-day chronometer.

Fig. 514 Chronometer movement.

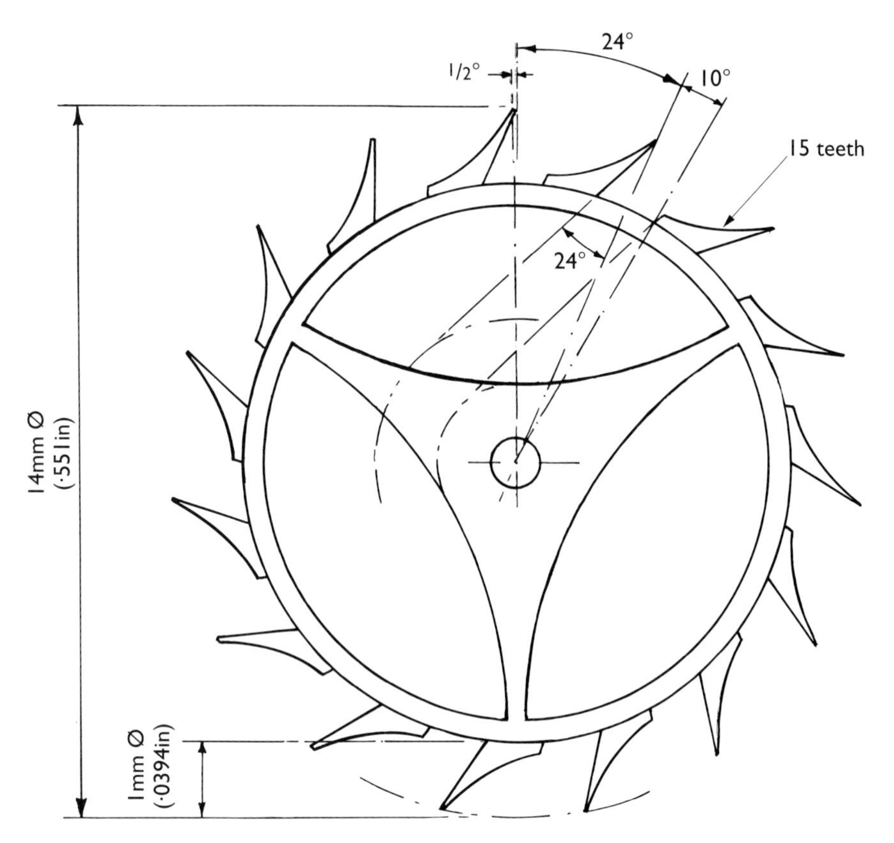

Fig. 515 Geometry of chronometer escape wheel.

wheel on the Hamilton 21 ship's chronometer was particularly susceptible to wear.

A chronometer escape wheel can be cut on the lathe or milling machine using the techniques previously described, though the geometry is more complicated than escape wheels covered before. If care is taken, however, and separate operations are broken down, a satisfactory job can be achieved.

A typical chronometer escape wheel would have fifteen teeth and an outside diameter of 14mm (0.562in), and a tooth height of 1mm (0.0394in); the undercut is 24 degrees, to ensure only the tip of the tooth is in engagement. This is a good average, but can vary between individual makers. The width of the tip of the tooth is equal to ½ degree. Generally two cutters are required, and these would be different for each escape wheel. Sometimes it is possible to use the radius cutter again, but the gashing cutter has to have a concave radius to match the rim, which is the root of the tooth. Fly cutters are ideal for the cutting of chronometer escape wheels. There are no commercial cutters available for this type of work, and individual companies made up tooling to suit their requirements; an escape wheel cutter used by Mercers to produce escape wheels for their own chronometers is shown.

Before commencing any machining of a replacement wheel, it is essential that a large-

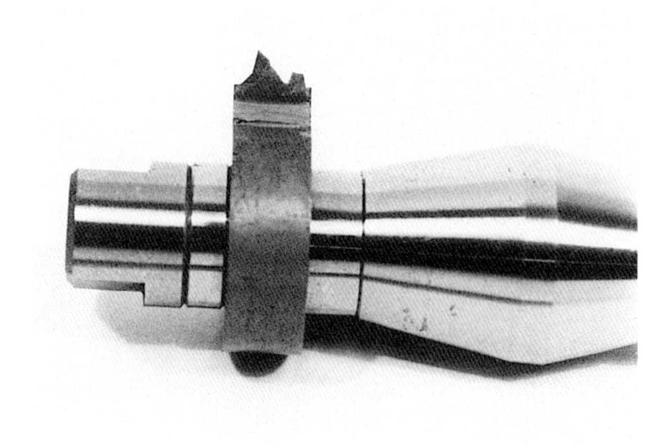

Fig. 516 Mercer fly cutter for cutting chronometer escape wheel.

scale drawing of the tooth form is produced; this will assist when shaping the two fly cutters. It is possible to grind the profile of the cutter to an existing wheel, but this is not easy, and when a cutter has been produced, trial cuts should be taken on a scrap wheel blank so that any errors can be corrected before attempting to cut the actual escape wheel. The lapping tool shown previously in the chapter on cutter making is ideal for shaping and finishing the cutter profile. The width of the cutter tip is determined by the 10 degree angle shown on the drawing of the wheel teeth. This is not important, but an angle that gives a nice proportion to the tooth should be used. It is essential that all cutting edges have relief and are extremely sharp and polished. This, in turn, will produce a good finish to the work. Once the cutters are complete, the escape wheel blank can be machined.

The material for the wheel should be hard brass. Mercers prepared their wheel blanks by work-hardening the metal with the use of a 20-ton press. Drawn brass bar gives good results as the manufacturing process work-hardens the material. The wheel blank is machined to the correct outside diameter and the recess bored; this profile is to form a wide, strong tooth tip and produce an escape wheel that is relatively light.

Once the blank is complete, the milling spindle and vertical slide is fitted to the lathe. The first cut is taken by the gashing cutter. The cutter is set central, and then moved over to produce the 24 degree undercut. For this dimension a calculation is required, as shown; desk-top or pocket calculators usually have functions to cover simple trigonometry. To calculate distance '*x*', the offset, the radius of the escape wheel blank is required, together with the angle:

Angle = 24°

radius = 7mm (2.76in)

Offset x = SINE 24° × 7mm

Offset x = 0.4067 × 7

Offset x = 2.85mm (0.112in)

Set the dividing head or division plate on the lathe spindle for fifteen divisions, and lower the cutter until it just touches the outer diameter of the escape-wheel blank. Whilst the depth of cut can be calculated, this is not necessary, as all that is required is to cut down until the cutter breaks through into the recess or counter-bore.

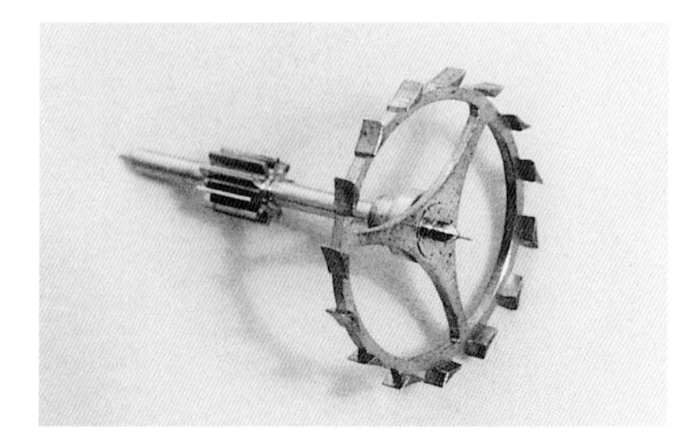

Fig. 517 Mercer chronometer escape wheel and pinion.

Fig. 518 Gashing fly cutter and arbor.

Fig. 519 Radiused fly cutter and arbor.

Index fifteen times; now the cutter can be changed for the second cut. This is to form the radius at the back of the tooth. It is necessary to move the cross-slide into a new position: take a number of trial cuts until a small land is produced, as shown in the drawing. The tip of the tooth is equal to ½ degree. A small land is also

Fig. 520 Machining the escape-wheel blank.

Fig. 522 Gashing the blank.

Fig. 521 Using the centring micrometer to set the gashing cutter.

Fig. 523 Fly-cutting the radius at the back of the tooth.

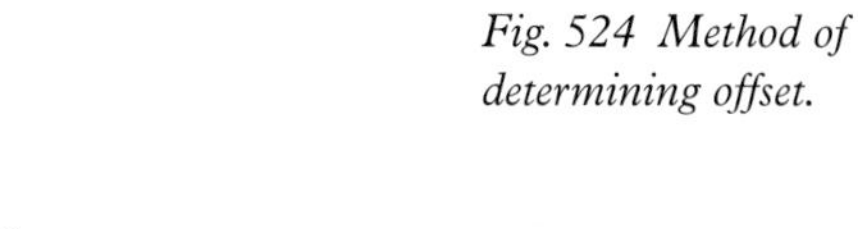

Fig. 524 Method of determining offset.

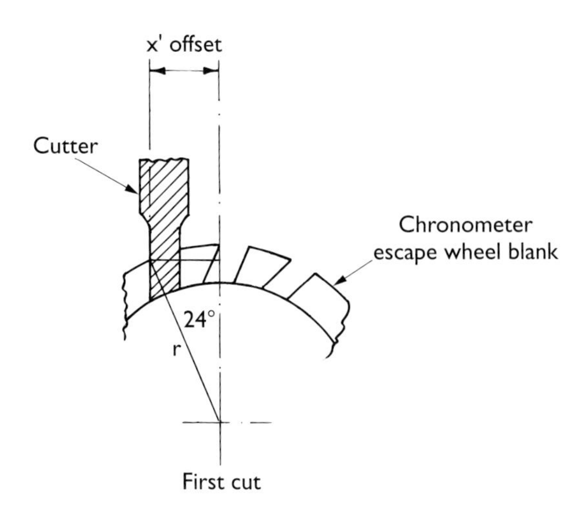

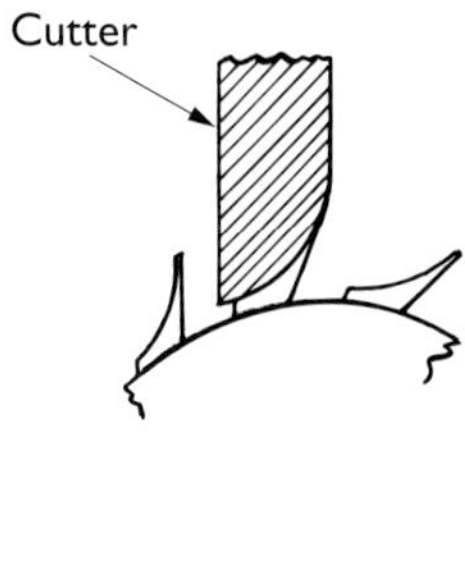

Fig. 525 Taking the last cut.

required at the root of the tooth adjacent to the rim.

An alternative method that will give added support when cutting the teeth is as follows. Prepare the blank without the recess, cut the escape-wheel teeth as before, then complete by forming the counter bore or recess. With this method it is more difficult to determine the depth of cut, and calculations are required. As the radius cutter removes a large amount of material, there is a tendency to bend the acting face of the tooth very slightly. To overcome this, take a light cut down the face of the tooth as a final operation.

Whilst the wheel is still mounted in the lathe drill for the collet, take care to ensure the hole is true, then part off. Normally the wheel has three crossings or spokes, see Fig. 336. The mounting of the wheel onto its collet is shown in Fig. 376.

Fig. 526 The completed escape wheel ready for parting off.

Finally all that is required is to cross out the spokes.

Duplex Escape Wheel

The duplex escape wheel has two sets of teeth: the long teeth are for locking, and the short teeth that are cut on the side of the wheel provide the main impulse. For constructional purposes this is two escape wheels in one, which means that it is therefore quite a difficult wheel to cut.

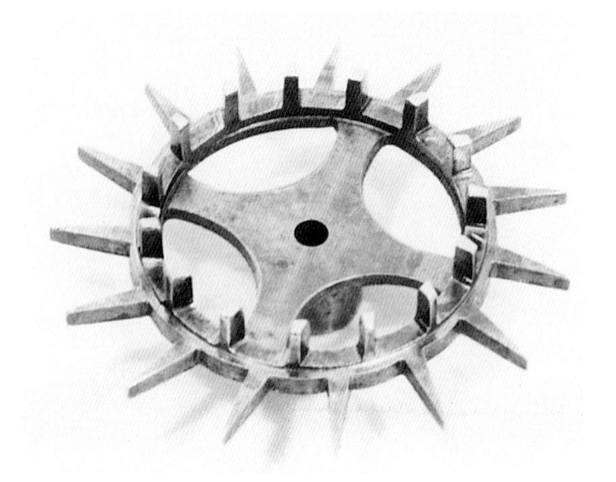

Fig. 527 Duplex escape wheel.

A blank is prepared to the correct size, assuming a sample is available to work to. *Watchmaking* by G. Daniels gives details and methods of calculating the various angles of the teeth; as with the chronometer escape wheel, there are generally fifteen teeth on a duplex wheel. Fly cutters are required to carry out the work. Only a brief description will be given, as it is unlikely that this type of escape wheel will be encountered very often.

As the teeth are very long and slender, it is essential that when cutting, two cuts are taken to produce the space, as a large amount of metal is being removed. Once the outer teeth are complete, the teeth on the side of the wheel can be attempted. Another fly cutter is required, as shown, and the vertical slide is employed to traverse the cutter through the work. The teeth are cut radially, and it is necessary to offset the cutter to produce the correct tooth form.

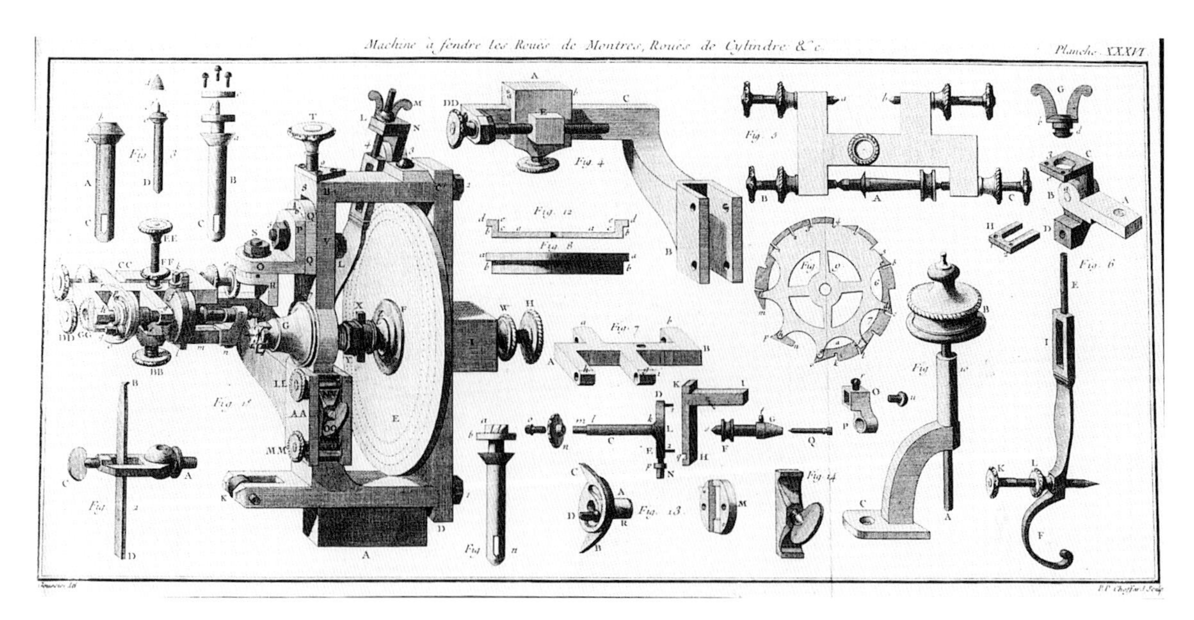

Fig. 528 Machine for cutting cylinder escape wheels – Ferdinand Berthoud, 1763.

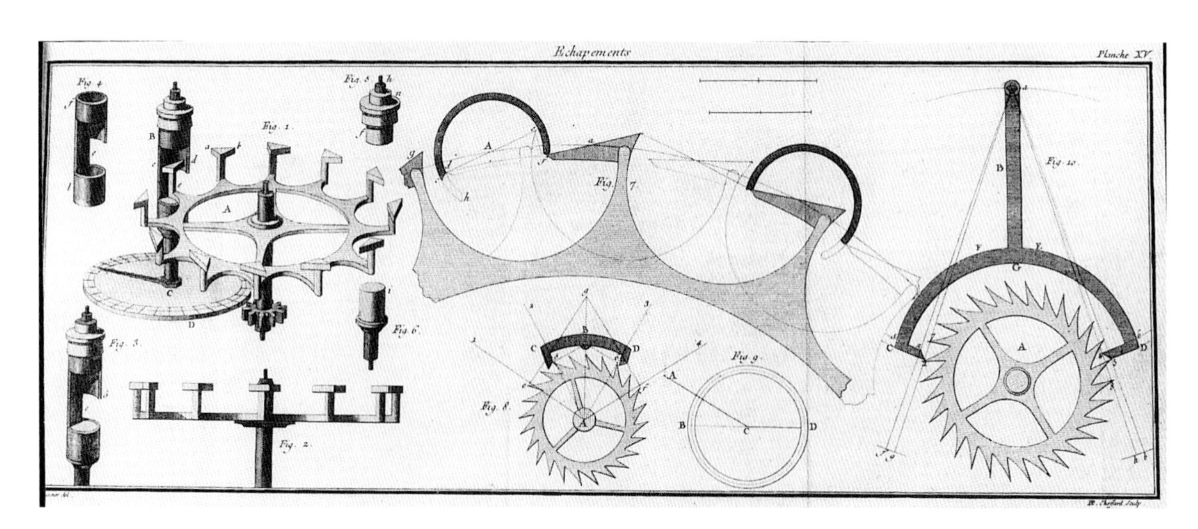

Fig. 529 Cylinder escape wheel and cylinder – Ferdinand Berthoud, 1763.

Cylinder Escape Wheel

Two plates are reproduced from Ferdinand Berthoud's *Essai sur l'horlogerie* of 1763: one shows the machine with the escape wheel mounted, the other shows the completed wheel. To produce a cylinder escape wheel such as this is outside the scope of this book – again, readers should be referred to G. Daniels' book *Watchmaking* for the geometry of this escapement.

CALCULATIONS FOR MISSING WHEELS AND PINIONS

Going Train

Sometimes it is necessary to replace a missing wheel or pinion in a train of wheels where one or the other is missing. Before proceeding with these calculations, it may be of interest to study the following examples of complete trains for various timepieces.

Fig. 530 Eight-day longcase clock.

Fig. 531 Simple longcase train – going side.

If we first consider a standard eight-day longcase-clock going train, the following is a typical count for this type of clock with a one-second pendulum, that is a pendulum with an effective length of 1m.

Note: for this calculation we do not include the great wheel or centre pinion.

Great wheel = 96 teeth

Centre wheel = 60 teeth/pinion of 8 leaves

Third wheel = 56 teeth/pinion of 8 leaves

Escape wheel = 30 teeth/pinion of 7 leaves

Each tooth of the escape wheel acts twice, once on each pallet face, therefore the equation is:

$$\frac{60 \times 56 \times 30 \times 2}{8 \times 7} = 3{,}600 \text{ beats per hour}$$

A 1m pendulum therefore makes one beat every second.

If we now take the train of an English dial clock, a typical train would be:

Great wheel = 96 teeth

Centre wheel = 84 teeth/pinion of 8 leaves

Fourth wheel = 80 teeth/pinion of 8 leaves

Escape wheel = 30 teeth/pinion of 7 leaves

$$\text{Then } \frac{84 \times 80 \times 30 \times 2}{8 \times 7} = 7{,}200 \text{ beats per hour}$$

The pendulum in this clock therefore beats half-seconds.

A further example is for a clock that instead of having a pendulum, is fitted with a balance or platform escapement, for example a carriage clock or a ship's chronometer.

Fig. 532 English dial wall clock.

Fig. 533 French striking carriage clock.

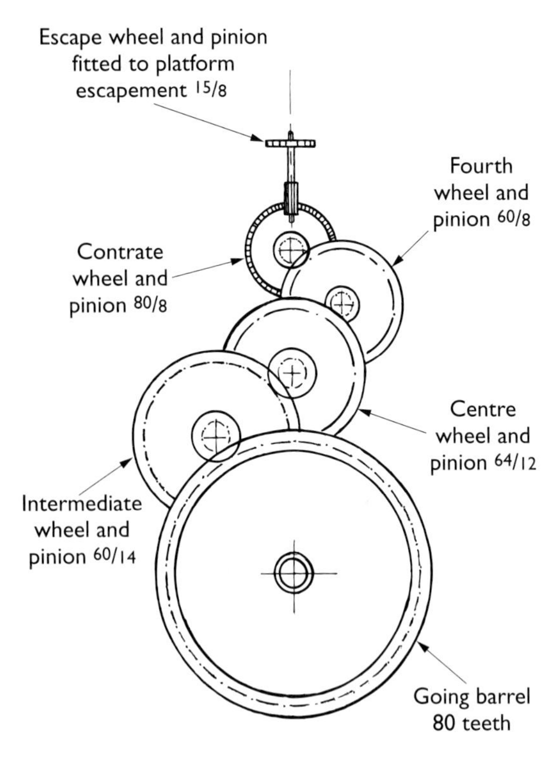

Fig. 534 Carriage clock going train.

We will now consider a train from a French carriage clock that normally has an additional wheel and pinion:

Barrel	= 80 teeth
Intermediate wheel	= 60 teeth/pinion of 14 leaves
Centre wheel	= 64 teeth/pinion of 12 leaves
Fourth wheel	= 60 teeth/pinion of 8 leaves
Contrate wheel	= 80 teeth/pinion of 8 leaves

The platform escapement has fifteen-tooth escape wheel and eight-leaf pinion, then:

$$\frac{64 \times 60 \times 80 \times 15 \times 2}{8 \times 8 \times 8}$$

$$= 18{,}000 \text{ beats per hour}$$

This is a fairly standard train for this type of clock. Many pocket watches also have the same number of beats per hour.

Ship's chronometers are different due to there being two beats of the balance for each tooth of

Fig. 535 French carriage clock showing platform escapement.

the escape wheel; this is classed as a single beat escapement. The escape wheel advances only with each alternate vibration of the balance. The fourth wheel revolves once per minute, and its pivot is extended through the dial to show dead half-seconds.

The following example is a typical train for a standard two-day ship's chronometer:

Great wheel	=	90 teeth
Centre wheel	=	90 teeth/pinion of 14 leaves
Third wheel	=	80 teeth/pinion of 12 leaves
Fourth wheel	=	80 teeth/pinion of 10 leaves
Escape wheel	=	15 teeth/pinion of 10 leaves

Then:

$$\frac{90 \times 80 \times 80 \times 15 \times 2}{12 \times 10 \times 10} = 14{,}400 \text{ beats per hour}$$

Note, if an electronic timer is used, due to the dummy vibration or passing tick, the instrument may show only half the number of beats per hour, i.e. 7,200.

Now that a basis has been laid down, it is possible to go ahead using the preceding formulae, and calculate for missing wheels and pinions.

If a longcase-clock movement is first considered with the following train, with the third wheel (w) missing:

$$\frac{60 \times w \times 30 \times 2}{8 \times 7} = 3{,}600$$

$$\text{Missing third wheel } w = \frac{3600 \times 8 \times 7}{60 \times 30 \times 2}$$

$$= 56 \text{ teeth}$$

In the second example, both wheel and the pinion (p) it drives are missing;
Then:

$$\frac{60 \times w \times 30 \times 2}{p \times 7} = 3{,}600$$

$$7p = \frac{60 \times w \times 30 \times 2}{3600}$$

$$\therefore \quad 7p = w$$

$$\therefore \quad \frac{w}{p} = 7 \quad \text{Ratio of } 7/1$$

Referring back now to the previous calculation, a typical wheel would be fifty-six teeth driving a pinion of eight leaves, or a forty-nine tooth wheel driving a seven-leaf pinion is also possible. Both these combinations would give the ratio of 7/1.

Where a complete arbor is missing, the pinion and wheel are considered separately. The wheel driving the missing pinion exists, as does the pinion driven by the missing wheel, and these are two separate calculations, as already covered.

MISSING PENDULUMS

We have seen that a longcase clock with a 1m pendulum will beat seconds for every tooth of the escape wheel. Then if the vibrations are known, the length of the pendulum can be calculated.

The formula for the law of the pendulum is as follows, for one single beat with a 1sec pendulum.

T = Time for one beat in seconds

l = Length of pendulum in inches

g = Force of gravity, taken as 32.2ft/sec^2

$$T = \pi\sqrt{\frac{l}{g}}$$

T = 1sec

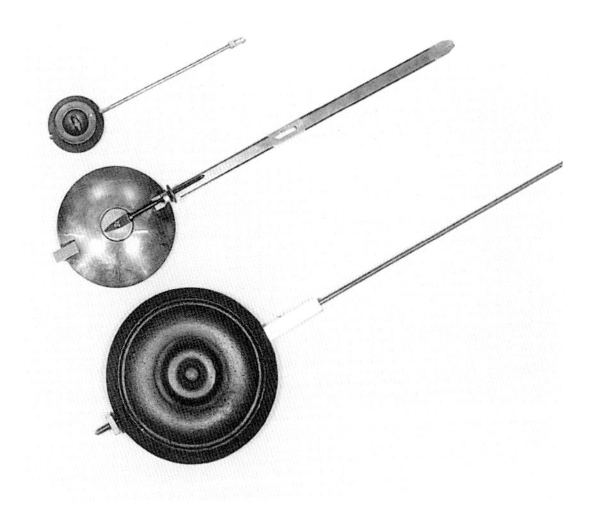

Fig. 536 Typical bob pendulums.

$$\therefore 1 = 3.142\sqrt{\frac{l}{32.2}}$$

$$\therefore 1^2 = \frac{(3.142)^2 l}{32.2}$$

$$\therefore l = \frac{32.2 \times 12}{9.872}$$

(Note the '12' in the formula is to convert feet to inches)

$$\therefore l = 39.14\text{in or } 1{,}000\text{mm (1m)}$$

To find the theoretical length of a pendulum that is missing, a count of the train is required, then apply the previous formula to find the vibrations or beats per hour. From this, the required length of pendulum can be calculated.

Using a constant of 140,904, divide this by the square of the number of beats per minute.

Example, to find the theoretical length of a pendulum beating 7,200 vibrations per hour, then

$$\frac{7200}{60} = \text{beats per minute} = 120$$

$$\therefore \frac{140904}{120 \times 120} = 9.78\text{in (248mm)}$$

From this information, a table has been prepared for various lengths of pendulum and their vibrations – see Appendix.

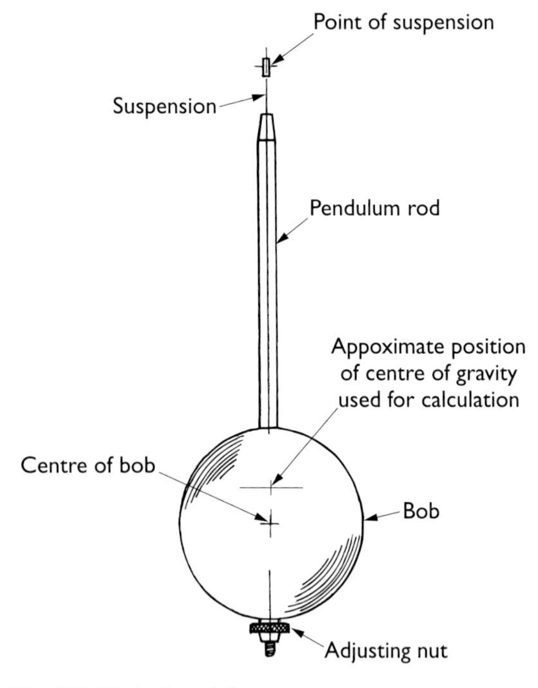

Fig. 537 Typical pendulum.

The information alongside the wheel-train calculations is most useful when bringing a clock to time using one of the modern clock-rate timers, such as the Microset or Timetrax machines.

STRIKING TRAINS – REPLACING MISSING WHEELS AND PINIONS

The following considerations refer to 'period' English clocks, but the basic principles apply to many other types.

Count-Wheel Striking

Count wheels are often referred to as locking plates, but here the term 'count wheel' is used mainly so as not to confuse it with the locking wheel – which, of course, is part of the main wheel train (also referred to as the 'hoop wheel' when applicable). The expression 'count wheel' often refers to both the count plate itself and the gear fixed to it to drive it as one unit.

Let us look first at thirty-hour longcase movements and lantern clocks. On very early balance-wheel lantern clocks, the count-wheel gear is driven by a four-leaf pinion formed on the end of the extended arbor of the great/pin wheel. Later a separate pinion was used, which was fitted on

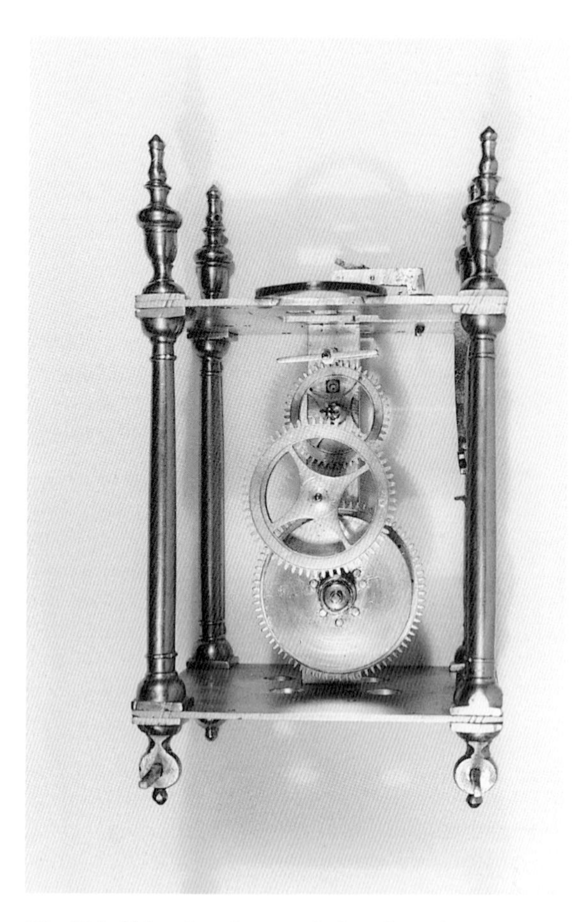

Fig. 538 Thirty-hour lantern clock strike train.

Fig. 539 Seventeenth-century lantern clock showing four-leaf pinion.

Fig. 540 Eight-leaf pinion fitted to a lantern clock.

the squared end of the arbor and held in place by the count-wheel plate, which is larger in diameter than the wheel fixed to it. One of the most common problems is that this pinion sometimes gets lost when the count wheel is removed when repairs are carried out. As discussed when covering the making of the locking plate, seventy-eight blows on the bell are required in twelve hours. For each pin on the great wheel, and therefore each blow on the bell, the count wheel must advance 1/78 revolution. If, as is frequently the case, the great wheel has thirteen pins, one revolution will represent thirteen blows on the bell, and we require one revolution of the count wheel for seventy-eight blows: so the great wheel has to turn six times to turn the count wheel once, that is a reduction ratio of 6:1 is required. In this instance a pinion of eight driving a count-wheel gear of forty-eight is the most common arrangement. Similarly, a great wheel with twelve pins requires a reduction ration of 78:12, or

Fig. 541 Locking plate or count wheel with matching pinion.

6.5:1, and a pinion of six driving a count-wheel gear of thirty-nine is normal.

Sometimes a count wheel has pins instead of slots, or is a casting with the slots projecting out sideways, in which case there are seventy-eight teeth on the outer edge of the count wheel, one tooth per blow on the bell. In this instance, the driving pinion will have the same number of leaves as there are pins in the great wheel. When a count wheel has been lost and a visit to the spares box produces another, it is vital to check the gear as well – it is not unknown for the replacement to have the wrongly numbered wheel or pinion, causing the count to get progressively more and more out of step and, on meeting this problem, it can be quite puzzling until it is realized that the clock has a non-original component. The following formula can be used for this arrangement of counting:

$$\frac{78}{\text{No. of pins}} = \frac{\text{Count-wheel teeth}}{\text{pinion}}$$

For example, assume both the small pinion and locking plate are missing:

The striking train has a pin wheel with thirteen pins:

$$\text{Then } \frac{78}{13} = 6 \qquad \text{Ratio of 6/1}$$

A suitable wheel would be sixty and a pinion of ten or, alternatively, a forty-eight tooth wheel with a matching pinion of eight. Either would suit, whichever fitted into the available space.

Moving on to eight-day clocks, the very earliest longcase clocks had a count wheel high up on the backplate, usually driven by a pinion of four formed on the rearwards extension of the pin wheel arbor, as is also found in early lantern clocks. Since this is a similar arrangement to the thirty-hour clock, the above formula can be used, substituting the pin wheel (usually with eight pins) for the thirty-hour great wheel. Shortly after this, the count plate (or locking plate as it is often called) was still outside the backplate but attached to an extension of the great wheel arbor. This meant that the arbor had to be in two halves, the great wheel being fixed rigidly to one

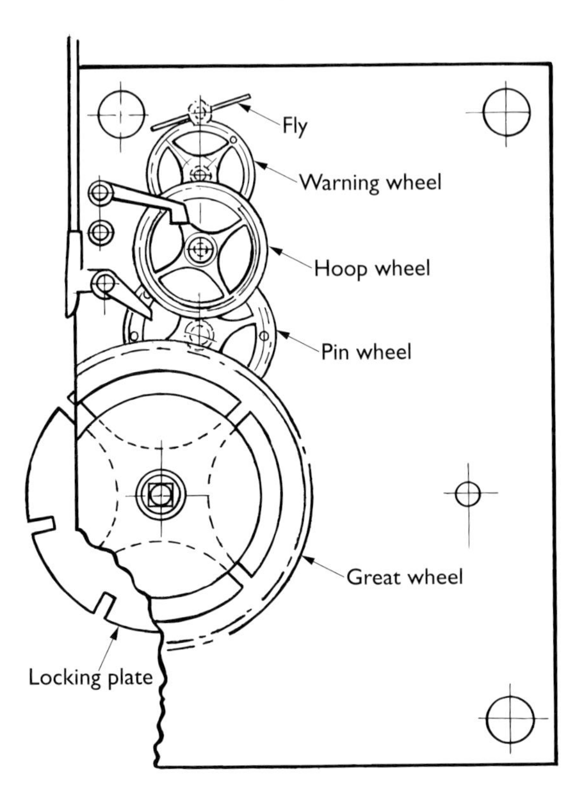

Fig. 542 Striking train with locking plate.

Fig. 543 Striking train from an eight-day longcase clock.

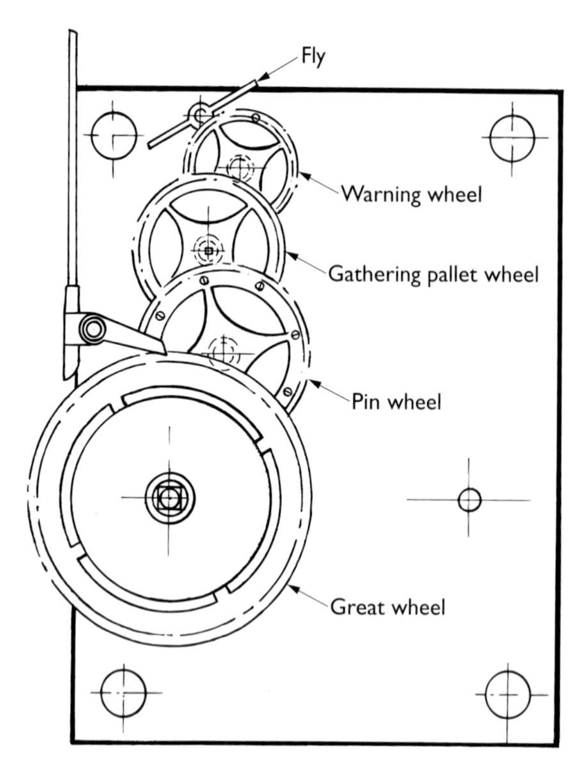

Fig. 544 Rack-striking train.

half that had an inward-facing stub rotating inside the barrel which, in turn, was mounted on the front half of the arbor. This complication was soon overcome by mounting the count plate directly on to the great wheel like a concentric ring which, in turn, was mounted in the manner with which we are familiar, loose on the arbor and held there with a keyhole disc. In either case, the great wheel itself has seventy-eight teeth to match the count plate, one tooth of the wheel then equalling one stroke of the bell, effectively removing the need for a pinion and wheel to drive the count plate.

On a month clock with a 4:1 reduction from the great wheel, the locking plate was mounted outside the backplate on an extension of the intermediate wheel and, thereafter, it was just like an eight-day clock. In other cases, it was again often driven from an extension of the pin wheel, and the above formula would apply.

MISSING ARBORS WITH THEIR WHEELS AND PINIONS

Irrespective of the going period or whether it is a count wheel or rack control, with a striking train it is vital that the locking and warn wheels always stop in the same place. This creates the golden rule that from the pin wheel up to the warning pinion, the wheel count must *always* be an exact multiple of the pinion it drives. For missing arbors, a little detective work is required. Since it is difficult to devise an easy formula for this, the following examples should help.

Assume that a thirty-hour clock is minus the next arbor up from the great wheel. Note with thirty-hour clocks that the pinions invariably have six leaves. This means that the great wheel must have six teeth between each pin to turn the hoop/locking wheel one exact revolution. At the other end, the warning-wheel pinion being six means that the hoop wheel must have a multiple of six teeth, that is, forty-two or forty-eight or fifty-six and so on. It will probably be forty-eight, but it can be calculated using the centre distance and the pinion dimensions as described elsewhere in this book. The wheels and pinions of old clocks are frequently theoretically incorrect, and an over- or undersized pinion could affect the calculations and result in a wheel count one or two teeth out from that required. However, since it is known that the wheel is a direct multiple of the pinion, it would be near enough to establish which wheel count is the correct one and to adjust accordingly, whilst keeping the same pitch diameter.

As another example, if the pin-wheel arbor of an eight-day count-wheel clock is missing, using centre distances and the known wheel count of the great wheel, it is possible to establish the pinion count (eight on most clocks). This will, in turn, tell us the number of pins on the pin wheel, as we know that one tooth of the great wheel is one count, so the number of pins must be equal to the number of leaves on the driving pinion.

Next, note the count of the next pinion up on the hoop/locking arbor (or gathering pallet arbor if a rack-striking clock). This will determine the pin-wheel count, as the number of teeth between the pins must be the same as the driven pinion in order to turn the hoop/locking arbor one complete turn, that is the tooth count equals the number of pins × the pinion count, so if there are eight pins and a pinion of seven on the next arbor, the pin wheel will have fifty-six teeth. With a rack-striking clock it is not vital that the pinion on the pin-wheel arbor matches the number of pins on the pin wheel, but it always does!

SOME COMMENTS ON STRIKING WHEEL TRAINS

Where there is more than one arbor missing, the rule about wheels being multiples of the driven pinion still applies, but more has to be done in the way of mathematics and speculation. The accuracy of speculation has to be based on experience with clocks of the period. Very few clock books contain precise details of trains, though one exception is *English 30-hour Clocks* by Darken and Hooper, which contains a complete list of combinations of all thirty-hour wheel counts for all periods, together with much other useful information for the repairer/restorer.

Consider now trains of longer duration clocks, and the following is offered as a guide. Working up the train, the wheels and pinions must get gradually smaller, and the earlier the clock, the more exaggerated the reduction. On early clocks, much more use is made of six-leaf pinions than on later ones. By the nineteenth century, six-leaf pinions had virtually disappeared on all but thirty-hour clocks (and cheap foreign imports). A comparison of typical eight-day longcase strike trains over two centuries is shown in the table.

The earlier trains, like the first one, tended to have more revolutions of the fly per blow on the bell to give a slower strike to bring out the qualities of early bells. Also, since this may have been the only clock in the house, it would give more time to count the blows if the listener was a little way from the clock.

The middle sample gives a quicker strike, which must have been more popular. To suit different tastes, the speed of the main train was fairly standard; adjustment of the strike speed was accomplished by varying the count of the wheel driving the fly, from around forty to fifty teeth or even more, forty-five being an average. A higher wheel count would obviously cause the fly to make more revolutions per blow, and so slow down the strike speed. Note the 'upgrade' to seven-leaf pinions.

The last example has a very slow strike, as it would have been used for gong striking, when a ponderous strike would have been more suitable.

Fig. 545 Month longcase clock, walnut case, c. *1710.*

A few makers were still using a seventy-eight-tooth great wheel, but with rack striking allowing repeating, an eighty-four-tooth wheel was more commonly used to give a longer going period to cover for this, as shown in the second two sample trains. It is most important, also, that the train of rack-striking clocks is not allowed to run down before the going train, as this could cause the strike to jam up and stop the clock. With a count wheel, all that would happen is that the striking would get out of phase with the hands.

Month and longer duration clocks have more wheels, but from the pin wheel upwards they are

Date	Great Wheel	Pin Wheel	Pins	Locking Wheel	Warning Wheel	Fly Pinion
1700	78	48/8	8	48/6	48/6	6
1800	84	56/8	8	49/7	45/7	7
1900	84	64/8	8	70/8	63/7	7

Fig. 546 Dial of Peter King – London, month-going longcase.

Fig. 547 Rear of month movement, showing position of the locking plate.

Fig. 548 Front of movement showing under-dial work.

Fig. 549 Intermediate wheel and pinion for a month-going longcase clock.

the same as a normal eight-day clock. In striking longcase clocks with a regulator-type going train, the striking train would often also have high count wheels and pinions.

With bracket clocks, except for the very early ones, they were almost always rack strike, and the trains did not vary much over the years throughout the eighteenth century. A typical, early eighteenth-century train would be:

84 / 8–60 (10 pins) / 6–54 / 6–48 / 6

Note that on these early bracket clocks the pin wheel usually has ten pins rather than the eight used on a longcase clock, and six-leaf pinions remained for a longer period before the use of seven- and eight-leaf pinions took over. By the nineteenth century, eight-pin wheels and seven- and eight-leaf pinions became more normal, and a typical train would be:

84 / 8–64 (8 pins) / 8–64 / 8–56 / 7

Obviously variations in wheel trains did occur, particularly for the better makers whose work was of high quality. They were great innovators who were always improving on their designs.

WORKSHOPS

At the beginning of this book, engravings show typical clockmakers' workshops of the seventeenth, eighteenth and nineteenth centuries. The tools used by the craftsmen were quite simple and crude by today's standards, but due to their skill, excellent work was achieved. Shown here are various views of a jobbing workshop with machines manufactured in the twentieth century.

Now that the twenty-first century is upon us and technology is advancing so rapidly, no doubt the machines shown will be completely outdated in another twenty-five years!

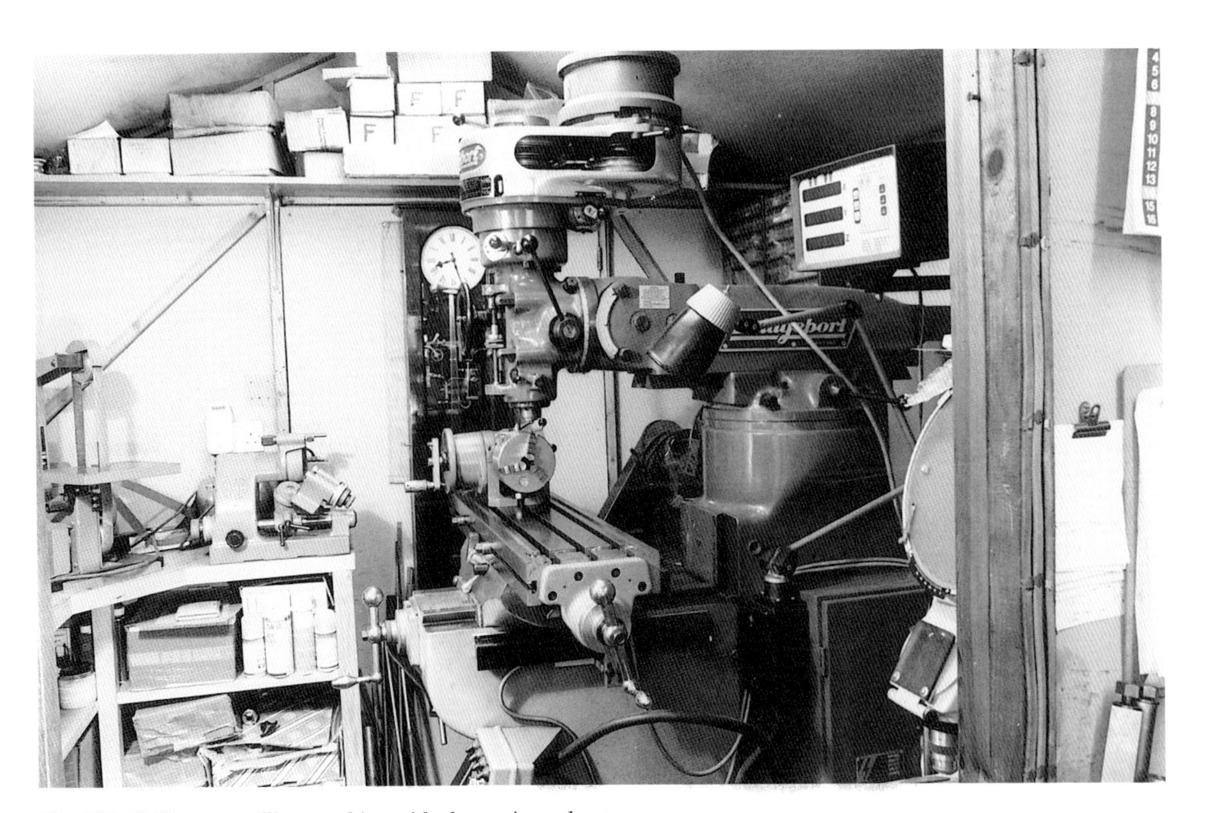

Fig. 550 Bridgeport milling machine with electronic read-out.

Fig. 551 General view showing instrument lathe, engraving machine and large centre lathe.

Fig. 552 In the foreground is the Myford Super 7 bench with lathe, 3½in centre.

Fig. 553 General view showing a Bergeon watchmaker's lathe, an American Watch Tool Co. wheel- and pinion-cutting machine, and an Aciera F1 precision milling machine.

Fig. 554 On right, a Boley Leinen instrument lathe, next to which is the Quorn tool and cutter grinder, and a Star wheel and pinion cutter.

Fig. 555 General view showing a Schaublin 70 precision instrument lathe with overhead drive.

BIBLIOGRAPHY

American Watch Tool Company *Catalogue of Tools*, 1980, Ken Roberts Publisher 1980 reprint.

Ashton, T. S. *An 18th Century Industrialist – Peter Stubs of Warrington*, 1939 and reprinted 1961.

Berthoud, Ferdinand *Essai sur l'horlogerie* 1763.

Britten, F. J. *Watch & Clock Makers Handbook* 15th edition, revised by J. W. Player 1955.

Cain, Tubal *Hardening & Tempering of Metals*, Argus Books, 1984.

Camus, M. *A Treatise on the Teeth of Wheels*, translated into English by John Isaac Hawkins, 1806.

Chaddock, D. H. *The Quorn Universal Tool & Cutter Grinder*, published by Argus Books 1984, reprinted TEE Publishing.

Craven, Maxwell *John Whitehurst, Clockmaker & Scientist 1713–1788*, Mayfield Books.

Crom, Theodore R. *Horological Wheel Cutting Engines 1700–1900*, published 1970.

Dane, E. Surrey, *Peter Stubs and the Lancashire Hand Tool Industry*, 1973.

Daniels, G. *Watchmaking*, publisher Philip Wilson, 1981.

Darken, N. J. & Hooper J. *English 30 Hour Clocks, Origin & Development 1600–1800*, Penita Books 1997.

Davis, W. O. *Gears for Small Mechanisms*, NAG Press 1953, reprinted by TEE Publishing.

De Carle, D. *Practical Clock Repairing*, NAG Press 1952.

De Carle, D. *Practical Watch Repairing*, NAG Press 1946.

De Carle, D. *The Watchmakers & Model Engineers Lathe*, 5th edition, Robert Hale 1997.

Encyclopédie de Diderot et d'Alembert Paris 1751–1772.

Evans, J. H. *Ornamental Turning*, 1886.

Gazeley, W. J. *Clock & Watch Escapements*, Newnes Butterworth 1956.

Gazeley, W. J. *Watch & Clockmaking & Repairing*, Heywood & Company Ltd 1958.

Gordon, G. F. C. *Clockmaking Past & Present*, Crosby Lockwood & Son 1928.

Haswell, J. Eric, *Horology*, Chapman & Hall 1928.

Hatton, Thomas *An Introduction to the Mechanical Part of Clock and Watch Work* 1773.

Holtzapffel's *Turning & Mechanical Manipulation Vol 5*, 1884.

Hooke, Robert *The Diary of Robert Hooke 1672–1680*, transcribed from the original in 1935, reprinted 1968.

Law, I. *Gears & Gear Cutting*, Argus Books 1988.

Les Sciences Les Arts Liberaux et Les Arts Mechaniques, 1765.

Loomes, Brian, *Watchmakers & Clockmakers of the World Volume 2*, NAG Press Ltd 1976.

Machinery's Handbook, 15th Edition, The Industrial Press 1954.

Penman, L. *Practical Clock Escapements*, Mayfield Books 1998.

Rees Cyclopaedia 1819/20, David & Charles 1970.

Reid, Thomas *A Treatise on Clock and Watch Making*, 1826.

Saunier, C. *The Watchmaker's Handbook*, 1891, reprinted 1924 Crosby Lockwood & Son.

Smiles, Samuel *Industrial Biography*, John Murray 1863.

Thiout, Antoine, *Trait de l'Horlogerie* 1741.

Thomas, Geo H. *Dividing & Graduating*, Argus Books Ltd, 1983. Now reprinted by TEE Publishing with new title *Workshop Techniques* and includes construction of the Universal Pillar Tool.

Timmins, A. *Making an Eight Day Longcase Clock*, TEE Publishing 1981.

Whitney, M. E. *The Ship's Chronometer*, American Watchmakers Institute 1985.

Wilding, J. H. *How to Repair Antique Clocks Vols 1–4*, Brant Wright Publishers.

Wyke, John *A Catalogue of Tools for Watch & Clockmakers*, edited by Alan Smith, University Press of Virginia.

PERIODICALS:

Antiquarian Horology, published by The Antiquarian Horological Society, issued four times per year. Excellent for anyone interested in the historical aspect of horology.

Engineering in Miniature, published by TEE Publishing monthly. Covers workshop practice and articles on building models and clocks.

Horological Journal, published monthly by The British Horological Institute. An excellent publication for anyone interested in horology.

Model Engineer, published by Nexus Special Interests twice per month. Covers workshop practice and articles on building models and clocks.

LIST OF SUPPLIERS

The British Horological Institute
Upton Hall
Upton
Newark
Notts
NG23 5TE
Tel. No. 01636 813795
Horological books

Chronos Ltd
Unit 8 Executive park
229/231 Hatfield Road
St Albans
Herts
AL1 4TA
Small tools, lathes, measuring equipment

Clock Spares of Dereham
The Yard
East Dereham
Norfolk
NR19 2BP
Materials and parts for clocks

Cousins Material House
Unit J Chesham Close
Romford
Essex
RM7 7PJ
Materials, tools and parts for clocks and watches

Cowells Small Machine Tools Ltd
Manor Workshops
Church Road
Little Bentley
Colchester
Essex
CO7 8SE
Small lathes and millers suitable for the clockmaker

DEJAY distribution Ltd
PO Box 195
9 The Business Centre
Molly Millars Lane
Wokingham
Berkshire
RG11 2GS
Special carbide drills

Drill Services (Horley) Ltd
Albert Road
Horley
Surrey
RH6 7HR
Special and carbide drills

Evenson Engineering
4 Duchy Crescent
Bradford
BD9 5NJ
Lathes and millers suitable for the model engineer and clockmaker

Fenn Tools
44 Springwood Drive
Springwood Industrial Estate
Braintree
Essex
CM7 2YN
Special carbide drills

G & M Tools
The Mill
Mill Lane
Ashington
West Sussex
RH20 3BX
Small Swiss machines suitable for the clock- and watchmaker

G K Hadfield
Beck Bank
Great Salkeld
Penrith
Cumbria
CA11 9LN
Horological books

Hemingway
Wadsworth House

Greens Lane
Burstwick
Hull
HU12 9EY
Kits for small machines and tools

Home & Workshop Machinery
144 Maidstone Road
Footscray
Sidcup
Kent
DA14 5HS
Small machines, lathes, millers, engraving machines, accessories

JMW Clocks
12 Norton Green Close
Sheffield
S8 8BP
Specialist clockmakers tools: milling spindle, division plates, wheel and pinion cutters, centring microscope, wheel and pinion cutting machine, cutter lapping tool, clock and watchmakers depthing tools, pinion steel

Millhill Supplies
66 The Street
Crowmarsh Gifford
Nr Wallingford
Oxon
OX10 8ES
Small lathes, chucks, cutting tools, measuring equipment

Model Engineering Services
Pipworth Farm
Pipworth Lane
Eckington
Sheffield
S21 4EY
Kits for tool and cutter grinders, Quorn & Kennet

Myford Machine Tools
Wilmot Lane
Chilwell Road
Beeston
Nottingham
NG9 1ER
New and reconditioned Myford lathes and millers

Project Machinery
43 Haw Lane
Bledlow Ridge
Buckinghamshire
HP14 4JH
Small machines for the model engineer and clockmaker

Repton Clocks
48 High Street
Repton
Derby
Microscopes for the horologist

Richards of Burton
Woodhouse Clock Works
Swadlincote Road
Woodville
Burton on Trent
Parts and materials for the clock restorer

Rita Shenton
148 Percy Road
Twickenham
Middlesex
TW2 6JG
Horological books

P P Thornton (Successors) Ltd
The Old Bakehouse
Upper Tysoe
Warwickshire
CV35 0TR
Cutters for clocks, wheels and pinions, escape wheel cutters

Tracy Tools Ltd
2 Mayors Avenue
Dartmouth
South Devon
TQ6 9NF
Cutting tools, end mills, taps and dies

H S Walsh & Sons Ltd
243 Beckenham Road
Beckenham
Kent
BR3 4RP
Tools, parts and spares for the clock- and watchmaker

John Wardle
(Prop Steve Bevan)
Express Works
Brailsford
Derby
DE6 3BY
Manufacturer and supplier of clock parts

APPENDIX I

SWISS STANDARD

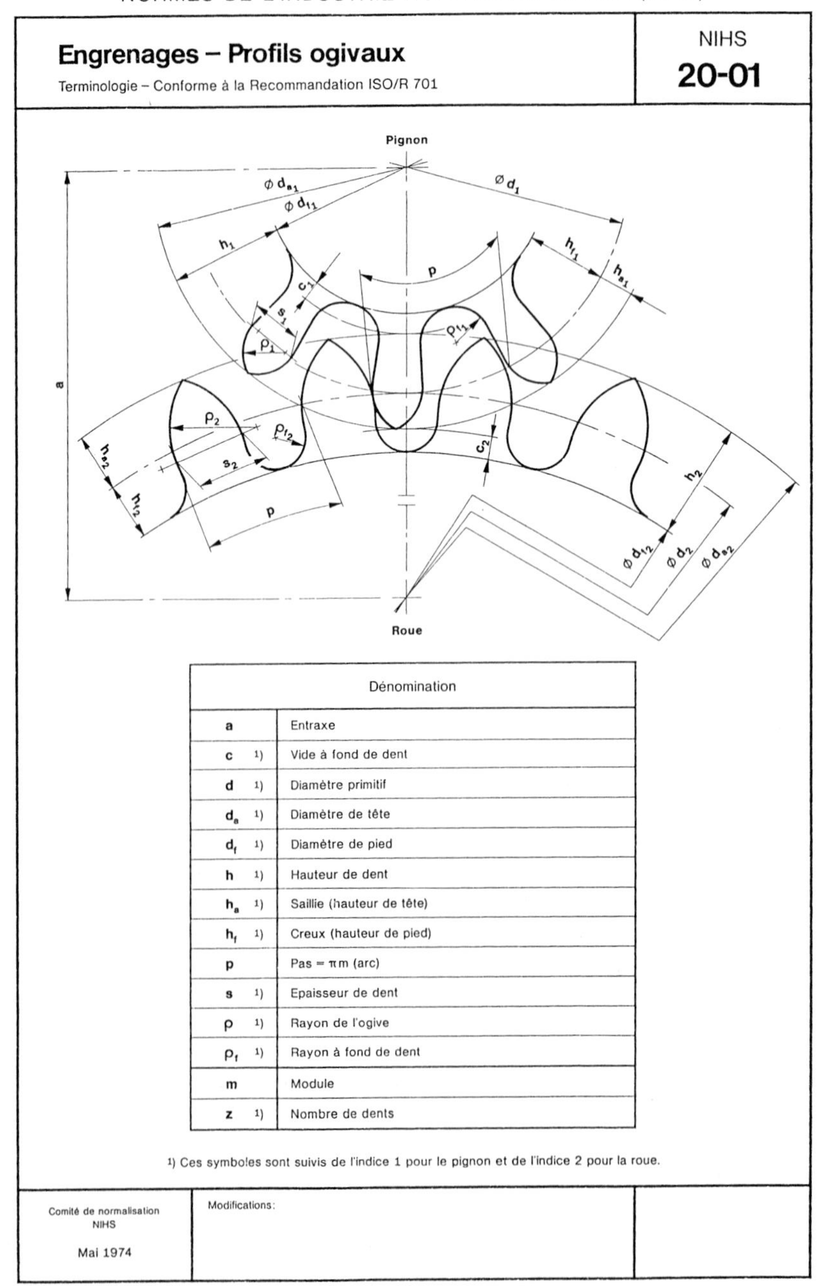

NORMES DE L'INDUSTRIE HORLOGÈRE SUISSE (NIHS)

Engrenages – Profils ogivaux

Terminologie – Conforme à la Recommandation ISO/R 701

NIHS **20-01**

Dénomination		
a		Entraxe
c	1)	Vide à fond de dent
d	1)	Diamètre primitif
d_a	1)	Diamètre de tête
d_f	1)	Diamètre de pied
h	1)	Hauteur de dent
h_a	1)	Saillie (hauteur de tête)
h_f	1)	Creux (hauteur de pied)
p		Pas = πm (arc)
s	1)	Epaisseur de dent
ρ	1)	Rayon de l'ogive
ρ_f	1)	Rayon à fond de dent
m		Module
z	1)	Nombre de dents

1) Ces symboles sont suivis de l'indice 1 pour le pignon et de l'indice 2 pour la roue.

Comité de normalisation NIHS Mai 1974	Modifications:	

NORMES DE L'INDUSTRIE HORLOGÈRE SUISSE (NIHS)

Engrenages multiplicateurs (roue menante)

Profils ogivaux – Dimensions

NIHS **20-02**

Caractéristiques:
- Ces dentures d'engrenages à profils ogivaux permettent d'obtenir une faible variation du couple transmis.
- Les parties rectilignes des flancs visent le centre, et sont tangentes aux arcs des ogives et aux fonds arrondis de la denture.
- Seul un engrenage roue-pignon est possible.
- La roue doit avoir au minimum 40 dents.

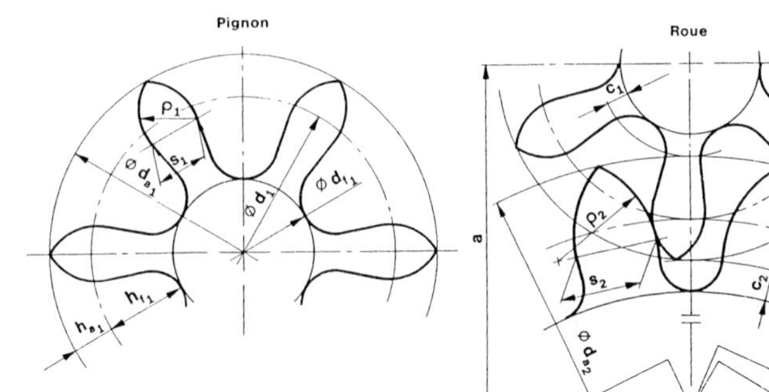

Pignon et roue	
$d = mz$ [1]	$a = m \frac{z_1 + z_2}{2}$
$d_a = m(z + 2h_a)$ [1]	
$d_f = m(z - 2h_f)$ [1]	

Facteurs pour module 1							
	Pour z_1 =	6	7	8	9	10	11 à 20
Pignon	$2h_{a_1}$	1,64		1,52			1,44
	$2h_{f_1}$	3,20	3,30	3,40	3,45	3,50	3,60
	ρ_1	1,05		0,95			0,85
	s_1	1,10					
	c_1	0,40					
Roue	$2h_{a_2}$	2,40	2,50	2,60	2,65	2,70	2,80
	$2h_{f_2}$	2,80	2,70	2,60	2,55	2,50	2,40
	ρ_2	2,00					
	s_2	1,60					
	c_2	0,58	0,53	0,54	0,515	0,49	0,48

[1] Ces symboles sont suivis de l'indice 1 pour le pignon et de l'indice 2 pour la roue.

Comité de normalisation NIHS

Mai 1974

Modifications:

NORMES DE L'INDUSTRIE HORLOGÈRE SUISSE (NIHS)

Engrenages

Profils ogivaux – Tracé pour pignon et roue

NIHS **20-10** F. 1

Eléments donnés

z $\mathbf{d_a}$ $\mathbf{d_f}$ $\boldsymbol{\rho}$ **s**

Tracé du profil

1. A partir du centre **O**, déterminer le sommet **Y** de la dent sur le diamètre $\mathbf{d_a}$.
2. A la distance **t** de part et d'autre de l'axe **OY**, tracer deux parallèles

$$t = \rho - \frac{s}{2}$$

3. A partir du sommet **Y**, tracer un arc de cercle de rayon **ρ**, puis à partir des centres **C**, les arcs de l'ogive.
4. Tangentiellement aux arcs de l'ogive, tracer les parties droites des flancs visant le centre **O**.
5. Tracer les axes **EO** de part et d'autre de **YO**.

$$\alpha = \frac{180°}{z}$$

6. Tracer le cercle de pied $\mathbf{d_f}$ et les tangentes aux points **F**, sur les axes **EO**.
7. Les intersections des bissectrices **B** et des axes **EO** donnent les centres des rayons à fond de dents ρ_f.

Remarque

Le cercle primitif n'intervient pas pour le tracé du profil.

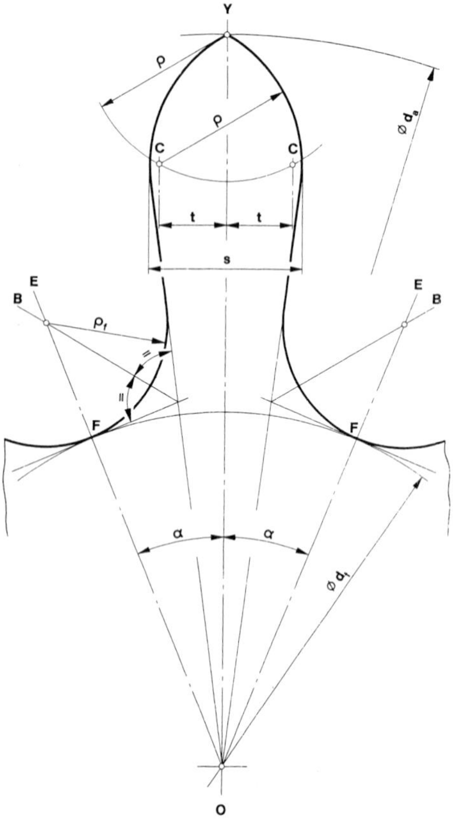

Comité de normalisation NIHS Mai 1974	Modifications:	

NORMES DE L'INDUSTRIE HORLOGÈRE SUISSE (NIHS)

Engrenages

Profils ogivaux – Calcul pour pignon et roue

NIHS **20-10** F. 2

Eléments donnés

$z \quad d_a \quad d_f \quad \rho \quad s$

Calcul du profil

1. $t = \rho - \frac{s}{2}$

2. $v_1 = \frac{d_a}{2} - \sqrt{\rho^2 - t^2}$

3. $r_1 = \sqrt{t^2 + v_1^2}$

4. $\alpha = \frac{180°}{z}$

5. $\beta = \alpha - \arcsin \frac{\rho}{r_1} + \arcsin \frac{t}{r_1}$

6. $\rho_f = \frac{d_f \sin \beta}{2\,(1 - \sin \beta)}$

7. $r_2 = \frac{d_f}{2} + \rho_f$

8. $u = r_2 \cdot \sin \alpha$

9. $v_2 = r_2 \cdot \cos \alpha$

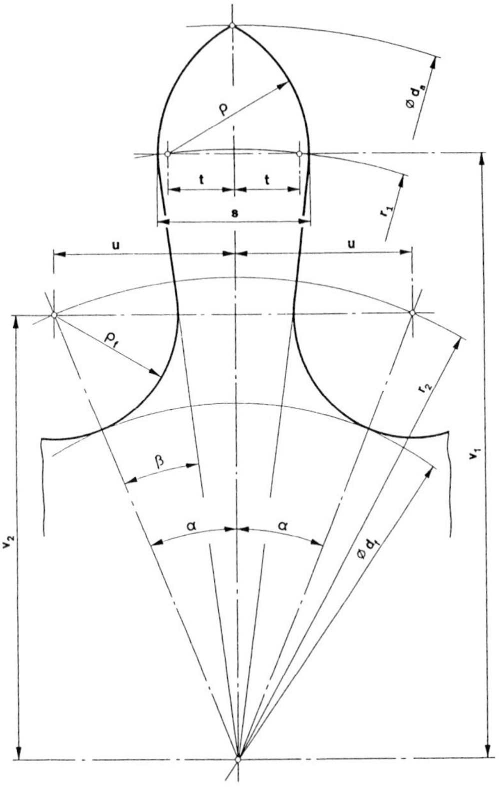

Comité de normalisation NIHS

Mai 1974

Modifications:

Schweizer Norm
Norme Suisse
Norma Svizzera **SN**

20 - 25 **282 025**

EINGETRAGENE NORM DER SCHWEIZERISCHEN NORMEN-VEREINIGUNG SNV NORME ENREGISTRÉE A L'ASSOCIATION SUISSE DE NORMALISATION

Engrenages

Remplace : NHS 56704
Octobre 1939

Mécanismes de remontoir et de mise à l'heure
Dentures épicycloïdales corrigées

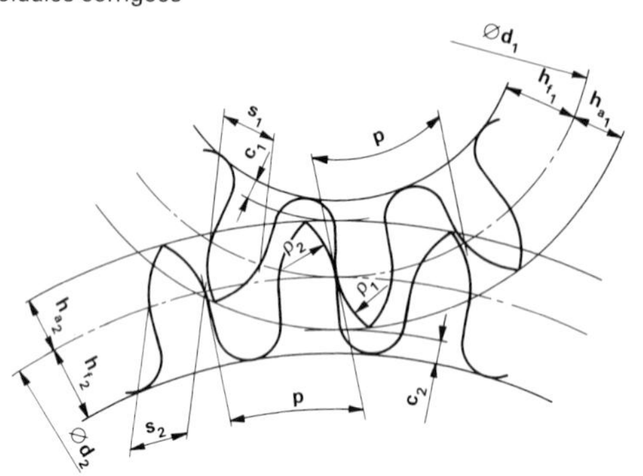

Figure 1 - Détails de la denture

Tableau 1 - Roues menantes et menées Dimensions en mm

m = module f = facteur théorique proportionnel	p = pas
Diamètre des cercles générateurs	3 m
Epaisseur des dents	$s_1 = s_2 = 1{,}41$ m
Hauteur des pieds des dents	$h_{f_1} = h_{f_2} = 1{,}75$ m
Doubles saillies	$2h_{a_1}$ ou $2h_{a_2} = f$ m
Rayons de courbure de l'ogive	ρ_1 ou $\rho_2 = 0{,}8$ f m
Vide à fond de dent	$c_1 = h_{f_1} - h_{a_2}$ $c_2 = h_{f_2} - h_{a_1}$
Jeu circonférentiel	1/10 p = 0,31 m

Nombre de dents z_1 ou z_2	$f = \frac{2h_{a_1}}{m}$ ou $\frac{2h_{a_2}}{m}$
8	2,32
9	2,34
10 - 11	2,38
12 - 13	2,40
14 - 16	2,44
17 - 20	2,48
21 - 25	2,52
26 - 34	2,54
35 - 54	2,58
55 - 134	2,62
135 - ∞	2,64

$$m = \frac{\text{Distance des centres des deux engrenages}}{\text{Demi somme du nombre de dents des deux engrenages}}$$

Les arcs de cercle de rayons ρ_1 ou ρ_2 sont prolongés au-dessous du cercle primitif et raccordés par des flancs radiaux tangents.
Un arc de cercle constitue le fond de denture.
Pratiquement arcs s_1 ou s_2 = cordes correspondantes.

Editeur/Distributeur:
Bureau des Normes NIHS
Rue d'Argent 6
2501 Bienne

No d'enregistrement SN
et année de parution: 282 025:1993

Modifications:

Approbation: Octobre 1993
Comité de Normalisation NIHS

Nombre de page: 1

APPENDIX II

BRITISH STANDARD FOR GEARS FOR INSTRUMENTS AND CLOCKWORK MECHANISMS PART 2. CYCLOIDAL TYPE GEARS

FOREWORD

This British Standard supersedes B.S. 978, 'Gears for clockwork mechanisms', which was published in 1941 as a war emergency measure to assist in the reduction of types and sizes of both gears and cutters.

As a result of experience gained in its use it became evident that a revision was necessary and it was considered desirable to divide the standard into four separate parts as follows:-

Part 1. Involute spur, helical and crossed helical gears.
Part 2. Cycloidal type spur gears.
Part 3. Bevel gears.
Part 4. Worm gears.

Each part will be published as it becomes available.

Suitable hobs and cutters will be dealt with in a later British Standard.

The provisions of this part differ substantially from those of the original B.S. 978, as a cycloidal type of gear has been substituted for the circular arc form. The standards of the Swiss and Black Forest watch and clock industries, particularly the former, have been used as a basis; this was considered desirable because of the prevalent use of those standards of horological work and other applications where a cycloidal form is preferred.

An explanation of this type of gearing, as applied to wheels driving pinions, is included in Appendix A.

Arising out of discussions which have taken place in connection with the work undertaken by the International Organization for Standardization on gears, there are indications that it may be possible to simplify the provisions of the standard in regard to tooth form and proportions. It is intended to give further attention to this subject when more evidence is available.

SECTION ONE : GENERAL

Scope

1. This part applies to cycloidal type spur gears for spring or weight-driven mechanisms in which:

a. the pinion is the driven member, or
b. the pinion is required to act sometimes as the driver and sometimes as the driven member.

For integrating electricity meters the particular form of gear specified herein is not necessary and it is recognised that other existing forms of gear will continue to be used.

Terminology and notation

2. The terms and notation shall be as defined in B.S. 2519, 'Glossary of terms and notation for gearing.' For the convenience of users of the standard, an abbreviated list of symbols is given in Appendix B.

Range of standard modules

3. The standard modules shall be as set out in Table 1, preferred sizes being shown in bold type.

* In course of preparation.

TABLE 1. RANGE OF MODULES (dimensions in millimetres)

Modules		
0.079	**0.150**	0.320
0.0725		0.340
0.075	0.155	0.360
0.0775	0.160	0.380
0.080	0.165	**0.400**
	0.170	
0.0825	0.175	0.420
0.085	0.180	0.440
0.0875	0.185	0.460
0.090	0.190	0.480
0.0925	0.195	**0.500**
0.095	**0.200**	
0.0975		0.550
0.100	0.210	**0.600**
	0.220	
0.105	0.230	0.650
0.110	0.240	**0.700**
0.115	**0.250**	
0.120		0.750
0.125	0.260	**0.800**
0.130	0.270	
0.135	0.280	0.850
0.140	0.290	0.900
0.145	**0.300**	0.950
		1.000

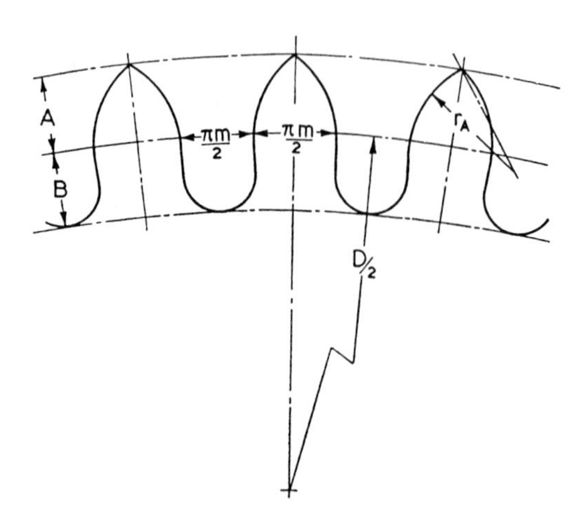

Fig. 1. Tooth proportions for wheels driving pinions in accordance with Clause 4

TABLE 2. PROPORTIONS OF DRIVING WHEELS

Tooth thickness along pitch circle		1.57 m
Addendum A		fm
Dedendum B		1.57 m
Radius of curvature of addendum r_A		mf_r
Pitch diameter	D	Tm
Outside diameter	D_a	D + 2fm
Root diameter	D_r	D – 3.14m

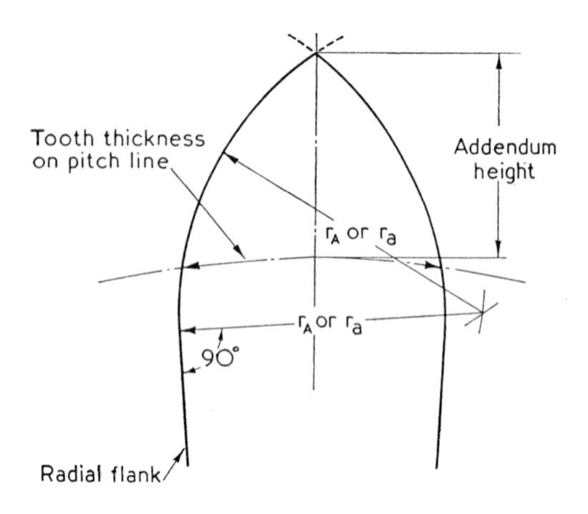

Fig. 2. Construction for position of centre of curvature of addendum

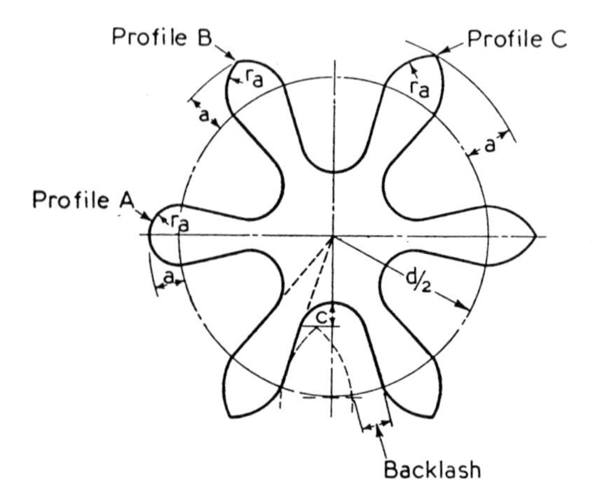

Fig. 3. Proportions for driven pinions with alternative addendum profiles

SECTION TWO : TOOTH FORM AND PROPORTIONS

Wheels driving pinions

4. *a. General.* The teeth of wheels driving pinions designed in accordance with Clause 5 shall have the proportions given in Table 2 and Fig. 1.

Values for f and f_r, appropriate to the gear ratio and the number of teeth in the pinion, shall be obtained from Table 3 (overleaf), or Charts 1 and 2.

The addendum height and the radius of curvature of the addendum shall be in accordance with Table 2. The centre of curvature for the addendum curve shall be chosen so that the curve and the centre line of the tooth intersect on the addendum circle; the radial flank is tangential to this curve (see Fig. 2). The bottom of the tooth spaces shall be approximately semi-circular (see Fig. 1).

Wheels designed in accordance with the provisions of this clause are only suitable to drive pinions designed in accordance with Clause 5.

The two addendum curves intersect at the addendum circle

TABLE 4. PROPORTIONS OF DRIVEN PINIONS (see Fig. 3)

			Number of teeth (t)	
			6–10	11 and over
Tooth thickness along pitch circle			1.05 *m.*	1.25 *m.*
Pitch diameter		d	tm	tm
Outside diameter		d_a	$d + 2a$	$d + 2a$
Root diameter		d_r	$d - 2b$	$d - 2b$
Dedendum		b	$(f + 0.4)m$	$(f + 0.4)m$
Bottom clearance		c	0.4 *m.*	0.4 *m.*
Minimum backlash			0.52 *m.*	0.32 *m.*
*Addendum	Profile A		0.525 *m.*	0.625 *m.*
	Profile B	a	0.67 *m.*	0.805 *m.*
	Profile C		0.855 *m.*	1.05 *m.*
*Radius of	Profile A		0.525 *m.*	0.625 *m.*
curvature of	Profile B	r_a	0.70 *m.*	0.84 *m.*
addendum	Profile C		1.05 *m.*	1.25 *m.*

* See Clause 5, 3rd and 4th paragraphs.

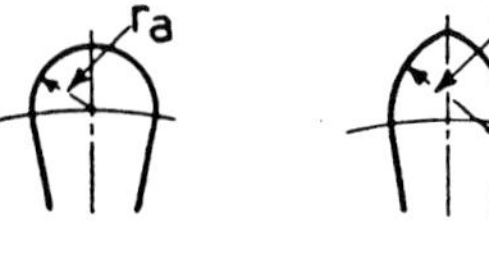

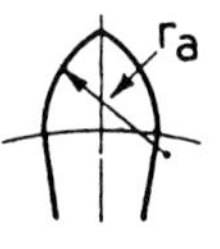

Fig. 4. Pinion addendum profiles

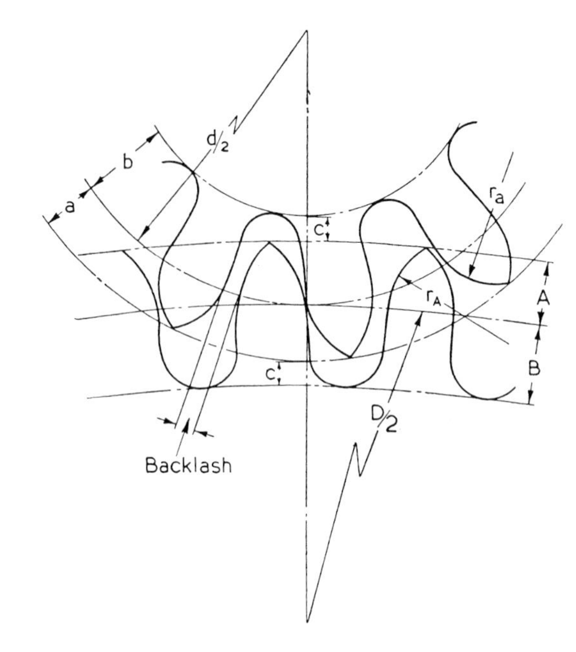

Fig. 5. Tooth proportions for gears in accordance with Clause 5

Driven pinions

5. The teeth of pinions to be driven by wheels designed in accordance with Clause 4 shall have the proportions given in Table 4, appropriate to the number of teeth.

The centres of curvature for the addendum curves shall be chosen so that the curves and the centre line of the tooth intersect in the addendum circle; the radial flank is tangential to this curve (see Fig. 2). The bottom of the tooth spaces shall be approximately semi-circular (see Fig. 3).

Three different addendum profiles are given, with different values for the addendum and the radius of curvature (see Fig. 4). In general these profiles are used under the following conditions:

Profile A	10 teeth and over
Profile B	8 and 9 teeth
Profile C	6 and 7 teeth

TABLE 3. VALUES OF ADDENDUM FACTOR (*f*) AND ADDENDUM RADIUS FACTOR (f_r)

Ratio	6 Teeth		7 Teeth		8 Teeth		9 Teeth		10 Teeth	
T/t	f	f_r	f	f_r	f	f_r	f	f_r	f	f_r
3	1.259	1.855	1.335	1.968	1.403	2.068	1.465	2.160	1.523	2.244
4	1.280	1.886	1.359	2.003	1.430	2.107	1.494	2.202	1.554	2.290
5	1.293	1.906	1.374	2.025	1.447	2.132	1.513	2.230	1.574	2.320
6	1.303	1.920	1.385	2.041	1.459	2.150	1.526	2.249	1.588	2.341
6½	1.307	1.926	1.389	2.048	1.464	2.157	1.531	2.257	1.594	2.349
7	1.310	1.930	1.393	2.053	1.468	2.163	1.536	2.263	1.599	2.356
7½	1.313	1.934	1.396	2.058	1.471	2.169	1.540	2.269	1.603	2.363
8	1.315	1.938	1.399	2.062	1.475	2.173	1.543	2.274	1.607	2.368
8½	1.318	1.942	1.402	2.066	1.478	2.177	1.547	2.279	1.610	2.373
9	1.320	1.944	1.404	2.069	1.480	2.181	1.549	2.283	1.613	2.377
9½	1.321	1.947	1.406	2.072	1.482	2.184	1.552	2.287	1.616	2.381
10	1.322	1.949	1.408	2.075	1.484	2.187	1.554	2.290	1.618	2.385
11	1.326	1.954	1.411	2.080	1.488	2.193	1.558	2.296	1.623	2.391
12	1.328	1.957	1.414	2.084	1.491	2.197	1.561	2.301	1.626	2.397

Ratio	12 Teeth		14 Teeth		15 Teeth		16 Teeth	
T/t	f	f_r	f	f_r	f	f_r	f	f_r
3	1.626	2.396	1.718	2.532	1.760	2.594	1.801	2.654
4	1.661	2.448	1.756	2.589	1.801	2.654	1.843	2.715
5	1.684	2.482	1.782	2.626	1.827	2.692	1.870	2.756
6	1.700	2.505	1.799	2.652	1.845	2.719	1.889	2.784
6½	1.707	2.516	1.807	2.662	1.853	2.730	1.897	2.795
7	1.712	2.523	1.812	2.671	1.859	2.739	1.903	2.804
7½	1.717	2.530	1.818	2.679	1.864	2.748	1.909	2.813
8	1.721	2.536	1.822	2.686	1.869	2.755	1.914	2.820
8½	1.725	2.542	1.827	2.692	1.874	2.761	1.919	2.827
9	1.728	2.547	1.830	2.697	1.878	2.767	1.923	2.833
9½	1.731	2.552	1.834	2.703	1.881	2.773	1.926	2.839
10	1.734	2.556	1.837	2.707	1.884	2.777	1.929	2.844
11	1.739	2.563	1.842	2.715	1.890	2.785	1.935	2.852
12	1.743	2.569	1.847	2.722	1.895	2.792	1.940	2.859

It should be noted, however, that profiles A and B may, in certain circumstances, be used with advantage for pinions with smaller numbers of teeth than those given above.

Wheels and Pinions for Trains in Which the Pinion may act Sometimes as the Driver and Sometimes as the Driven Member

6. The proportions for wheels and pinions for trains in which the pinion may act sometimes as the driver and sometimes as the driven member shall be in accordance with Table 5 and Fig. 5.

The centres of curvature for the addendum curves shall be chosen so that the curves and the centre line of the tooth intersect on the addendum circle; the radial flank is tangential to this curve (see Fig. 2).

The roots of the teeth shall be approximately semi-circular.

For minute to hour reduction trains, winding trains and hand setting trains, involute teeth in accordance with Part 1 of this standard, 'Involute spur, helical and crossed helical gears,' should be used.

The values of f and f_r appropriate to the number of teeth in the gear shall be obtained from Table 6.

TABLE 5. PROPORTIONS OF WHEELS AND PINIONS FOR TRAINS IN WHICH THE PINION MAY ACT SOMETIMES AS THE DRIVER AND SOMETIMES AS THE DRIVEN MEMBER

Tooth thickness along pitch circle		1.41 *m.*
Pitch diameter	D or *d*	T*m* or *tm*
Addendum	A or *a*	*fm*
Dedendum	B or *b*	1.75 *m.*
Radius of curvature of addendum	r_A or r_a	$f_r m$
Bottom clearance	*c*	B – *a* or *b* – A
Minimum backlash		0.32 *m.*

TABLE 6. VALUES OF ADDENDUM FACTOR (*f*) AND ADDENDUM RADIUS FACTOR (f_r) FOR NORMAL DRIVES

Number of teeth in gear (T or t)	Addendum factor (f)	Addendum radius factor (f_r)
8	1.16	1.85
9	1.17	1.87
10, 11	1.19	1.90
12, 13	1.20	1.92
14 to 16	1.22	1.95
17 to 20	1.24	1.98
21 to 25	1.26	2.01
26 to 34	1.27	2.03
35 to 54	1.29	2.06
55 to 134	1.31	2.09
135	1.32	2.11

APPENDIX III

LANTERN PINIONS

Lantern pinion Data Sheet from *Gears for Small Mechanisms* by W O Davis reproduced by kind permission of Robert Hale & NAG Press.

LANTERN PINIONS AND EPICYCLOIDAL DRIVING WHEELS

Lantern pinions must always be the driven member, and therefore, lantern pinion gearing is primarily suitable for the going trains of timepieces.

The tooth forms for driving wheels are rather taller than the teeth of wheels which engage radial flanked pinions. The gears are often formed by stamping, and a small tip rounding aids this operation. Where the pinion has 8 or fewer pins, this rounding should be very slight, in order to retain the maximum path of contact. Where the pinion has 12 or more pins, the rounding may be generous, such that the working height of the addendum is reduced to $1 \times m$. Teeth abbreviated in this manner are illustrated.

PINIONS

Pin dia. = 1.05 where t is 10 or less

Pin dia. = 1.25 where t is more than 10.

WHEELS

$s_2 = 1.57$

B = 1.2 for heavily stressed Teeth

B = 1.4 for lightly stressed Teeth

A and R are given in the table below.

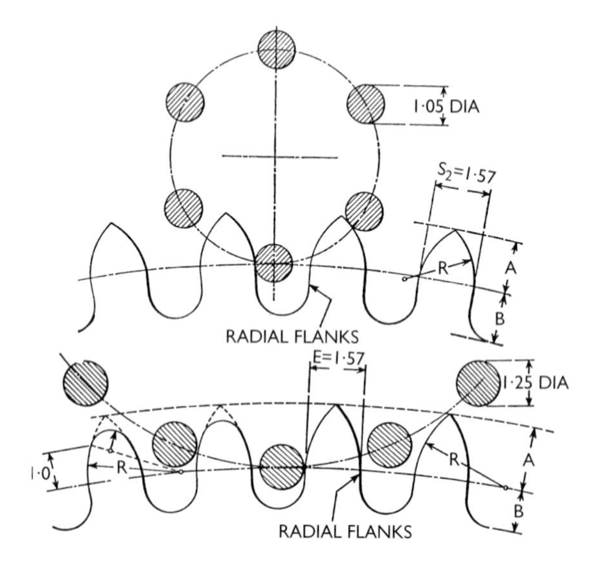

EPICYCLOIDAL DRIVING WHEELS FOR LANTERN PINIONS

	No. of pins in pinions													
	6		7		8		10		12		14		16	
Ratio T/t	A	R	A	R	A	R	A	R	A	R	A	R	A	R
3	1.31	1.62	1.39	1.83	1.45	1.92	1.56	2.12	1.62	2.19	1.70	2.34	1.78	2.48
4	1.37	1.75	1.45	1.96	1.52	2.07	1.63	2.27	1.69	2.34	1.78	2.52	1.87	2.67
6	1.45	1.97	1.53	2.20	1.60	2.32	1.73	2.55	1.79	2.64	1.88	2.82	1.97	2.99
8	1.48	2.01	1.56	2.25	1.64	2.38	1.77	2.62	1.85	2.73	1.94	2.93	2.04	3.11
10	1.50	2.03	1.58	2.28	1.65	2.41	1.79	2.65	1.88	2.78	1.97	2.97	2.08	3.17
12	1.51	2.05	1.60	2.30	1.67	2.44	1.82	2.70	1.90	2.82	2.00	3.02	2.10	3.21
Rack or Contrate	1.64	2.35	1.72	2.49	1.79	2.62	1.93	2.99	2.02	3.24	2.13	3.70	2.25	4.19

APPENDIX IV

CARPANO TABLES

Caractéristiques des Fraises "L. CARPANO" pour ROUES a Profils Epicycloidaux

MODULE	Numéro CARPANO	Diamètre PITCH	Pas circonférentiel	Épaisseur de la Fraise
0,0400	0,22	635	0,126	0,063
0,0425	0,21	598	0,134	0,067
0,0450	0,20	564	0,141	0,070
0,0475	0,19	534	0,149	0,074
0,0500	0,18	508	0,147	0,078
0,0525	0,17	484	0,165	0,082
0,0550	0,16	462	0,173	0,086
0,0575	0,15	442	0,181	0,090
0,0600	0,14	423	0,188	0,094
0,0625		406	0,196	0,098
0,0650	0,13	391	0,204	0,102
0,0675	0,12	376	0,202	0,106
0,0700	0,11	363	0,220	0,110
0,0725	0,10	350	0,228	0,114
0,0750	0,9	339	0,236	0,118
0,0775	0,8	328	0,243	0,121
0,0800	0,7	318	0,251	0,125
0,0825		308	0,259	0,129
0,0850	0,6	299	0,267	0,133
0,0875	0,5	290	0,275	0,137
0,0900	0,4	282	0,283	0,141
0,0925	0,3	275	0,291	0,145
0,0950	0,2	267	0,298	0,149
0,0975	0,1	261	0,306	0,153
0,1000	0	254	0,314	0,157
0,1025	½	248	0,322	0,161
0,1050		242	0,330	0,165
0,1075	1	236	0,338	0,169
0,1100	1 ½	231	0,346	0,173
0,1125	2	226	0,353	0,176
0,1150		221	0,361	0,180
0,1175	2 ½	216	0,369	0,184
0,1200	3	212	0,377	0,188
0,1225	3 ½	207	0,385	0,192
0,1250	4	203	0,393	0,196
0,1275	4 ½	199	0,401	0,200

MODULE	Numéro CARPANO	Diamètre PITCH	Pas circonférentiel	Épaisseur de la Fraise
0,1300	5	195	0,408	0,204
0,1325	5½	192	0,416	0,208
0,1350		188	0,424	0,212
0,1375	6	185	0,432	0,216
0,1400	6½	181	0,440	0,220
0,1425	7	178	0,448	0,224
0,1450	7½	175	0,456	0,228
0,1475	8	172	0,463	0,231
0,1500	8½	169	0,471	0,235
0,1525		167	0,479	0,239
0,1550	9	164	0,487	0,243
0,1575		161	0,495	0,247
0,1600	10	159	0,503	0,251
0,1625	10½	156	0,511	0,255
0,1650	11	154	0,518	0,259
0,1675	11½	152	0,526	0,263
0,1700	12	149	0,534	0,267
0,1725	12½	147	0,542	0,271
0,1750		145	0,550	0,275
0,1775	13	143	0,558	0,279
0,1800	13½	141	0,565	0,282
0,1825	14	139	0,573	0,286
0,1850	14½	137	0,581	0,290
0,1875	15	135	0,589	0,294
0,1900	15½	134	0,597	0,298
0,1925	16	132	0,605	0,302
0,1950		130	0,613	0,306
0,1975	16½	129	0,620	0,310
0,2000	17	127	0,628	0,314
0,2050	18	124	0,644	0,322
0,2100	19	121	0,660	0,330
0,2150	20	118	0,675	0,337
0,2200	21	115	0,691	0,345
0,2250		113	0,707	0,353
0,2300	22	110	0,723	0,361
0,2350	23	108	0,738	0,369

MODULE	Numéro CARPANO	Diamètre PITCH	Pas circonférentiel	Épaisseur de la Fraise	MODULE	Numéro CARPANO	Diamètre PITCH	Pas circonférentiel	Épaisseur de la Fraise
0.2400	24	106	0,754	0,377	0,4300	58	59,1	1,351	0,675
0,2450	25	104	0,770	0,385	0,4350		58,4	1,367	0,683
0.2500	26	102	0,785	0,392	0,4400	59	57.7	1,382	0,691
0,2550	27	100	0,801	0,400	0,4450	60	57,1	1,398	0,699
0,2600	28	97,7	0,817	0,408	0,4500	61	56,4	1,414	0,707
0,2650		95,8	0,833	0,416	0,4550	62	55,8	1,429	0,714
0,2700	29	94,11	0,848	0,424	0,4600	63	55,2	1,445	0,722
0,2750	30	92,4	0,864	0,432	0,4650	64	54,6	1,461	0,730
0,2800	31	90,7	0,880	0,440	0,4700	65	54	1,477	0,738
0,2850	32	89,2	0,895	0,447	0,4750		53,4	1,492	0,746
0,2900	33	87,6	0,911	0,455	0,4800	66	52,9	1,508	0,754
0,2950	34	86,1	0,927	0,463	0,4850		52,4	1,524	0,762
0,3000	35	84,7	0,942	0,471	0,4900	67	51,8	1,539	0,769
0,3050		83,3	0,958	0,479	0,4950	68	51,3	1,555	0,777
0,3100	36	81,9	0,974	0,487	0,5000	69	50,8	1,571	0,785
0,3150	37	80,6	0,990	0,495	0,5050	70	50,3	1,586	0,793
0,3200	38	79,4	1,005	0,502	0,5100	71	49,8	1,602	0,801
0,3250	39	78,2	1,021	0,510	0,5150	72	49,3	1,618	0,809
0.3300	40	77	1,037	0,518	0,5200	73	48,8	1,634	0,817
0,3350		75,8	1,052	0,526	0,5250		48,4	1,649	0,824
0,3400	41	74,7	1,068	0,534	0,5300	74	47,9	1,665	0,832
0,3450	42	73,6	1,084	0,542	0,5350	75	47,5	1,681	0,840
0,3500	43	72,6	1,110	0,550	0.5400	76	47	1,696	0,848
0,3550	44	71,5	1,115	0,557	0,5450	77	46,6	1,712	0,856
0,3600	45	70,6	1,131	0,565	0,5500		46,2	1,728	0,864
0,3650	46	69,6	1,147	0,573	0,5550	78	45,8	1,744	0,872
0,3700	47	68,6	1,162	0,581	0,5600	79	45,4	1,759	0,879
0,3750	48	67,7	1,178	0,589	0,5650	80	45	1,775	0,887
0,3800		66,8	1,194	0,597	0,5700	81	44,6	1,791	0,895
0,3850	49	66	1,210	0,605	0,5750		44,2	1,806	0,903
0,3900	50	65,1	1,225	0,612	0 5800	82	43,8	1,822	0,911
0,3950	51	64,3	1,241	0,620	0 5850	83	43,4	1,838	0,919
0,4000	52	63,5	1,257	0,628	0,5900	84	43,1	1,853	0,926
0,4050	53	62,7	1,272	0,635	0,5950	85	42,7	1,869	0,934
0,4100	54	62	1,288	0,644	0,6000	86	42,3	1,885	0,942
0,4150	55	61,2	1,304	0,652	0,6050		42	1,901	0,950
0,4200	56	60,5	1,319	0,659	0,6100	87	41,6	1,916	0,958
0,4250	57	59,8	1,335	0,667	0,6150	88	41,3	1,932	0,966

MODULE	Numéro CARPANO	Diamètre PITCH	Pas circonférentiel	Épaisseur de la Fraise
0,6200	89	41	1,948	0,974
0,6250	90	40,6	1,963	0,981
0.6300	91	40,3	1,979	0,989
0,6350		40,0	1,995	0,997
0,6400	92	39,7	2,011	1,005
0,6450	93	39,4	2,026	1,013
0,6500	94	39,1	2,042	1,021
0,6550	95	38,8	2,058	1,029
0,6600	96	38,5	2,073	1,036
0,6650	97	38,2	2,089	1,044
0,6700	98	37,9	2,105	1,052
0,6750	99	37 6	2,121	1,060
0,6800	100	37,4	2,136	1,068
0,6850	101	37,1	2,159	1,076
0,6900	102	36,8	2,168	1,084
0,6950		36,5	2,183	1.091
0,7000	103	36,3	2,199	1,099
0,7050	104	36	2,215	1,107
0,7100	105	35,8	2,231	1,115
0,7150	106	35,5	2,246	1,122
0,7200	107	35,3	2,262	1,131
0,7250	108	35	2,278	1,139
0,7300	109	34,8	2,293	1,146
0,7350	110	34,6	2,309	1,154
0.7400	111	34,3	2,325	1,162
0,7450	112	34,1	2,340	1,170
0,7500	113	33,9	2,356	1,178
0,7550	114	33.6	2,372	1,186
0,7600	115	33,4	2,388	1,194
0,7650	116	33,2	2,403	1,201
0,7700	117	33	2,419	1.209
0,7750	118	32,8	2,435	1,217
0,7800	119	32,6	2,450	1,225
0,7850	120	32,4	2,466	1,233
0,7900	121	32,2	2,482	1,241
0,7950	122	32,0	2,498	1,249
0,8000	123	31,8	2,513	1,256
0,8050	124	31,6	2,529	1.264
0,8100	125	31,4	2,545	1.272
0,8150	126	31,2	2,560	1,280
0,8200	127	31	2.576	1,288
0,8250	128	30,8	2,592	1,296
0,8300	129	30,6	2,607	1,303
0,8350	130	30,4	2,623	1.311
0,8400	131	30,2	2,639	1,319
0,8450	132	30,1	2,655	1,327
0,8500	133	29.9	2,670	1,335
0,8550	134	29,7	2,686	1,343
0,8600	135	29,5	2,702	1,351
0,8650	136	29,4	2,717	1 358
0,8700	137	29,2	2,733	1,366
0,8750	138	29,0	2,749	1,374
0,8800	139	28,9	2,765	1,382
0,8850	140	28,7	2,780	1,390
0,8900	141	28.5	2,796	1,398
0,8950	142	28,4	2,812	1,406
0.9000	143	28.2	2.827	1,413
0,9050	144	28,1	2,843	1,421
0,9100	145	27,9	2,859	1,429
0,9150	146	27,8	2,875	1,437
0,9200	147	27,6	2,890	1,445
0,9250	148	27,5	2,906	1,453
0,9300	149	27,3	2,922	1,461
0,9350	150	27,2	2,937	1,468
0,9400	151	27.0	2,953	1,476
0,9450	152	26,9	2,969	1,484
0,9500	153	26,7	2,984	1,492
0,9550	154	26,6	3,000	1,500
0,9600	155	26,5	3,016	1,508
0,9650	156	26,3	3,032	1,516
0,9700	157	26,2	3.047	1,523
0,9750	158	26,1	3,063	1,531
0,9800	159	25,9	3,079	1,539
0,9850	160	25,8	3,094	1,547
0,9900	161	25,7	3,110	1,555
0,9950	162	25,5	3,126	1,563
1,0000	163	25,4	3,142	1,571

APPENDIX V

PINION TOOTH & CUTTER DIMENSIONS

Dimensions are given in millimetres and inches.

Module 'm'	**No. of Leaves**	**Addendum 'a'**	**Dedendum 'd'**	**Addendum Radius 'r'**	**Tooth Thickness 't'**	**Tooth Space 's'**
	6	0.17(.007)	0.32(.013)	0.21(.008)	0.21(.008)	0.42(.016)
	7	0.17(.007)	0.37(.015)	0.21(.008)	0.21(.008)	0.42(.016)
0.2	8	0.17(.007)	0.38(.015)	0.21(.008)	0.21(.008)	0.42(.016)
	10	0.16(.006)	0.41(.016)	0.16(.006)	0.25(.009)	0.38(.015)
	12	0.16(.006)	0.42(.017)	0.16(.006)	0.25(.009)	0.38(.015)
	16	0.16(.006)	0.42(.017)	0.16(.006)	0.25(.009)	0.38(.015)
	6	0.21(.008)	0.40(.016)	0.26(.010)	0.26(.010)	0.52(.020)
	7	0.21(.008)	0.46(.018)	0.26(.010)	0.26(.010)	0.52(.020)
0.25	8	0.21(.008)	0.48(.019)	0.26(.010)	0.26(.010)	0.52(.020)
	10	0.20(.008)	0.51(.020)	0.21(.008)	0.31(.012)	0.47(.018)
	12	0.20(.008)	0.53(.021)	0.21(.008)	0.31(.012)	0.47(.018)
	16	0.20(.008)	0.53(.021)	0.21(.008)	0.31(.012)	0.47(.018)
	6	0.26(.010)	0.47(.019)	0.32(.013)	0.32(.013)	0.64(.025)
	7	0.26(.010)	0.56(.022)	0.32(.013)	0.32(.013)	0.64(.025)
0.3	8	0.26(.010)	0.57(.022)	0.32(.013)	0.32(.013)	0.64(.025)
	10	0.24(.009)	0.62(.024)	0.25(.010)	0.38(.015)	0.56(.022)
	12	0.24(.009)	0.63(.025)	0.25(.010)	0.38(.015)	0.56(.022)
	16	0.24(.009)	0.63(.025)	0.25(.010)	0.38(.015)	0.56(.022)
	6	0.30(.012)	0.55(.022)	0.37(.015)	0.37(.015)	0.74(.029)
	7	0.30(.012)	0.65(.026)	0.37(.015)	0.37(.015)	0.74(.029)
0.35	8	0.30(.012)	0.67(.026)	0.37(.015)	0.37(.015)	0.74(.029)
	10	0.28(.011)	0.72(.028)	0.29(.011)	0.44(.017)	0.66(.026)
	12	0.28(.011)	0.74(.029)	0.29(.011)	0.44(.017)	0.66(.026)
	16	0.28(.011)	0.74(.029)	0.29(.011)	0.44(.017)	0.66(.026)
	6	0.34(.013)	0.63(.025)	0.42(.017)	0.42(.017)	0.84(.033)
	7	0.34(.013)	0.74(.029)	0.42(.017)	0.42(.017)	0.84(.033)
0.4	8	0.34(.013)	0.76(.030)	0.42(.017)	0.42(.017)	0.84(.033)
	10	0.32(.012)	0.82(.032)	0.33(.013)	0.50(.020)	0.75(.030)
	12	0.32(.012)	0.84(.033)	0.33(.013)	0.50(.020)	0.75(.030)
	16	0.32(0.12)	0.84(.033)	0.33(.013)	0.50(.020)	0.75(.030)
	6	0.38(.015)	0.71(.028)	0.47(.019)	0.47(.019)	0.94(.037)
	7	0.38(.015)	0.83(.033)	0.47(.019)	0.47(.019)	0.94(.037)
0.45	8	0.38(.015)	0.86(.034)	0.47(.019)	0.47(.019)	0.94(.037)
	10	0.36(.014)	0.92(.036)	0.37(.015)	0.56(.022)	0.84(.033)
	12	0.36(.014)	0.95(.037)	0.37(.015)	0.56(.022)	0.84(.033)
	16	0.36(.014)	0.95(.037)	0.37(0.15)	0.56(.022)	0.84(.033)

Root Land 'w'	Full Tooth Depth 'h'	Cutter Angle in degrees	Ogive Type	Cutter Form Thickness	Pinion Root Dia.	Pinion O/Dia.
.207(.008)	0.49(.019)	40	'C'	2.0(.078)	0.568(.022)	1.54(.061)
.204(.008)	0.54(.021)	34	'C'	2.0(.078)	0.66(.026)	1.74(.069)
.225(.009)	0.55(.022)	30	'C'	2.0(.078)	0.84(.033)	1.94(.076)
.225(.009)	0.57(.022)	21.5	'B'	2.0(.078)	1.18(.046)	2.32(.091)
.247(.010)	0.58(.023)	18	'B'	2.0(.078)	1.56(.061)	2.72(.107)
.279(.011)	0.58(.023)	13.5	'B'	2.0(.078)	2.36(.093)	3.52(.139)
.258(.010)	0.61(.024)	40	'C'	2.0(.078)	0.71(.028)	1.93(.076)
.255(.010)	0.68(.027)	34	'C'	2.0(.078)	0.83(.032)	2.18(.086)
.281(.011)	0.69(.027)	30	'C'	2.0(.078)	1.05(.041)	2.43(.096)
.281(.011)	0.71(.028)	21.5	'B'	2.0(.078)	1.48(.058)	2.90(.114)
.309(.012)	0.73(.029)	18	'B'	2.0(.078)	1.95(.077)	3.40(.134)
.349(.014)	0.73(.029)	13.5	'B'	2.0(.078)	2.95(.116)	4.40(.173)
.310(.012)	0.73(.029)	40	'C'	2.0(.078)	0.85(.034)	2.31(.091)
.306(.012)	0.81(.032)	34	'C'	"	0.99(.039)	2.61(.103)
.338(.013)	0.83(.033)	30	'C'	"	1.26(.050)	2.91(.115)
.338(.013)	0.86(.034)	21.5	'B'	"	1.77(.070)	3.48(.137)
.371(.015)	0.87(.034)	18	'B'	"	2.34(.092)	4.08(.161)
.419(.016)	0.87(.034)	13.5	'B'	"	3.54(.139)	5.28(.208)
.362(.014)	0.85(.036)	40	'C'	2.0(.078)	0.99(.039)	2.70(.106)
.356(.014)	0.95(.037)	34	'C'	"	1.16(.045)	3.05(.120)
.394(.016)	0.96(.038)	30	'C'	"	1.47(.058)	3.40(.134)
.394(.016)	1.00(.039)	21.5	'B'	"	2.07(.081)	4.06(.160)
.432(.017)	1.02(.040)	18	'B'	"	2.73(.107)	4.76(.188)
0.489(.019)	1.02(.040)	13.5	'B'	"	4.13(.163)	6.16(.243)
0.413(.016)	0.97(.038)	40	'C'	3.25(.127)	1.14(.045)	3.08(.121)
0.407(.016)	1.08(.043)	34	'C'	"	1.32(.052)	3.48(.137)
0.450(.018)	1.10(.043)	30	'C'	"	1.68(.066)	3.88(.153)
0.450(.018)	1.14(.045)	21.5	'B'	"	2.36(.093)	4.64(.183)
0.494(.019)	1.16(.046)	18	'B'	"	3.12(.123)	5.44(.214)
0.559(.022)	1.16(.046)	13.5	'B'	"	4.72(.186)	7.04(.277)
0.465(.018)	1.10(.043)	40	'C'	3.25(.127)	1.28(.050)	3.47(.137)
0.458(.018)	1.22(.048)	34	'C'	"	1.49(.058)	3.92(.154)
0.506(.020)	1.24(.049)	30	'C'	"	1.89(0.74)	4.37(.172)
0.506(.020)	1.28(.050)	21.5	'B'	"	2.66(.105)	5.22(.206)
0.556(.022)	1.31(.051)	18	'B'	"	3.51(.138)	6.12(.241)
0.628(.025)	1.31(.051)	13.5	'B'	"	5.31(.209)	7.92(.312)

Module 'm'	No. of Leaves	Addendum 'a'	Dedendum 'd'	Addendum Radius 'r'	Tooth Thickness 't'	Tooth Space 's'
	6	0.43(.017)	0.79(.031)	0.53(.021)	0.53(.021)	1.05(.042)
	7	0.43(.017)	0.93(.037)	0.53(.021)	0.53(.021)	1.05(.042)
0.5	8	0.43(.017)	0.95(.037)	0.53(.021)	0.53(.021)	1.05(.042)
	10	0.40(.016)	1.03(.040)	0.41(.016)	0.63(.025)	0.94(.037)
	12	0.40(.016)	1.05(.041)	0.41(.016)	0.63(.025)	0.94(.037)
	16	0.40(.016)	1.05(.041)	0.41(.016)	0.63(.025)	0.94(.037)
	6	0.47(.019)	0.87(.034)	0.58(.023)	0.58(.023)	1.16(.046)
	7	0.47(.019)	1.02(.040)	0.58(.023)	0.58(.023)	1.16(.046)
0.55	8	0.47(.019)	1.05(.041)	0.58(.023)	0.58(.023)	1.16(.046)
	10	0.44(.017)	1.13(.044)	0.45(.018)	0.69(.027)	1.04(.041)
	12	0.44(.017)	1.16(.046)	0.45(.018)	0.69(.027)	1.04(.041)
	16	0.44(.017)	1.16(.046)	0.45(.018)	0.69(.027)	1.04(.041)
	6	0.51(.020)	0.95(.037)	0.63(.025)	0.63(.025)	1.26(.050)
	7	0.51(.020)	1.11(.044)	0.63(.025)	0.63(.025)	1.26(.050)
0.6	8	0.51(.020)	1.14(.045)	0.63(.025)	0.63(.025)	1.26(.050)
	10	0.48(.019)	1.23(.048)	0.49(.019)	0.75(.030)	1.13(.044)
	12	0.48(.019)	1.26(.050)	0.49(.019)	0.75(.030)	1.13(.044)
	16	0.48(.019)	1.26(.050)	0.49(.019)	0.75(.030)	1.13(.044)
	6	0.56(.032)	1.03(.041)	0.68(.027)	0.68(.027)	1.36(.054)
	7	0.56(.022)	1.20(.047)	0.68(.027)	0.68(.027)	1.36(.054)
0.65	8	0.56(.022)	1.24(.049)	0.68(.027)	0.68(.027)	1.36(.054)
	10	0.52(.020)	1.33(.052)	0.53(.021)	0.81(.032)	1.22(.048)
	12	0.52(.020)	1.37(.054)	0.53(.021)	0.81(.032)	1.22(.048)
	16	0.52(.020)	1.37(.054)	0.53(.021)	0.81(.032)	1.22(.048)
	6	0.60(.024)	1.11(.044)	0.74(.029)	0.74(.029)	1.47(.058)
	7	0.60(.024)	1.30(.051)	0.74(.029)	0.74(.029)	1.47(.058)
0.7	8	0.60(.024)	1.33(.052)	0.74(.029)	0.74(.029)	1.47(.058)
	10	0.56(.022)	1.44(.057)	0.57(.022)	0.88(.034)	1.31(.052)
	12	0.56(.022)	1.47(.058)	0.57(.022)	0.88(.034)	1.31(.052)
	16	0.56(.022)	1.47(.058)	0.57(.022)	0.88(.034)	1.31(.052)
	6	0.64(.025)	1.19(.047)	0.79(.031)	0.79(.031)	1.58(.062)
	7	0.64(.025)	1.39(.055)	0.79(.031)	0.79(.031)	1.58(.062)
0.75	8	0.64(.025)	1.43(.056)	0.79(.031)	0.79(.031)	1.58(.062)
	10	0.60(.024)	1.54(.061)	0.62(.024)	0.94(.037)	1.41(.056)
	12	0.60(.024)	1.58(.062)	0.62(.024)	0.94(.037)	1.41(.056)
	16	0.60(.024)	1.58(.062)	0.62(.024)	0.94(.037)	1.41(.056)
	6	0.68(.027)	1.26(.050)	0.84(.033)	0.84(.033)	1.68(.066)
	7	0.68(.027)	1.48(.058)	0.84(.033)	0.84(.033)	1.68(.066)
0.8	8	0.68(.027)	1.52(.060)	0.84(.033)	0.84(.033)	1.68(.066)
	10	0.64(.025)	1.64(.065)	0.66(.026)	1.00(.039)	1.50(.059)
	12	0.64(.025)	1.68(.066)	0.66(.026)	1.00(.039)	1.50(.059)
	16	0.64(.025)	1.68(.066)	0.66(.026)	1.00(.039)	1.50(.059)

ot Land 'w'	Full Tooth Depth 'h'	Cutter Angle in degrees	Ogive Type	Cutter Form Thickness	Pinion Root Dia.	Pinion O/Dia.
517(.020)	1.22(.048)	40	'C'	3.25(.127)	1.42(.056)	3.86(.152)
509(.020)	1.35(.053)	34	'C'	"	1.65(.065)	4.36(.171)
563(.022)	1.38(.054)	30	'C'	"	2.10(.083)	4.86(.191)
563(.022)	1.43(.056)	21.5	'B'	"	2.95(.116)	5.81(.229)
618(.024)	1.45(.057)	18	'B'	"	3.90(.154)	6.81(.268)
698(.027)	1.45(.057)	13.5	'B'	"	5.90(.232)	8.81(.347)
569(.022)	1.34(.053)	40	'C'	3.25(.127)	1.56(.061)	4.24(.167)
560(.022)	1.49(.059)	34	'C'	"	1.82(.072)	4.80(.189)
619(.024)	1.52(.060)	30	'C'	"	2.31(.091)	5.34(.210)
619(.024)	1.57(.062)	21.5	'B'	"	3.25(.128)	6.39(.251)
679(.027)	1.60(.063)	18	'B'	"	4.29(.169)	7.49(.295)
768(.030)	1.60(.063)	13.5	'B'	"	6.49(.256)	9.69(.381)
620(.024)	1.46(.058)	40	'C'	3.25(.127)	1.70(.067)	4.63(.182)
611(.024)	1.62(.064)	34	'C'	"	1.98(.078)	5.23(.206)
.675(.027)	1.65(.065)	30	'C'	"	2.52(.099)	5.83(.229)
.675(.027)	1.71(.067)	21.5	'B'	"	3.54(.139)	6.97(.274)
.741(.029)	1.74(.069)	18	'B'	"	4.68(.184)	8.17(.321)
.838(.033)	1.74(.069)	13.5	'B'	"	7.08(.279)	10.57(.416)
.672(.026)	1.58(.062)	40	'C'	3.25(.127)	1.85(.073)	5.01(.197)
.662(.026)	1.76(.069)	34	'C'	"	2.15(.084)	5.66(.223)
.732(.029)	1.79(.071)	30	'C'	"	2.73(.107)	6.31(.248)
.732(.029)	1.86(.073)	21.5	'B'	"	3.84(.151)	7.55(.297)
.803(.032)	1.89(.074)	18	'B'	"	5.07(.200)	8.85(.348)
.908(.036)	1.89(.074)	13.5	'B'	"	7.67(.302)	11.45(.451)
.724(.028)	1.70(.067)	40	'C'	3.50(.138)	1.99(.078)	5.40(.212)
.713(.028)	1.89(.075)	34	'C'	"	2.31(.091)	6.10(.240)
.788(.031)	1.93(.076)	30	'C'	"	2.94(.116)	6.80(.268)
.788(.031)	2.00(.079)	21.5	'B'	"	4.13(.163)	8.13(.320)
.865(.034)	2.03(.080)	18	'B'	"	5.46(.215)	9.53(.375)
.978(.038)	2.03(.080)	13.5	'B'	"	8.26(.325)	12.33(.485)
.775(.031)	1.83(.072)	40	'C'	4.50(.178)	2.13(.084)	5.78(.228)
.764(.030)	2.03(.080)	34	'C'	"	2.48(.097)	6.53(.257)
.844(.033)	2.07(.081)	30	'C'	"	3.15(.124)	7.28(.287)
.844(.033)	2.14(.084)	21.5	'B'	"	4.43(.174)	8.71(.343)
.927(.036)	2.18(.086)	18	'B'	"	5.85(.230)	10.21(.402)
.047(.041)	2.18(.086)	13.5	'B'	"	8.85(.348)	13.21(.520)
.827(.033)	1.95(.077)	40	'C'	4.50(.178)	2.27(.089)	6.17(.243)
.815(.032)	2.16(.085)	34	'C'	"	2.64(.104)	6.97(.274)
.900(.035)	2.20(.087)	30	'C'	"	3.36(.132)	7.77(.306)
.900(.035)	2.28(.090)	21.5	'B'	"	4.72(.186)	9.29(.366)
.988(.039)	2.32(.091)	18	'B'	"	6.24(.246)	10.89(.429)
.117(.044)	2.32(.091)	13.5	'B'	"	9.44(.372)	14.09(.555)

Module 'm'	No. of Leaves	Addendum 'a'	Dedendum 'd'	Addendum Radius 'r'	Tooth Thickness 't'	Tooth Space 's'
	6	0.73(.029)	1.34(.053)	0.89(.035)	0.89(.035)	1.78(.070)
	7	0.73(.029)	1.57(.062)	0.89(.035)	0.89(.035)	1.78(.070)
0.85	8	0.73(.029)	1.62(.064)	0.89(.035)	0.89(.035)	1.78(.070)
	10	0.68(.027)	1.74(.069)	0.70(.028)	1.06(.042)	1.59(.063)
	12	0.68(.027)	1.79(.070)	0.70(.028)	1.06(.042)	1.59(.063)
	16	0.68(.027)	1.79(.070)	0.70(.028)	1.06(.042)	1.59(.063)
	6	0.77(.030)	1.42(.056)	0.95(.037)	0.95(.037)	1.90(.075)
	7	0.77(.030)	1.67(.066)	0.95(.037)	0.95(.037)	1.90(.075)
0.9	8	0.77(.030)	1.71(.067)	0.95(.037)	0.95(.037)	1.90(.075)
	10	0.72(.028)	1.85(.073)	0.74(.029)	1.13(.044)	1.70(.067)
	12	0.72(.028)	1.89(.074)	0.74(.029)	1.13(.044)	1.70(.067)
	16	0.72(.028)	1.89(.074)	0.74(.029)	1.13(.044)	1.70(.067)
	6	0.81(.032)	1.50(.059)	1.00(.039)	1.00(.039)	2.00(.079)
	7	0.81(.032)	1.76(.069)	1.00(.039)	1.00(.039)	2.00(.079)
0.95	8	0.81(.032)	1.81(.071)	1.00(.039)	1.00(.039)	2.00(.079)
	10	0.76(.030)	1.95(.077)	0.78(.031)	1.19(.047)	1.78(.070)
	12	0.76(.030)	2.00(.079)	0.78(.031)	1.19(.047)	1.78(.070)
	16	0.76(.030)	2.00(.079)	0.78(.031)	1.19(.047)	1.78(.070)
	6	0.86(.034)	1.58(.062)	1.05(.041)	1.05(.041)	2.10(.083)
	7	0.86(.034)	1.85(.073)	1.05(.041)	1.05(.041)	2.10(.083)
1.0	8	0.86(.034)	1.90(.075)	1.05(.041)	1.05(.041)	2.10(.083)
	10	0.81(.032)	2.05(.081)	0.82(.032)	1.25(.049)	1.88(.074)
	12	0.81(.032)	2.10(.083)	0.82(.032)	1.25(.049)	1.88(.074)
	16	0.81(.032)	2.10(.083)	0.82(.032)	1.25(.049)	1.88(.074)

…ot Land 'w'	Full Tooth Depth 'h'	Cutter Angle in degrees	Ogive Type	Cutter Form Thickness	Pinion Root Dia.	Pinion O/Dia.
879(.035)	2.07(.081)	40	'C'	4.50(.178)	2.41(.095)	6.55(.258)
866(.034)	2.30(.091)	34	'C'	"	2.81(.110)	7.40(.291)
957(.038)	2.34(.092)	30	'C'	"	3.57(.141)	8.25(.325)
957(.038)	2.43(.096)	21.5	'B'	"	5.02(.197)	9.87(.389)
050(.041)	2.47(.097)	18	'B'	"	6.63(.261)	11.57(.455)
187(.047)	2.47(.097)	13.5	'B'	"	10.03(.395)	14.97(.590)
930(.037)	2.19(.086)	40	'C'	4.50(.178)	2.56(.101)	6.94(.273)
917(.036)	2.43(.096)	34	'C'	"	2.97(.117)	7.84(.309)
013(.040)	2.48(.098)	30	'C'	"	3.78(.149)	8.74(.344)
013(.040)	2.57(.101)	21.5	'B'	"	5.31(.209)	10.45(.411)
112(.044)	2.61(.103)	18	'B'	"	7.02(.276)	12.25(.482)
257(.040)	2.61(.103)	13.5	'B'	"	10.62(.420)	15.85(.624)
982(.039)	2.31(.091)	40	'C'	4.50(.178)	2.70(.106)	7.32(.288)
967(.038)	2.57(.101)	34	'C'	"	3.14(.123)	8.27(.326)
069(.042)	2.62(.103)	30	'C'	"	3.99(.151)	9.22(.363)
069(.042)	2.71(.107)	21.5	'B'	"	5.61(.221)	11.03(.434)
174(.046)	2.76(.109)	18	'B'	"	7.41(.292)	12.93(.509)
327(.052)	2.76(.109)	13.5	'B'	"	11.21(.441)	16.73(.659)
034(.041)	2.44(.096)	40	'C'	4.50(.178)	2.84(.112)	7.71(.304)
018(.040)	2.71(.107)	34	'C'	"	3.30(.130)	8.71(.343)
125(.044)	2.76(.109)	30	'C'	"	4.20(.165)	9.71(.382)
125(.044)	2.86(.113)	21.5	'B'	"	5.90(.232)	11.61(.457)
240(.049)	2.91(.115)	18	'B'	"	7.80(.307)	13.61(.536)
397(.055)	2.91(.115)	13.5	'B'	"	11.80(.465)	17.61(.693)

APPENDIX VI

WHEEL CUTTER DIMENSIONS

Wheel Tooth & Cutter Dimensions in mm & inches

Module 'M'	Addendum 'A'	Dedendum 'D'	Tooth Thickness 'T'	Addendum Radius 'R'	Tooth Height 'H'	Cutter Blank Thickness
	mm ins	mm ins	mm ins	mm ins	mm ins	mm ins
0.20	0.28(0.011)	0.31(0.012)	0.31(0.012)	0.39(0.015)	0.59(0.023)	0.94(0.037)
0.25	0.35(0.014)	0.39(0.015)	0.39(0.015)	0.48(0.019)	0.74(0.029)	1.14(0.045)
0.30	0.41(0.016)	0.47(0.019)	0.47(0.019)	0.58(0.023)	0.89(0.035)	1.40(0.055)
0.35	0.48(0.019)	0.55(0.022)	0.55(0.022)	0.68(0.026)	1.03(0.041)	1.62(0.064)
0.40	0.55(0.022)	0.63(0.025)	0.63(0.025)	0.77(0.030)	1.18(0.046)	1.85(0.073)
0.45	0.62(0.024)	0.71(0.028)	0.71(0.028)	0.87(0.034)	1.33(0.052)	2.11(0.083)
0.50	0.69(0.027)	1.00(0.039)	0.79(0.031)	0.97(0.038)	1.69(0.067)	2.34(0.092)
0.55	0.76(0.030)	1.10(0.043)	0.86(0.034)	1.06(0.042)	1.86(0.073)	2.54(0.100)
0.60	0.83(0.033)	1.20(0.047)	0.94(0.037)	1.16(0.046)	2.03(0.080)	2.79(0.110)
0.65	0.90(0.035)	1.30(0.051)	1.02(0.040)	1.25(0.049)	2.20(0.086)	3.02(0.119)
0.70	0.97(0.038)	1.40(0.055)	1.10(0.043)	1.35(0.053)	2.37(0.093)	3.12(0.123)
0.75	1.04(0.041)	1.50(0.059)	1.18(0.046)	1.45(0.057)	2.54(0.100)	3.48(0.137)
0.80	1.10(0.043)	1.60(0.063)	1.26(0.049)	1.54(0.061)	2.70(0.106)	3.68(0.145)
0.85	1.17(0.046)	1.70(0.067)	1.33(0.053)	1.64(0.065)	2.87(0.113)	3.93(0.155)
0.90	1.24(0.049)	1.80(0.071)	1.41(0.056)	1.74(0.068)	3.04(0.120)	4.14(0.163)
0.95	1.31(0.052)	1.90(0.075)	1.49(0.059)	1.83(0.072)	3.21(0.126)	4.37(0.172)
1.0	1.38(0.054)	2.00(0.079)	1.57(0.062)	1.93(0.076)	3.38(0.133)	4.59(0.181)
1.1	1.52(0.060)	1.73(0.068)	1.73(0.068)	2.12(0.084)	3.25(0.128)	5.07(0.200)
1.2	1.66(0.065)	1.88(0.074)	1.88(0.074)	2.32(0.091)	3.54(0.139)	5.53(0.218)
1.3	1.79(0.071)	2.04(0.080)	2.04(0.080)	2.51(0.099)	3.84(0.151)	5.96(0.235)
1.4	1.93(0.076)	2.20(0.087)	2.20(0.087)	2.70(0.106)	4.13(0.163)	6.45(0.254)
1.5	2.07(0.081)	2.36(0.093)	2.36(0.093)	2.90(0.114)	4.43(0.174)	7.00(0.276)

APPENDIX VII

LENGTHS OF PENDULUM AND NUMBER OF VIBRATIONS

Vibrations of Pendulum Per Hour	Length of Pendulum		Vibrations of Pendulum Per Hour	Length of Pendulum	
	in mm	in inches		in mm	in inches
3600	1000	39.37	7218	247	9.73
4080	774	30.49	7320	240	9.46
4200	730	28.75	7488	229	9.02
4320	690	27.17	7500	228	9.01
4500	636	25.05	7560	225	8.87
4680	588	23.15	7680	218	8.59
4800	559	22.01	7758	214	8.42
5040	507	19.97	7800	212	8.34
5160	484	19.06	7920	205	8.08
5280	462	18.19	8022	200	7.90
5346	450	17.72	8100	196	7.73
5400	441	17.39	8292	187	7.38
5616	408	16.08	8400	182	7.18
5640	405	15.94	8520	177	6.99
5730	392	15.45	8616	176	6.93
5760	388	15.28	8700	170	6.69
5880	372	14.66	8826	165	6.50
5934	366	14.41	9000	159	6.26
6000	358	14.09	9096	155	6.10
6120	345	13.54	9120	154	6.09
6144	341	13.44	9240	151	5.94
6300	324	12.78	9360	147	5.78
6348	320	12.59	9600	140	5.50
6420	312	12.30	9630	139	5.47
6480	307	12.08	9894	131	5.15
6552	300	11.82	9900	130	5.17
6600	295	11.64	10164	124	4.88
6684	288	11.35	10200	123	4.87
6720	285	11.22	10428	118	4.65
6756	282	11.11	10500	117	4.60
6840	275	10.82	10560	115	4.55
6900	270	10.65	10698	112	4.43
6954	266	10.49	10800	110	4.35
7200	248	9.78	11280	101	3.99

REFERENCES

(1) *The Diary of Robert Hooke 1672–80*, transcribed from the original in 1935, re-printed 1968.
(2) Bodleian Library M S Carte 264 fol. 44r.
(3) Jeremy Evans, *Antiquarian Horology, Vol. 24 No. 6*, 1999.
(4) Jose A. Garcia-Diego – Juanelo Torriano – Charles V's clockmaker.
(5) Henry Hindley 1701–71, R. J. Law, *AHS Vol. 7 No. 3*, 1971.
(6) By kind permission of Professor Alan Smith.
(7) Early wheel-cutting engine signed Humphrey Marsh; Science Museum collection, SSTL.
(8) Wheel-cutting machine signed Wyke & Green; Prescot Museum Collection.
(9) The Stradanus design of a clockmaker's workshop *c.* 1600, courtesy of the British Museum.
(10) Courtesy Leicester Museum.
(11) Thomas Tompion movement by kind permission of Christie's, London.
(12) John Griffiths: 'Methods of Cutting Wheels and Pinions in Horology' – *Horological Journal*, March 1992.
(13) Pinion-cutting machine from the Prescot Museum collection.
(14) American Watch Tool Co. *Catalogue of Tools* 1980.
(15) C.N.C. rotary indexer: photo by kind permission of Sherline Products and Bryan Mumford.
(16) Article by David Burton 'Care and Use of Wheel and Pinion Cutters' *Horological Journal* 1993.
(17) Duplex: 'Gear Cutting in the Lathe' *Model Engineer* commencing June 1949.
(18) 'The Eureka, A Continuous Form-Relieving Tool for Gear Cutting', Prof. D. H. Chaddock and Ivan Law, *Model Engineer*, February 1987.
(19) 'Simple Form-Relieved Milling Cutters', D. J. Unwin, *Model Engineer* commencing August 1970.

INDEX